Nursery Manual for Native Plants

A Guide for Tribal Nurseries

Volume One
Nursery Management

United States Department of Agriculture
Forest Service

AGRICULTURE HANDBOOK 730
MARCH 2009

Edited by

R. Kasten Dumroese
National Nursery Specialist and Research Plant Physiologist
USDA Forest Service, Rocky Mountain Research Station
Moscow, ID

Tara Luna
Botanist and Native Plant Horticulturist
East Glacier, MT

Thomas D. Landis
National Nursery Specialist, USDA Forest Service (retired)
Native Plant Nursery Consulting
Medford, OR

Pesticides used improperly can be injurious to humans, animals, and plants. Follow the directions and heed all precautions on the labels. Store pesticides in original containers under lock and key—out of the reach of children and animals—and away from food and feed. Apply pesticides so that they do not endanger humans, livestock, crops, beneficial insects, fish, and wildlife. Do not apply pesticides when there is danger of drift, when honey bees or other pollinating insects are visiting plants, or in ways that may contaminate water or leave illegal residues. Avoid prolonged inhalation of pesticide sprays or dusts; wear protective clothing and equipment if specified on the container. If your hands become contaminated with a pesticide, do not eat or drink until you have washed. In case a pesticide is swallowed or gets in the eyes, follow the first-aid treatment given on the label, and get prompt medical attention. If a pesticide is spilled on your skin or clothing, remove clothing immediately and wash skin thoroughly. Do not clean spray equipment or dump excess spray material near ponds, streams, or wells. Because it is difficult to remove all traces of herbicides from equipment, do not use the same equipment for insecticides or fungicides that you use for herbicides. Dispose of empty pesticide containers promptly. Have them buried at a sanitary land-fill dump, or crush and bury them in a level, isolated place. NOTE: Some States have restrictions on the use of certain pesticides. Check your State and local regulations. Also, because registrations of pesticides are under constant review by the Federal Environmental Protection Agency, consult your county agricultural agent or State extension specialist to be sure the intended use is still registered.

Nomenclature for scientific names follows the U.S. Department of Agriculture, Natural Resources Conservation Service PLANTS (Plant List of Accepted Nomenclature, Taxonomy, and Symbols) database (2008). http://plants.usda.gov

CONTENTS

Preface

In 2001, the Forest Service, U.S. Department of Agriculture (USDA), through its Virtual Center for Reforestation, Nurseries, and Genetics Resources (RNGR), invited Native Americans from across the United States to attend the Western Forest and Conservation Nursery Association annual meeting. About 25 tribal members, representing 20 tribes, attended the meeting at Fort Lewis College in Durango, Colorado. The following year, a similar meeting was held in Olympia, Washington, and tribal members initiated a Tribal Nursery Council and requested that RNGR facilitate the organization. During 2003, RNGR requested information from 560 tribes across the United States, seeking specific information on tribes' needs for native plants, facilities, training, and so on. Results from the responding 77 tribes were incorporated into the *Tribal Nursery Needs Assessment*. Based on the results of that questionnaire, and input from tribal members attending the 2003 Intertribal Nursery Council meeting in Coeur d'Alene, Idaho, it was agreed that a nursery handbook was needed. That fall, planning began for writing the manual, loosely based on Agriculture Handbook 674, *The Container Tree Nursery Manual*, but with special attention to the uniqueness of Native American cultures.

The team consisted of Thomas D. Landis, recently retired after 30 years with the Forest Service (the last 25 of those years as Western and National Nursery Specialist), lead author of the seven-volume *Container Tree Nursery Manual*, author of numerous other technology transfer publications, and currently principal of Native Plant Nursery Consulting in Medford, Oregon; Tara Luna, a botanist, who has grown native plants 20+ years in the Western United States and worked extensively with the Confederated Salish and Kootenai Tribes (Montana), the Blackfeet Nation (Montana), the Hopi Tribe (Arizona), and the Confederated Tribes of the Umatilla Indian Reservation (Oregon) in their native plant nurseries; Kim M. Wilkinson, who founded a nursery for native and culturally important plants on the Island of Hawai'i and managed it for 10 years, and

has authored several books on ecological restoration and cultural renewal; Douglass F. Jacobs, Associate Professor of Regeneration Silviculture with the Hardwood Tree Improvement and Regeneration Center in the Department of Forestry and Natural Resources at Purdue University, who has extensive research in nursery production and seedling quality of forest tree species for reforestation and restoration; and R. Kasten Dumroese, Research Plant Physiologist in the Forest Service's Rocky Mountain Research Station, who with 24 years experience growing and researching native plants is the current National Nursery Specialist and editor of the *Native Plants Journal*.

The vision was to provide a handbook that covered all aspects of managing a native plant nursery, from initial planning through crop production to establishing trials to improve nursery productivity into the future. The handbook is divided into four main sections: Getting Started, Developing a Nursery, Growing Plants, and Problem Solving.

During development of the handbook, and with input from the Intertribal Nursery Council, the production team decided to split the handbook into two volumes. This first volume, Nursery Management, contains 17 chapters devoted to that topic, whereas the second volume will include nearly 300 protocols for propagating native plants important to, and identified by, the tribes for cultural, medicinal, and restoration purposes. Together, these two volumes should provide a solid foundation for Native Americans and others interested in producing native plants to do so.

Acknowledgments

Photos were contributed by Terrence Ashley, Confederated Salish and Kootenai Tribes; Michael A. Castellano, USDA Forest Service, Pacific Northwest Research Station; Efren Cazares, Oregon State University; Kingsley Dixon, Botanic Parks and Garden Authority, Australia; John L. Edson, Hawai'i Reforestation Nursery; Mike Evans, Tree of Life Nursery; Richard Hannan, USDA Agricultural Research Service; Bev Hills, Ktnuaxa-Kinbasket First Nation; J. Chris Hoag, USDA Natural Resources Conservation Service; JFNew Nursery; Kate Kramer, USDA Forest Service, Region 5; Joyce Lapp, U.S. Department of the Interior, National Park Service; Ben Luoma, Oregon State University; Charles Matherne, State of Louisiana Dept. of Agriculture, retired; Terry McGrath Photography; USDA Forest Service, Missoula Technology and Development Center; Joseph F. Myers, USDA Forest Service, Region 1; Ronald P. Overton, USDA Forest Service, Northeast Area; William Pink, Temecula Band of Luiseno Indians; Jeremiah R. Pinto, Navajo Nation and USDA Forest Service, Rocky Mountain Research Station; William Sayward, ITASCA Greenhouses; Nancy Shaw, USDA Forest Service, Rocky Mountain Research Station; David Steinfeld, USDA Forest Service, Region 6; Stuewe and Sons, Inc.; Dawn Thomas, Confederated Salish and Kootenai Tribes; University of Idaho Library, Special Collections and Archives; and Chuck Williams, Redwood Valley Band of Pomo Indians. Illustrations were prepared by Jim Marin Graphics and Steve Morrison. Early drafts were reviewed by Peggy Adams, Jamestown S'Klallam Tribe; Haley McCarty, Makah Indian Tribe; Jeremiah R. Pinto, Navajo Nation; Max Taylor and Priscilla Pavatea, Hopi Tribe; and Gloria Whitefeather-Spears, Red Lake Band of Chippewa Indians. Timber Press, Inc. and Bruce McDonald allowed use of their copyrighted material. Design and layout was prepared by Grey Designs. Sonja Beavers and Candace Akins edited and proofed various versions of this volume. Karl Perry reviewed images. Richard Zabel and the Western Forest and Conservation Association, through an agreement with the Southern Research Station, were instrumental in producing this handbook. We thank everyone for their generous and professional support.

Funding for this volume came from the Forest Service. Primary funding was provided by State and Private Forestry, Cooperative Forestry, Larry Payne, former Director, through the Virtual Center for Nurseries, Reforestation, and Genetics Resources. Additional support provided by State and Private Forestry, Joel Holtrop, former Deputy Chief; Southern Research Station, Peter J. Roussopoulus, former Director; and Rocky Mountain Research Station, George "Sam" Foster, Director.

Nursery Manual for Native Plants

A Guide for Tribal Nurseries

Nursery Management

Planning a Native Plant Nursery

Kim M. Wilkinson and Thomas D. Landis

Every nursery is unique. The environmental, social, and economic context is different for each nursery. A wide variety of species and outplanting environments contributes to nursery diversity (figure 1.1). In addition, each nursery has a distinct vision and purpose. The methods a nursery will use to bring people together, produce high-quality plants for the community, and share knowledge about those plants will also be unique. With so many diverse factors to consider, no standard blueprint for how to design a particular nursery exists. On the contrary, the very best nursery design will be matched to a particular situation, resources, and objectives. Although outside resources may be consulted during the planning phase, ultimately it is the nursery team that best understands the place, the plants, and the community.

Planning involves both strategic and tactical thinking. Strategic planning addresses questions about why, what, when, and where. Why start (or expand) a nursery? What kind of nursery will best reach the goals? When and where will be best to perform certain tasks? This kind of strategic thinking is introduced in this chapter and expanded on in the first six chapters of this handbook. The rest of the handbook will help with tactical planning, addressing questions of who, how, and how much. How will plants be propagated and the nursery managed? When are goals met?

Ideally, read this chapter, browse through this entire handbook, and read relevant sections before making any big decisions or investments in a nursery.

Wilbert Fish of the Blackfeet Nation in Montana by Terry McGrath Photography.

Why? A nursery is a web of interrelated factors. Each aspect of the nursery affects every other aspect. For example, consider the seemingly simple act of choosing what kind of containers to use for growing plants. Containers come in many sizes and shapes, but, hopefully, container selection will be based on which containers will yield the best plant size and type to meet needs on the outplanting sites. Container type will dictate what kinds of nursery layout and benches will be needed, what types of irrigation systems and growing media will be used, how seeds will be sown, and so on. Container type and size will also impact scheduling, fertilization practices, product costs, and so forth. These factors are only some examples of the interconnectedness involved in planning a nursery. Imagine how other single factors can affect other aspects. Interrelated factors also include management aspects, such as relating with customers and the public, budgeting, and scheduling timelines for plant production.

OVERVIEW OF THE PLANNING PROCESS

The start-up phase of many successful nurseries involves thoughtfulness, research, discussion, and careful planning on paper. Too often, this crucial planning phase is rushed to "get things done" and begin development on the ground. Sometimes people involved in planning or funding the nursery have a preconceived idea of how a nursery should look (for example, "All nurseries should have a big greenhouse") or what the nursery should do (for example, "We need to grow conifer seedlings for reforestation"). It would be a mistake to rush into making these tactical decisions without investing in some strategic planning first. Making tactical decisions prematurely would impose predetermined ideas on the nursery before the goals are clarified and the actual situation is assessed. The initial planning phase is an opportunity to step back and clarify the vision and goals of the nursery and strategically coordinate all components that will affect the nursery. For example, the first idea may be to build a large propagation structure, such as a big greenhouse with a very controlled climate. Assessment of the nursery's goals and the actual needs of the species grown, however, may lead to a site-appropriate design that instead creates several different, smaller scale environments that are ultimately more economical, efficient, and effective for producing plants.

One good way to work through the planning phase is to begin by clarifying the vision and objectives for the nursery. Objectives likely include producing plants that will survive and thrive on a client's outplanting sites; therefore, background research into the best kinds of plants to meet the needs of clients and outplanting sites will be part of this process (figure 1.2). After the vision and objectives are defined, the practical considerations for reaching the objectives are assessed. Are these practical considerations economically possible? Are necessary resources available? Starting a small pilot nursery can be an excellent way to gain an overall, holistic understanding of what will be involved in nursery development and management while minimizing risk. After goals and financial realities are understood, additional factors are assessed to see if starting the nursery is a realistic and achievable undertaking. The planning phase then moves into practicalities: selecting a nursery site, determining species and stock types to grow, designing structures and facilities, purchasing containers and growing media supplies, and so forth, as detailed in other chapters in this handbook. Figure 1.3 illustrates many of the factors that go into nursery design.

DEFINE NURSERY VISION AND OBJECTIVES

Most nurseries are founded on the vision of a person or group of people. The vision may involve an intuitive sense of how the landscape and community could be 10, 50, or more than 100 years from now as a result of efforts expended today. This vision of the nursery will be a guiding force that adapts to meet the needs of the community and environment, and translates into practical objectives (figure 1.4). Some community and ecology objectives of the nursery might include the following:

— Bringing people together.
— Perpetuating local heritage.
— Making culturally important plants more available.
— Providing employment and economic opportunities.
— Filling a community need for native plants for landscaping.
— Renewing resources for important food plants or other useful species.
— Educating children to pass on traditional ecological knowledge about plants to future generations.
— Restoring degraded land.
— Propagating rare species.

Visions are translated into practical objectives through interactions with the community. The nursery may have a vision, for example, of perpetuating native plants throughout the landscape. The ultimate hope may be to see many kinds of people using a wide diversity of native and culturally important plants provided by the nursery. This hope, however, must be tempered by reality. An approach that says, "If we grow it, they will plant it" may result in wasted initial effort if the community has little desire or knowledge about how to use the nursery's plants. To help bridge the gap between a vision and practical objectives, the nursery may start by asking questions such as:

— What is truly needed and wanted in our community?
— Who are the potential clients of our plant materials at this time?
— Who might be potential clients in the future (if we engage in outreach and education)?
— What are the needs and priorities of the potential clients?

Community needs and priorities can be assessed in many ways. Formal and informal avenues should be used to gather as much information as possible. Existing trade groups, guilds, elders, and instructors that work with the products of various plant materials are often tremendous sources of key information. Politics may also play a large role in shaping the nursery. Holding a public gathering, discussing hopes for the future of the local environment and community, interviewing people, publishing an article in the local newspaper and asking for responses, or conducting formal market research can be invaluable in this phase (figure 1.5).

In cases in which a high demand for certain species clearly exists, this step of end-user assessment is key to helping a nursery determine not only what species to grow but also the stock type and size clients might prefer. For plants used for land restoration or forestry, the planting sites determine the optimal specifications. With culturally important plants, particularly those used for medicine, clients often have very exacting specifications. It is essential to assess these expectations during the planning phase to ensure that the nursery will provide what clients actually need and want (figure 1.6). In addition, an assessment will help determine how long the existing demand is likely to

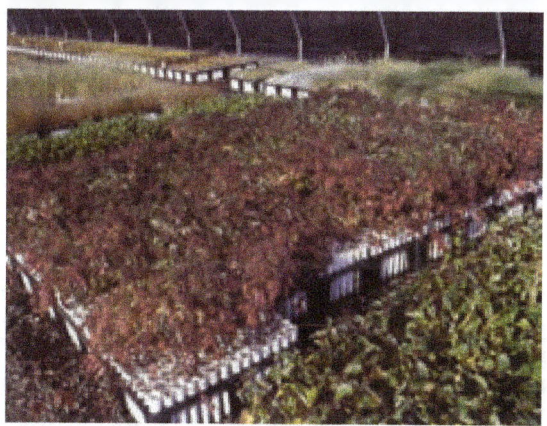

Figure 1.1—*Native plant nurseries have many unique characteristics, including growing a wide diversity of species.* Photo by R. Kasten Dumroese.

Figure 1.2—*Understanding the challenges of outplanting sites, such as those on this project near the Flathead River by the Confederated Salish and Kootenai Tribes in Montana, helps determine the optimal sizes and types of plant materials to grow, which affects all aspects of a nursery's design.* Photo by Dawn Thomas.

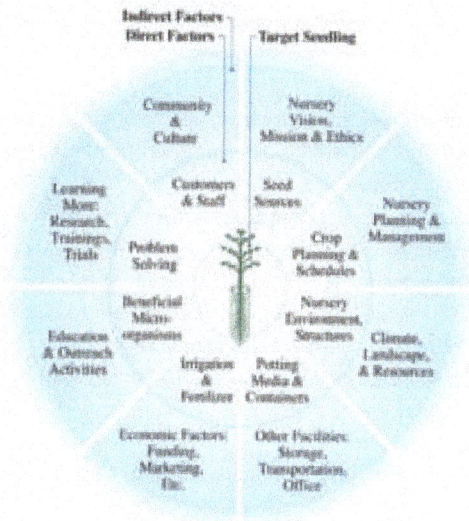

Figure 1.3—*Getting a good overview of direct and indirect factors for plant production in any situation will help design the best nursery to meet local needs.* Illustration by Jim Marin.

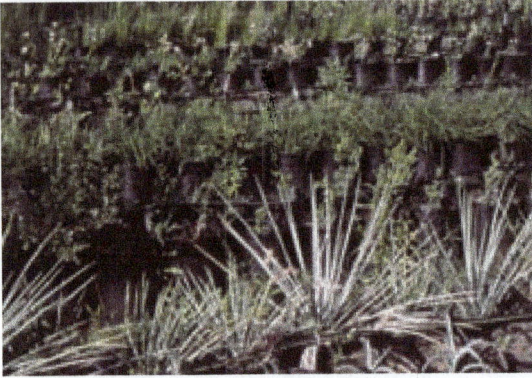

Figure 1.4—*The perpetuation of culturally important plant materials is a key objective for the Pechanga Band of the Luiseno Indians native plant nursery in California, which is cultivating many native plant species to provide basketry materials.* Photo by Tara Luna.

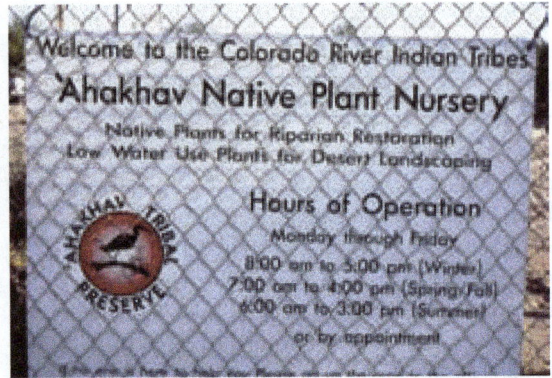

Figure 1.5—*Gathering information about the needs of the community and local ecosystems helps clarify the nursery's purpose. At the Colorado River Indian Tribes Reservation in Arizona, the 'Ahakhav Native Plant Nursery sign clearly communicates the nursery's objectives: to provide native plants for riparian restoration and low-water-use plants for desert landscaping.* Photo by Tara Luna.

last, which may indicate whether the nursery will be viable in the long term. On the other hand, the nursery may also use its own vision to avoid being swept up into meeting the short-term demands of a changing market. For example, if public interest in planting a certain exotic ornamental species becomes high, a nursery might expect financial gain from meeting this demand. Would meeting that demand, however, fulfill the mission of the nursery? Perhaps it would and perhaps not; deciding whether to meet a certain demand depends on the circumstances and the nursery's objectives.

Some topics to think about when assessing community needs include the following:

— The species the nursery is capable of growing.
— The types of environments in which plants will be outplanted.
— Specific end-user requirements for species (for example, seed source, special properties).
— The size and age of stock preferred.
— The season during which people prefer to plant.
— The quantities of species people may plant.
— The distance people are willing to travel to obtain the plant materials.

This information is invaluable in designing the nursery. It helps determine the species to grow and the "target plant" (optimal size, age, and type to thrive on your client's various outplanting sites) for each of those species. See Chapter 2, *The Target Plant Concept,*

for more discussion about this topic. The target criteria will be different for each species and will vary by outplanting site. The production of target plants will shape all other aspects of nursery design: location, structures, container types, species grown, scheduling, management practices, propagule collection, and so forth. The needs assessment responses can help the nursery identify its unique market niche and develop tactics to produce the best quality plant materials effectively and efficiently.

Keep in mind that nursery–client communication goes both ways. Although the nursery must listen to the stated needs and wants of its community, it can also share the visions and goals of the nursery with the community and potential clients. The nursery can engage in education and outreach to share information with clients and the community about the benefits and attributes of its plant materials (figure 1.7). For example, if the community thinks only a certain ornamental plant works well as a boundary hedge, it may be because people are unaware of a native species that can be planted for the same purpose. A good understanding of local ecology, environmental issues, history, soil types, and site needs for outplanting materials is important for good nursery design and will facilitate the production of high-quality, site-adapted plants that have high survival rates after outplanting. A little outreach to the community by the nursery can go a long way to overcome preconceived ideas (figure 1.8).

ASSESS RESOURCES AND COSTS

Nurseries differ greatly in terms of their financial objectives. Some nurseries may be funded through grants or government programs. Some may have startup money but are expected to be financially self-sufficient in the future. Private, for-profit nurseries must earn enough income from the sale of plants to at least pay for development, infrastructure, production costs, and staff time. Whatever the circumstances, finance is a key part of nursery planning. It determines:

— How much money can be invested in the nursery at the outset.
— What staff can be hired.
— In what timeframe the nursery can start to produce plants for sale.
— How many plants can be produced.
— What price can be charged for the plant materials.

Even if plants are to be distributed freely and not sold, it is still essential to know the cost to produce each kind of plant. Knowing production costs is crucial for planning, assessing feasibility, and ensuring the financial viability of the nursery. Will the nursery be able to meet its production expenses? Is the price the nursery will need to charge for plant materials a cost that the market can bear? For a new nursery, predicting the cost of plants is complicated; predicting costs depends a great deal on infrastructure, nursery size, staff skills, knowledge base, and many other factors. Nurseries that have gone through a pilot phase probably have a good grasp of the costs involved in producing plants (figure 1.9). These costs can be revised to reflect production on a larger scale. Without a pilot phase, estimating product costs prior to production is difficult to do accurately, and high crop losses may be expected during the first few seasons as successful propagation methods are developed.

Visiting other nurseries to get an idea of similar production processes can be very helpful as resources and finances are assessed. Government nurseries are a great source of information because staff members usually share production details openly, and their production costs are public information. Visiting private nurseries is also useful in assessing all the stages that go into plant production, although financial disclosures cannot be expected from private nurseries.

Figure 1.6—*Bringing people together to gather information about community needs is a key aspect of starting a nursery.* Photo by Thomas D. Landis.

Figure 1.7—*Community outreach and education is often an important part of working with native plants. This demonstration garden, part of the Santa Ana Pueblo Garden Center in New Mexico, helps visitors understand the beauty and the applications of plants grown in the tribe's nursery.* Photo by Tara Luna.

Figure 1.8—*The educational component of a native plant nursery can help connect people with traditional uses of plants, such as these ancestral structures made from native plant materials at the Pechanga Band of the Luiseno Indians native plant nursery in California.* Photo by Tara Luna.

Figure 1.9—*Starting with a smaller scale pilot phase reduces the number of unknown factors and risks and is a key part of developing a successful nursery.* Photo by Dawn Thomas.

STARTING WITH A SMALL PILOT NURSERY

Because of the enormous value of understanding the full range of processes that go into successful native plant production, it is often wise to start with a small pilot nursery. Instead of developing on a large scale, a pilot phase allows you to try out production on a smaller scale with less risk. The design of the pilot nursery is essentially your "best educated guess" on what type of set-up would be optimal to produce plants, based on this handbook and personal experiences. A few seasons of growth and observation in a pilot nursery can eliminate many of the unknowns regarding plant production and provide enough detailed information to effectively plan a larger facility.

Many unknowns are associated with nurseries growing native or culturally important plants. Therefore, a very large learning curve should be expected as production methods are developed for new species. High crop losses may occur during the first few seasons, and more losses may take place for the sake of experimentation. Starting with a small pilot nursery can be an effective strategy to decrease risks, preclude unnecessary expenditures, and develop viable propagation strategies. A pilot nursery can also be invaluable in estimating costs and making a more accurate feasibility assessment when the time comes to expand.

A small-scale pilot nursery can help accurately assess:

— What infrastructure is truly necessary for production (for example, perhaps a big greenhouse isn't needed at all).
— Which container types and propagation strategies are optimal for the plants.
— How to take crops through all phases of development, from germination through storage and distribution.
— What labor and material costs are involved.
— How clients respond to the plants produced.
— How to develop realistic timelines and budgets for future production.
— What aspects of the nursery vision are feasible to carry out at this time.

Many ideas are tested during the pilot phase of nursery development. The smaller-scale phase may last a few seasons or even a few years. Keep in mind that no one who works with plants will ever feel as if they have learned everything they need to know; even very established nurseries are always learning more and refining their propagation techniques. At some point, however, you will be confident that it is time to expand on the successes of the pilot nursery and continue development on a larger scale.

These production details can then be used to estimate costs in the planned nursery.

Estimate crop production costs by considering all phases of production, from seed procurement to delivery (figure 1.10). To improve the accuracy of the estimated costs, consider the timeframe for growing the crop, the size of the stock, the labor and materials required, and the fact that some crop losses will take place during production. Remember to account for the following factors:

— Material production costs (for example, growing media, water, fertilizer, seeds, pest control).
— Labor costs for production, maintenance, and delivery.
— Labor for customer relations (for example, answering e-mail messages and phone calls, handling correspondence, bookkeeping).
— Inventory required (for example, the time, space, and materials each crop will require, such as greenhouse benches, containers, trays).
— Structures (for example, greenhouse, shadehouse, storage).
— Overhead costs (for example, rent, insurance, water, utilities).
— Taxes.
— Time and funds for outreach, advertising, or educational programs.

It might be wise to estimate a range of best-case (most economical) and worst-case production scenarios. After production is under way and actual numbers are available, it is imperative to revisit the price structure of the plants to ensure they are in line with actual costs. In some cases, the costs of producing plants on a larger scale will be lower per plant than during the smaller-scale pilot phase. But it is also possible that costs related to rent, utilities, labor, and so on may be higher. Once the costs for crop production are understood, it is time to assess the feasibility of the overall nursery plan.

ASSESS FEASIBILITY

After assessing resources and costs that will be involved in carrying out the nursery's vision and objectives, it is time to take a good hard look at whether starting the nursery is a realistic and achievable undertaking. Again, starting with a small pilot phase is a good way to determine the feasibility of starting a nursery. The feasibility assessment should look at the species potentially available for the nursery to grow and match those species with the nursery site, goals, client needs, and nursery capabilities. Of course, the cost of plant materials and market price also must be considered. Will the emphasis be on growing plants from seeds or cuttings? How long does it take to grow these species to target specifications? What size plants should be produced? Several scenarios should be examined, including a variety of facility designs, sizes, and locations, so that the best conditions to meet projected needs are identified.

Necessary questions that should also be asked include, "Can the vision and objectives be fulfilled without starting a new nursery? Do alternatives exist? Can existing suppliers provide the desired plants that the nursery team could distribute for local outplanting needs?" Acting as a distributor instead of as a producer may be an economical alternative to starting a new nursery, but this has its benefits and drawbacks (table 1.1).

The final decision about whether to proceed with nursery development is ultimately in the hands of the people who had the vision for the nursery in the first place. Many nurseries have been developed to meet unique needs despite evidence that it would not be

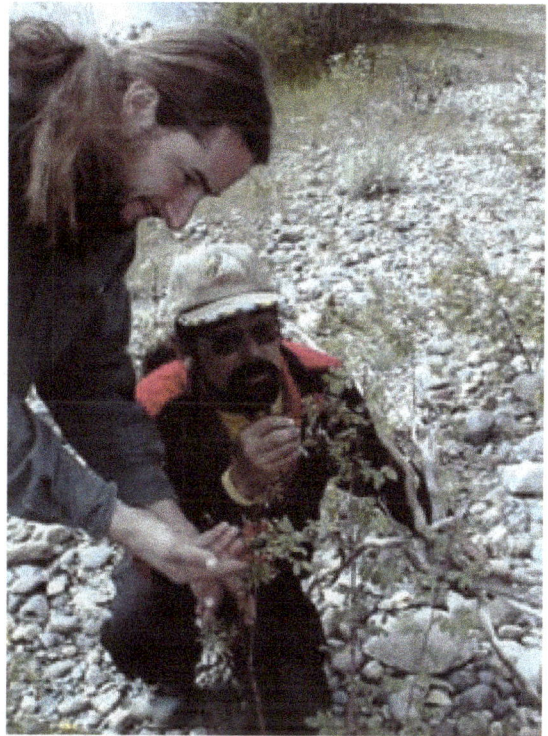

Figure 1.10—*To accurately estimate the time and expenses required for growing plant materials in a nursery, examine all aspects of plant production, including seed collecting and processing. Here Michael Keefer and Pete McCoy of Ktnuaxa-Kinbasket First Nation, British Columbia, gather seeds.* Photo by Bev Hills.

Table 1.1—The benefits and drawbacks of either starting a nursery or distributing plants from another supplier (after Landis and others 1994)

Purchase Plants

Benefits

—Time and capital available for other uses

—No nursery staff needed

—More long-term flexibility

—Plants grown by supplier in large quantities may be less expensive than plants grown in small, local nurseries

—Short-term or no commitment required

Drawbacks

—No control over growing process

—Less control over plant quality and availability

—Plants may not be adapted to local environment

—Unique needs of local clients may not be met

Start Own Nursery

Benefits

—High control over quality and availability of plants

—Can develop local expertise on plant growing and handling

—Can use traditional or culturally appropriate methods if applicable

—Plants will be adapted to local environment

—No reliance on other individuals or organizations

—Create job opportunities

—Others?

Drawbacks

—Large initial investment, capital, and time

—Long-term professional and economic commitment

—Must hire and maintain staff

—Native plant markets are notorious for year-to-year fluctuations

—Others?

practical to do so. If the decision is made to proceed, it is time to select a nursery site and think about nursery infrastructure.

SELECT A SITE FOR THE NURSERY

After the decision has been made to develop a container nursery, an appropriate site must be selected. This is a chance to be observant and think about working with nature, rather than against it, for the most effective, efficient, and economical design. The less the natural environment has to be modified to produce high-quality plants, the less expense the nursery will have to incur to create optimal crop conditions. Again, an understanding of the target plants to produce, as described in the following chapter, will help match the site to the needs of the crops. Careful observation of site conditions and an assessment of past and present climatic records are important.

Critical nursery site selection factors include the following:

— Access to good-quality, affordable, abundant water.
— Unobstructed solar access.
— Inexpensive and reliable energy.
— Easy access by staff.
— Adequate land area.
— Freedom from problematic ecological concerns (for example, free from neighboring chemical pollution, unmanageable noxious weeds, and so on).
— Freedom from problematic political concerns (for example, zoning restrictions, historical land use issues).

Climatic and biological attributes top the list for importance in site selection; an abundance of good-quality, reliable, affordable water is the number one factor, and water quality should always be tested when a site is being considered for nursery construction. See Chapter 10, *Water Quality and Irrigation*, for more information about this topic. Unobstructed solar access is also essential. Access to affordable electricity is very important. At least one person should have quick access to the nursery in case of emergency; if the nursery site will be far away from human dwellings, it may be advisable to construct a caretaker residence on site. The amount of land selected for the nursery must be large enough for the production areas and support buildings

and also allow for the efficient movement of equipment and materials. In addition to immediate needs, potential nursery sites should be evaluated on the basis of available space for possible expansion. Ecopolitical site selection factors, notably land use zoning and concerns about pesticide use and potential groundwater contamination, have severely reduced the number of sites suitable for nursery development.

Desirable attributes include those site selection criteria that are not absolutely necessary but will increase the efficiency of the nursery operation. If possible, choose a site with these desirable attributes:

— Protected microclimate.
— Gentle topography.
— Good labor supply.
— Easy access for staff and customers.
— Close proximity to markets.

Protected microclimate can make dramatic improvements in nursery productivity and reduce cost expenditures. A site with an equitable and sheltered climate and gentle topography is ideal. Access to the nursery by staff and clients is also important for economical nursery production.

Sometimes one or two factors are so important that the choice of site is obvious, but more commonly each site has good and bad attributes. If desired, make a list of potential nursery sites and compare them using a decision matrix. The decision matrix (table 1.2) is constructed by listing the potential nursery sites across the top and the significant site selection criteria down the left side. The next step is to assign each site selection criterion an importance value or weight on a scale from 1 to 10, with the most critical factors receiving the highest scores and the less important ones receiving progressively lower scores. Next, the suitability of each potential nursery location is evaluated and rated, again on a scale of 1 to 10, based on the information that has been gathered. After this task is accomplished, the score for each cell in the matrix is calculated by multiplying the weights for each site selection factor by the rating for each site. Finally, the weighted scores are totaled for each site, and, if the weights and rankings have been objectively assigned, then the potential nursery site with the highest total ranking should be the best choice. If all the potential sites are close in score, then the process should be repeated

Table 1.2—Decision matrix for evaluating potential container nursery sites. In this example, Site A received the highest score and is therefore considered the best choice for a nursery site (Landis and others 1994)

Site Selection Criteria	Weight Value*	Site A		Site B		Site C	
		Rating	Weighted Score	Rating	Weighted Score	Rating	Weighted Score
Critical Factors							
Good solar access	10	9	90	7	70	9	90
Water quality	9	9	81	7	63	4	36
Water quantity	8	10	80	8	64	9	72
Available energy	8	9	72	9	72	10	80
Adequate land area	7	8	56	8	56	10	70
Zoning restrictions	7	10	70	6	42	8	56
Pollution concerns	6	9	54	7	42	9	54
Secondary Factors							
Microclimate	6	9	54	8	48	9	54
Topography	5	10	50	9	45	10	50
Labor supply	4	9	36	8	32	10	40
Accessibility	4	8	32	6	24	8	32
Shipping distances	3	9	27	7	21	10	30
Totals			702		579		664
Site Suitability			#1		#3		#2

* Weights are relative importance values from 1 to 10, with 10 being highest

and careful attention paid to the relative weights and the ratings of the factors. If the scores are still close, the sites are probably equally good.

PLAN STRUCTURES AND FLOW OF WORK AROUND THE TARGET PLANT CONCEPT

All nursery planning pivots on producing target plant materials that match nursery stock type with the characteristics of the outplanting site to ensure plant survival and growth. See Chapter 2, *The Target Plant Concept,* for more discussion about this topic. The target plant must also meet the needs of clients; plants should produce the materials or products (medicine, wood, food, and so on) that clients expect. These needs dictate the plant's target size, age, seed source, container type, and management in the nursery. The requirements to produce target plants then guide all other aspects of nursery design. What propagation environments are best to produce these stock types efficiently? Good site selection and a sound knowledge of the plants and their needs as they go through the three phases of growth are important for creating the most appropriate structures for the crop's needs (figure 1.11). See Chapter 3, *Planning Crops and Developing Propagation Protocols,* for more information about plant growth phases. Commonly, rather than a single, large structure, a diversity of smaller structures is used and is tailored to meet the needs of the crops as they go through their development. See Chapter 4, *Propagation Environments,* for more details about this topic. Structural design is also affected by container types (chapter 6) and growing media (chapter 5). For example, bench layout must be planned to accommodate the container sizes. In turn, container type, growing media, and bench layout impact the design of the irrigation system. All these elements then come into play for management practices: fertilizing and watering, working with beneficial microorganisms and pests, and managing the overall flow of work in the nursery.

Although crop production is the core of nursery activities, it is only part of the whole picture. Preparation, cleanup, and storage must also be well planned. Where will seeds be cleaned, stored, treated, and tested? Where will containers be cleaned, sterilized, and stored when

Figure 1.11—*Good site assessment and an understanding of your crop's needs will help you determine the best structures for your nursery. Shown here is the greenhouse at the Browning High School on the Blackfeet Reservation in Montana.* Photo by Tara Luna.

Figure 1.12—*Good planning examines the flow of work and sets up effective workstations for efficient and enjoyable plant production during all phases of growth.* Photo by Tara Luna.

not in use? If crops are to be stored during winter, it is essential they are in an appropriate environment to ensure their survival. As nursery activities are planned, think about the flow of work and the design structures that facilitate the movement of people and plants in an efficient and safe way.

Good planning takes place in time and space (figure 1.12) and examines the flow of work and materials through all seasons and phases of growth. Time must be allotted for important activities such as outreach and educational programs; conducting trials and experiments to improve plant quality; and learning more through attending field days, meetings, and other events. See Chapter 16, *Nursery Management*, and Chapter 17, *Discovering Ways to Improve Crop Production and Plant Quality*, for more discussion about these topics.

Other environmental issues and risks should be considered. For example, the design should look at not just where good-quality water will come from for irrigation, but where the water will go after nursery use. The water may contain fertilizers and be a potential source of pollution, possibly creating legal issues for the nursery. With good planning, that same water may be used as a resource, directed to other crops (figure 1.13), or recycled. Thoughtful irrigation design and application minimizes the amount of water used, provides for the needs of plants, and deals with runoff appropriately, as discussed in Chapter 10, *Water Quality and Irrigation*. What other risks can be precluded by good design? In areas at risk of high winds, making use of a natural windbreak (figure 1.14) or having the ability to quickly

remove the plastic from the roof of a greenhouse may save a structure in a bad storm. A backup water supply ensures crop survival through periods of drought or uncertainty. Firebreaks or a site selected to minimize fire risks can preclude disaster. Knowing the site and thinking far into the future will help in planning for contingencies and increase the likelihood of long-term success. In short, think about the "big picture" during the planning process and build a strong foundation for producing high-quality plants efficiently.

PLANNING AS AN ONGOING PROCESS

The initial planning phase is a crucial part of successful nursery development, but the planning process does not stop after the nursery is operational. Instead, planning is an ongoing process. See Chapter 16, *Nursery Management*, for more discussion about this topic. The vision of the nursery should be revisited regularly. Time should be set aside to assess the progress the nursery is making in fulfilling its objectives, visualizing new possibilities, and adapting to changing circumstances. Following up with clients and revising the target plant specifications are essential steps for building a good nursery.

SUMMARY

No standard blueprint for designing a native plant nursery exists. On the contrary, each nursery will have a unique design based on distinct needs, resources, and requirements. Many factors go into the planning process, from defining the nursery vision and objectives to determining the kinds of outplanting materials that best meet the needs of the community and the outplanting sites. Strategic and tactical planning is necessary to develop a successful nursery efficiently. Starting with a small pilot nursery is a valuable step toward gaining a good grasp of all phases of plant production and delivery. The successes of the pilot phase can then be expanded as the nursery grows in size. A levelheaded assessment of the feasibility of running a nursery is important to carry out before launching into full-scale production.

The importance of the planning phase cannot be overemphasized. Resist the urge to start development on the ground until the big-picture strategy of the nursery operation is understood. A nursery is a web of interrelated factors, and every aspect of the nursery will affect everything else. Nursery development will be centered on the production of target plant materials. Proper site selection; design of propagation structures; and choice of containers, benches, growing media, irrigation systems, and so on will all be guided by target plant criteria. All these factors will be combined to produce the best quality plants in the least time at an acceptable cost. After production is under way, management practices, such as fertilization and scheduling, are also shaped by target plant criteria.

It is important, however, not to become so focused on planning crop production that other key aspects of nursery design are neglected. Space and time for "before" and "after" crop production must be planned: storage for seeds and propagation materials, overwintering and storage for crops, cleaning and sterilizing containers, shipping and delivery practices, and so forth. Equally important, time for training and education for nursery staff, outreach and education to clients and the public, and in-house research and trials to improve plant productivity should be planned.

The chapters in this handbook discuss each of these aspects of nursery design to give you an overview of the factors to consider when planning your nursery. Ultimately, the design of a nursery is personal, because you are the person who can best understand the unique needs and resources of your nursery and its crops, your community, and the environment around you.

Figure 1.13—*Good planning means making efficient use of resources and space. This small greenhouse at the Blackfeet High School in Browning, Montana, saves water and labor by growing sedges and wetland grasses on benches beneath those for dryland prairie forbs.* Photo by Tara Luna.

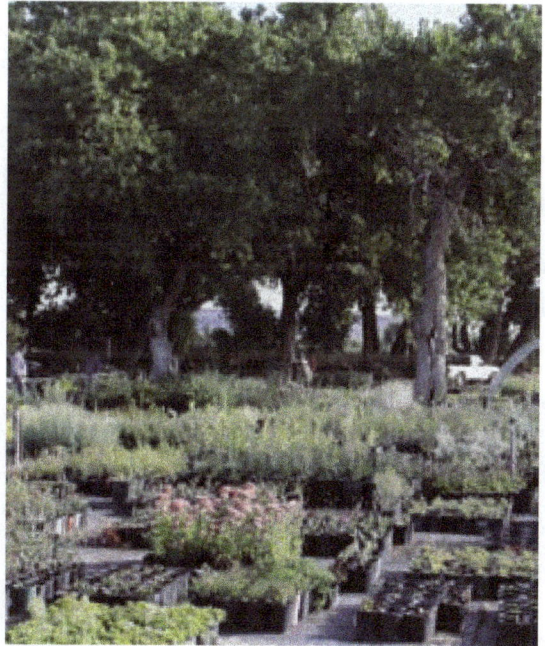

Figure 1.14—*The Santa Ana Pueblo nursery uses natural windbreaks to reduce water use and add protection to the nursery site. Planning for local environmental conditions and for risk is essential during the process of nursery design.* Photo by R. Kasten Dumroese.

LITERATURE CITED

Landis, T.D.; Tinus, R.W.; McDonald, S.E.; Barnett, J.P. 1994. The container tree manual: volume 1, nursery planning, development, and management. Agriculture Handbook 674. Washington, DC: U.S. Department of Agriculture, Forest Service. 188 p.

The Target Plant Concept
Thomas D. Landis

2

The first native plant nurseries in North America were gardens of plants transplanted from the wild by indigenous people. Specific plants were irrigated and otherwise cultured in these gardens to produce seeds, leaves, roots, or other desirable products (figure 2.1). As native people collected seeds from the largest or most productive plants, they were making the first genetic selections that resulted in new cultivated varieties. Plants that could not be easily domesticated were cultured at their natural sites by pruning to increase seed or fruit production.

Among the first contemporary native plant nurseries were forest tree nurseries that were established in the early 1900s. The objective of the nurseries was to reforest and restore forests and protect watersheds after timber exploitation or forest fires. The process was very simple: nurseries produced seedlings that were shipped for outplanting. Foresters took what they got without much choice. Tree planting was a mechanical process of getting seedlings into the ground in the quickest and least expensive manner. Not much thought was given to seedling quality, different stock types, or the possibility of matching seedlings to outplanting site conditions.

THE TARGET PLANT CONCEPT

The Target Plant Concept is a new way of looking at nurseries and the uses of native plants. After working with forest and native plant nurseries for almost

William Pink of the Pechanga Band of the Luiseno Indians in California by Kate Kramer.

Figure 2.1—*Native Americans cultured many native plants, such as camas, for nutritional, medicinal, and other cultural uses.* Image PG38-1062 courtesy of Special Collections and Archives, University of Idaho Library, Moscow, Idaho.

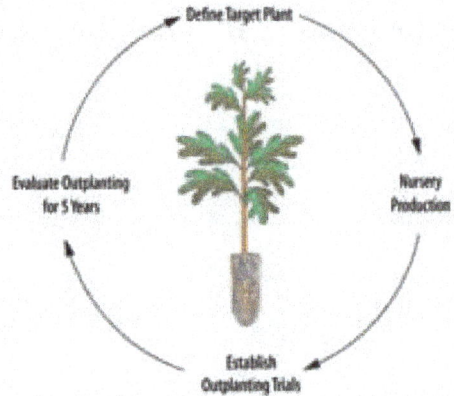

Figure 2.2—*Information from the client defines the initial target plant and outplanting performance then fine-tunes these specifications.* Illustration by Jim Marin.

30 years, I believe this concept is one of the critical aspects to understand when starting a new nursery or upgrading an existing one. The Target Plant Concept is based on three simple but often overlooked ideas.

1. Start at the Outplanting Site

Without the concept, a nursery produces a crop of plants that are provided to clients. In this one-way system, clients have little control over the type or quality of plants they receive. With the concept, the process is approached in a completely different manner: starting with the characteristics of the outplanting site, clients specify exactly what type of plant material would be best.

2. The Nursery and Client Are Partners

Without the concept, clients seek plants solely by price or availability. With the concept, the client specifies the ideal type of plant for the project, the nursery grows the plants, and they are outplanted. Based on performance of this first crop, the client and nursery manager work together to make necessary changes to improve survival and growth. Using these revised target plant characteristics, the nursery grows another crop that is again evaluated on the project site (figure 2.2). This feedback system fosters good communication between native plant customers and nursery managers, builds partnerships, and ensures the best possible plants for the project.

3. Emphasis on Seedling Quality

Without the concept of quality, inexperienced growers and clients think there are cheap, all-purpose native plants that will thrive just about anywhere. With the concept, it is clear that *plant quality cannot be described at the nursery, but can only be proven on the outplanting site.* A beautiful crop of plants in the nursery may perform miserably if they are inappropriate for conditions on the outplanting site.

DEFINING TARGET PLANT MATERIALS

The Target Plant Concept consists of six sequential but interrelated steps, which are illustrated in figure 2.3 and described below.

Step 1 — What Are the Project Objectives?

This step may seem obvious but it is all too often overlooked. Exactly what is trying to be accomplished? Native plants are grown for a variety of reasons and project objectives have an overriding effect on the best types of plants to produce. Native American tribes want native plants for many purposes, including reforesting after timber harvest, ensuring local supplies of cultural or medicinal plants, restoring plant communities, controlling invasive species, creating wildlife habitat, producing native foods, educating young people, and developing small businesses.

A few tribes have productive forestland and grow seedlings of commercial trees to replant after timber harvest. For example, the Red Lake Band of the Chippewa

Indians grows red pine and eastern white pine in their greenhouse in Minnesota to reforest after timber harvest (figures 2.4A and B).

One of the most mentioned objectives for Native American tribes is to produce native plants having cultural or spiritual significance. Often, these plants are getting harder to find and grow only in remote locations that are difficult to access, especially for elders. For example, many Indians use sweetgrass in their ceremonies. Unfortunately, some do not have access to this species, so tribes could grow sweetgrass in their nursery or purchase it from another tribe. The Confederated Tribes of the Umatilla Indian Reservation use tule for floor mats and other utilitarian purposes (figure 2.5A). Basketry is an important cultural heritage and Native Americans use many different native plants for basket materials. The Mohawk and other tribes of the Northeastern United States use black ash (Benedict and David 2003), whereas the Pechanga Band of the Luiseno Indians of southern California use dogbane. At the Redwood Valley Rancheria in northern California, white-root sedge was propagated and then outplanted in gardens to make collecting for baskets much easier and more convenient (figure 2.5B). Establishing these cultural plant gardens also protects wild plants from the stresses of overcollection.

Healing the earth is a subject that is very important to Native Americans. Restoring plant communities along streams and rivers is particularly important because many have been damaged by overgrazing, and several tribes have expressed an interest in growing native

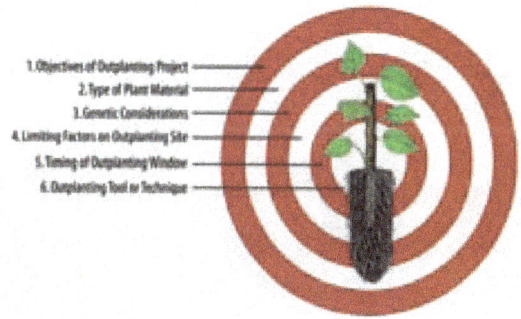

Figure 2.3—*The process of defining the target plant materials for a specific outplanting site consists of six steps.* Illustration by Jim Marin.

Figure 2.4—*(A) The Red Lake Band of Chippewa Indians grow commercial conifers in their greenhouse. (B) The conifers are used to reforest lands after timber harvest.* Photos by Ronald Overton.

Figure 2.5—*(A) Damon McKay of the Confederated Tribes of the Umatilla Indian Reservation in Oregon collects tule for a variety of purposes. (B) Women collecting roots of white-root sedge for basket making at the Redwood Valley Rancheria in California.* Photo A by Tara Luna, B by Chuck Williams.

Figure 2.6— *(A) Dawn Thomas-Swaney collects hardwood willow cuttings. (B) The cuttings are used by the Confederated Salish and Kootenai Tribes for riparian restoration projects along the Flathead and Jocko rivers in Montana.* Photo A by Joyce Lapp, B by Dawn Thomas.

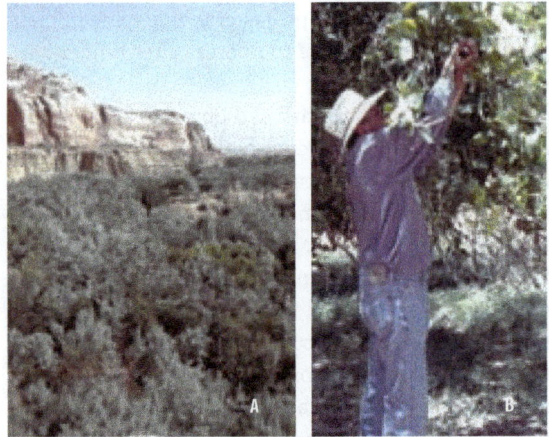

Figure 2.7— *(A) Many riparian areas on the Hopi Reservation in Arizona are overgrown with the invasive weed trees Russian-olive and saltcedar. (B) Chuck Adams of the Hopi tribe is collecting seeds of native cottonwoods and willows to grow plants that will be used to reintroduce these natives after the exotics have been removed.* Photos by Thomas D. Landis.

plants to restore these communities. The Confederated Salish and Kootenai Tribes of Montana have established a native plant nursery to produce rooted cuttings and seedlings for 200 acres (80 ha) of riparian restoration projects along the Flathead and Jocko rivers (figure 2.6).

One common use of native plants is for revegetation after the removal of invasive weeds. On the Hopi Reservation in northern Arizona, riparian areas have been overtaken by Russian-olive and saltcedar. These aggressive woody exotics have replaced the native willows, cottonwoods, and other riparian plants along streams,

washes, and springs. In some locations, these invasive trees have grown so thick that the tribe's cattle and horses cannot pass through them (figure 2.7A). The Hopi are collecting seeds and cuttings of native cottonwoods and willows (figure 2.7B) so that plants can be grown for outplanting after the removal of the Russian-olive and saltcedar.

Many tribes have voiced concern about the loss of plants and animals because of habitat degradation. Native plants can be used to restore or create new habitat on which other organisms depend. A good example is a project on the Paiute Reservation to restore wetlands that are critical to the survival of the northern leopard frog. Native wetland plants were grown in a nursery and outplanted (figure 2.8A), and, after only a few years, a viable wetland community was reestablished (figure 2.8B).

From both a cultural and health standpoint, Native Americans are trying to reintroduce traditional foods into their diets. Historically, native plants were a principal food source (see figure 2.1). Besides being the State flower, bitterroot (figure 2.9A) was an important food staple of tribes in Montana. After being dug, cleaned, and dried, the root provided a lightweight, nutritious food that could be stored. Tremendous potential exists for tribes to grow other traditional food plants in nurseries for outplanting into food gardens.

In the Pacific Northwest region of the United States, salmon forms the central part of the diet of many tribes and much of the recent decline in salmon runs

Figure 2.8—*(A) To create breeding habitat for the northern leopard frog, the Paiute Tribe of Nevada grew wetland plants for outplanting on degraded riparian areas. (B) Native plants quickly improved the habitat.* Photos by J. Chris Hoag.

can be attributed to habitat destruction. The Stillaguamish Tribe of Washington State has started a native plant nursery to produce stock for restoring streams and rivers that are essential to the survival of young salmon and steelhead (figure 2.9B).

One of the most rewarding objectives for growing and outplanting native plants is the education of young people. Many tribes have expressed interest in starting a native plant nursery with a primary objective of environmental education. The Blackfeet Nation and Confederated Salish and Kootenai Tribes have nurseries as part of their schools and use them to teach young people the names and uses of native plants. The Hopi have established a cultural plant propagation center at the Moencopi Day School in Tuba City, Arizona (figure 2.10). Science teachers at the school use the greenhouse in their classes. The plants are used for many purposes including restoration projects on surrounding Hopi and Navajo lands.

Of course, one objective of a native plant nursery is to sell plants for profit and some tribes are doing this. The Santa Ana Pueblo of New Mexico grows a variety of desert native plants in containers suitable for landscaping (figure 2.11). This practice would be most appropriate for tribes close to large population centers but a market for selling to tourists would be possible in any location.

Figure 2.9—*(A) Traditional food plants such as bitterroot could be produced in native plant nurseries. (B) Plants are also needed to help restore the riparian habitat that is critical for salmon, a cultural food staple of many tribes of the Northwestern United States.* Photo A by Tara Luna; B by Jeremy R. Pinto.

Figure 2.10—*One of the most exciting uses of native plants is environmental education. Discussing the tribal plant names and uses is an excellent way to get young people enthused about nature. Many tribes have located their nurseries at schools to foster this education. Steven Lomadafkie teaches children at Moencopi Day School on the Hopi Reservation in Arizona.* Photo by Tara Luna.

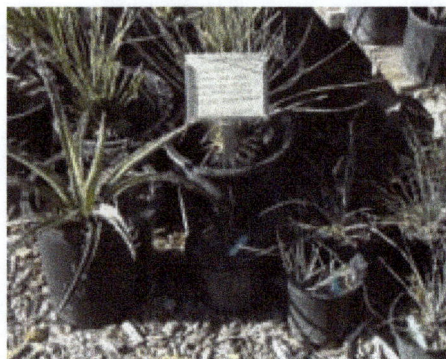

Figure 2.11—*The Santa Ana Pueblo of New Mexico operates a retail native plant nursery where it sells native plants in large containers for landscaping and other ornamental purposes.* Photo by Jeremy R. Pinto.

Step 2 — What Type of Plant Material Will Meet Project Objectives and Site Characteristics?

After the best plant species have been determined, the next decision is what type of plant material would best meet project objectives. The term "plant material" is commonly used to describe any plant part that can be used to establish new plants on the project site. Common plant materials include seeds, root stock, and nonrooted cuttings as well as traditional nursery stock types, such as rooted cuttings, bareroot seedlings, and container seedlings (table 2.1). Native plant nurseries are currently growing, or could grow, any or all of these categories of plant materials.

Seeds are an ideal type of plant material that are easy to handle, store, and outplant (figure 2.12A). The effectiveness of direct seeding on the project site varies with the species of plants, the harshness of the site, the objectives of the project, and the project timeframe. Directly broadcasting seeds offers three principal advantages: (1) seeds are inexpensive compared with other plant materials, (2) spreading seeds is relatively easy, and (3) seedlings from broadcast seeds develop a natural root system (table 2.1). Many drawbacks, however, exist as well. Native plant seeds from the proper species and origin are often difficult to obtain or are very expensive; some species do not produce adequate seed crops each year; and seeds of other species, such as the white oaks, do not store well.

Seeds of many diverse species require special cleaning and processing before they can be sown. Even if the proper seeds can be obtained and properly distributed over the site, predation by birds and rodents, competition from weed species, and unpredictable weather often reduce establishment success (Bean and others 2004). Finally, with direct seeding, it is difficult to control species composition and plant spacing over the project area (Landis and others 1993).

Direct seeding is most successful for grasses, forbs, and some woody shrubs, the seeds of which can be easily produced in bareroot beds in nurseries (figure 2.12B). Seeding with native grass species after wildfires is standard procedure to stabilize soils and prevent erosion (figure 2.12C). In California, the direct seeding of native oaks has been quite successful and the Department of Fish and Game has direct seeded cosmopolitan bulrush for the restoration of wetland wildlife habitat in the Sacramento River Delta (Landis and others 1993). Wild rice is an important food and cultural plant for the Ojibwa and other tribes from the northern Great Lakes region and is traditionally propagated by seeds (Luna 2000).

Root stock refers to specialized roots, such as bulbs and corms, and to modified underground stems, such as rhizomes and tubers (figure 2.13). Root stock can be used for the vegetative propagation of certain grasses and wetland plants. Grass and sedge rhizomes and root sections have been successfully used for wildland outplantings, such as a restoration project at Jepson Prairie in California (Landis and others 1993). Because of difficulties with seed dormancy, the Mason State

THE TARGET PLANT CONCEPT

Table 2.1—Many different types of native plant materials can be grown in nurseries

Plant Materials	Examples	Characteristics
Seeds	Wild rice	Small and easy to outplant Seeds of many native plants can be stored for long periods Plants develop natural root structure
Root stock	Camas	Can be stored under refrigeration Excellent survival after outplanting
Nonrooted cuttings	Willows Wormwood	Used for live stakes and structures for soil stabilization along streams Efficiently and economically produced in nursery stooling beds or from container stock plants
Rooted cuttings	Willows Cottonwoods Redosier dogwood	Can be grown in a variety of different container types and sizes Good option when seeds are unavailable or have complex dormancy
Bareroot seedlings	Most species, including: Ponderosa pine Turkey oak Bitterbrush	Nurseries require good soil Efficient way to produce large numbers of plants Not as practical for small native plant nurseries Shorter lifting window Store best under refrigeration
Container seedlings	All species	Require high quality and pure seeds Nurseries can be on harsher soils and climates Use artificial growing media Handling and storage are less demanding

Nursery in Illinois produces rooted cuttings and root divisions of several species of prairie forb, woodland understory, and wetland plants (Pequignot 1993).

Many riparian and wetland species can be successfully propagated on the project site by collecting cuttings and planting them immediately without roots (nonrooted cuttings). Under ideal conditions, planting nonrooted cuttings can be a very cost-effective means for establishing certain vegetation types. Nonrooted cuttings are prepared from long whips collected from dormant shrubs or trees on the project site or from stock plants at a nursery. If a large number of cuttings will be needed for several years, plants can be established at a local nursery and cuttings collected each year. Whips should be collected during the dormant season when the potential for new root formation is highest. Whips are cut into cuttings that range from 12 to 24 in (30 to 61 cm) in length and 0.4 to 0.75 in (10 to 19 mm) in diameter (figure 2.14A). When outplanted properly in moist soil and under favorable conditions, these cuttings will form new roots that follow the receding water table down as the young plant develops during the first growing season. Nonrooted cuttings of willow or cottonwood are often used as "live stakes" in riparian restoration projects (figure 2.14B). Often, however, nonrooted cuttings initially produce leaves but eventually die because of moisture stress or canker diseases.

Pole cuttings are an interesting type of nonrooted cutting that are sometimes used in riparian restoration projects (Hoag and Landis 2001). These cuttings are

Figure 2.12—*(A) Seeds are the best plant material type for many grasses and forbs. (B) Native grass production fields. (C) These fields produce source-identified seeds that can be applied immediately after wildfires to stop soil erosion.* Photos by Thomas D. Landis.

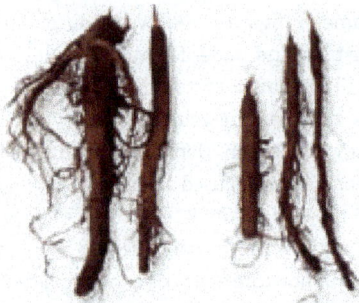

Figure 2.13—*An uncommon type of plant material, known as root stock, is used to establish some grasses, sedges, forbs, and wetland plants that cannot be direct seeded or outplanted as seedlings.* Photo by Thomas D. Landis.

often 6 ft (1.8 m) in length and 8 to 12 in (20 to 30 cm) in diameter and are obtained by cutting the major branches or stems of cottonwood or willow trees (figure 2.14C). The key to success is to outplant the poles deep enough so that the butt ends remain in contact with the water table. Pole cuttings are very effective in stabilizing stream or riverbanks because they resist erosion (figure 2.14D). When large numbers of poles are required, they should be grown in stooling beds in nurseries to avoid the negative impact of collecting from wild "donor" plants on the project site.

Often, it is more effective to root cuttings in a nursery before outplanting them on the project site. The type and size of cuttings used in nursery propagation to produce **rooted cuttings** is much different from those used as nonrooted cuttings. A much shorter stem section can be used (2 to 4 in [5 to 10 cm]; figures 2.15A and B) but it should have a healthy bud near the top (Dumroese and others 2003). See Chapter 9, *Vegetative Propagation*, for more information about collecting and culturing rooted cuttings.

Bareroot seedlings are grown in the ground and harvested without soil around their roots (figure 2.16). Because they require a considerable amount of high-quality soil and often take longer to reach shippable size, fewer species of native plants are grown as bareroot stock. One serious drawback of bareroot stock is that seedlings need much better storage and postharvest care than container stock does.

Container seedlings are a newer stock type than bareroot and continue to increase in popularity. Many different types of containers are being used, and all require artificial growing medium. The distinguishing feature of container seedlings is that, because the roots are restricted, they bind growing media into a cohesive "plug" (figure 2.17A). "Single-cell" containers are more popular for growing native plants than "block" containers because individual seedlings can be sorted or consolidated (see Chapter 6, *Containers*).

Container seedlings are the stock type of choice for most tribal nurseries because of lower land requirements and startup costs. When small amounts of many different native plants are desired, container propagation is the best option. Another advantage is that container stock is more tolerant and hardy during handling, shipping, storage, and outplanting. Compared with bareroot stock, container seedlings can be

Figure 2.14—*(A) Nonrooted cuttings of willows and cottonwoods. (B) These cuttings are a popular and inexpensive type of plant material that is used for live stakes in riparian restoration. (C) Poles are very large cuttings. (D) Poles are used along stream and riverbanks where water erosion would damage conventional nursery stock.* Photo by Thomas D. Landis, illustrations by Steve Morrison.

harvested at almost any time of the year, which creates a wider outplanting window. See Chapter 13, *Harvesting, Storing, and Shipping*, for further discussion about this topic. The compact root systems of container plants make outplanting easier, especially on harsh sites, and the cylindrical plugs offer more surface area for root egress (figure 2.17B).

Step 3 — What about Genetic or Sexual Considerations?

The third component of the Target Plant Concept concerns the question of genetic and sexual diversity.

Local Adaptation

Native plants have a "sense of place" and so, when collecting seeds, cuttings, or other plant materials, it is important to identify their origin. We know that plants are genetically adapted to local environmental conditions and, for that reason, plant materials should always be collected within the same area where the plants will be outplanted. "Seed zone" and "seed source" are terms that foresters use to identify their seed collections. A seed zone is a three-dimensional geographic area that has relatively similar climate and soil type (figures 2.18A and B). Native plant nurseries grow plants by seedlots and might have several lots of the same species that they identify and keep separate.

Local adaptation can affect outplanting survival and growth in a couple of ways: growth rate and cold tolerance. In general, plants grown from seeds or cuttings collected from higher latitudes or elevations will grow

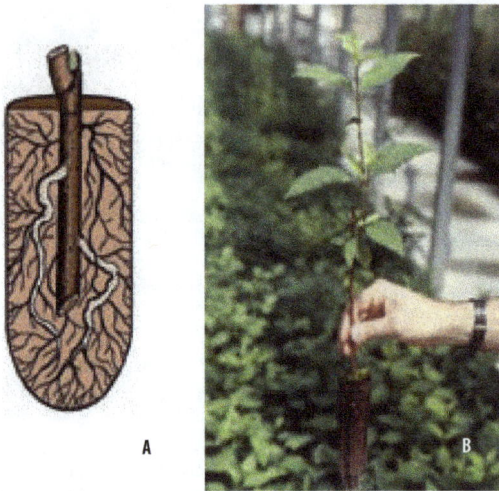

Figure 2.15—*(A) Rooted cuttings use a shorter section of stem with a bud. (B) Cuttings quickly grow into large plants under nursery culture, such as this redosier dogwood.* Illustration by Steve Morrison, photo by Thomas D. Landis.

Figure 2.16—*Bareroot seedlings are grown in raised seedbeds in fields and are harvested and shipped with no soil surrounding the roots. Because of the many limitations of bareroot stock, tribes prefer to grow their plants in containers.* Photo by Thomas D. Landis.

slower but tend to have more cold hardiness during winter than those collected from lower elevations or more southerly latitudes. This concept has been proven with research trials on commercial conifers (St. Clair and Johnson 2003). Although tests have not been done on other native plants, it makes sense that the same concepts may apply. Therefore, always collect plant materials from the same geographic area and elevation in which the nursery stock is to be outplanted.

Genetic and Sexual Diversity

Target plant materials should also represent all the genetic and sexual diversity present on the outplanting site. To maximize genetic diversity in the resultant plants, seeds should be collected from as many plants as possible. The same principal applies to plants that must be propagated vegetatively. Cuttings must be collected on or near the outplanting site to make sure they are properly adapted. On restoration projects with widely separated sites, be sure to collect plant materials from each location to ensure a good mix of genetic attributes (figure 2.19). Of course, collecting costs must be kept within reason and so the number of collections will always be a compromise. Guinon (1993) provides an excellent discussion of all factors involved in preserving biodiversity when collecting plant materials, and he suggests a general guideline of 50 to 100 donor plants. See Chapter 7, *Collecting, Processing, and Storing Seeds*, for a complete discussion on this topic.

Dioecious species, such as willows and cottonwoods, are challenging because they have male and female plants. Therefore, all vegetatively propagated plants will be the same sex as their parent (figure 2.19), which can be particularly important on sites where populations of plants are geographically separated. Therefore, when collecting cuttings at the project site, care must be taken to ensure that both male and female plants are equally represented. Rooted cuttings or other vegetatively propagated stock types should be labeled by sex so that males and females can be outplanted in a mixed pattern to promote seed production.

Step 4 — What Factors on the Project Site Could Limit Survival and Growth?

The fourth aspect of the Target Plant Concept is based on the ecological "principle of limiting factors" that states that any biological process will be limited

A

B

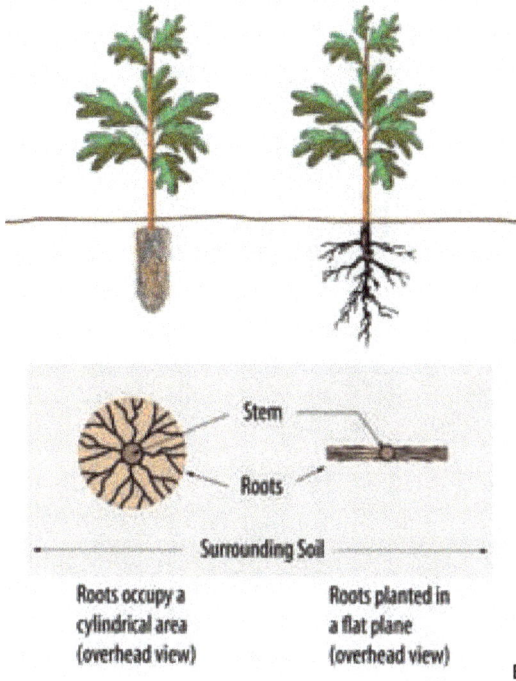

Figure 2.17—*(A) Container plants come in many types and sizes but all form their roots into a plug. (B)This tendency allows container stock to be harvested and outplanted with less disturbance to the root system and increases the area of root-to-soil contact.* Illustrations by Jim Marin.

A

B

Figure 2.18—*(A) In forest management, seed zones are used to make certain that seeds are collected from the same geographic area and elevation zone. (B) The size of seed zones varies with plant species. For example, in western Oregon, Douglas-fir seed zones are smaller and more numerous compared with western redcedar.* Modified from St.Clair and Johnson (2003).

by the factor present in the least amount. Therefore, each outplanting site should be evaluated to identify the environmental factors that are most limiting to survival and growth (figure 2.20A). Foresters do this procedure when they write "prescriptions" for each harvest unit specifying which tree species and stock type would be most appropriate. This same procedure should also be done for any project using native plants.

On most outplanting sites, native plants must quickly establish root contact with the surrounding soil to obtain enough water to survive and grow. For example, water is usually the most limiting factor on southwest slopes where sunlight is intense (figure 2.20B). On northern aspects or at higher elevations or latitudes, however, cold soil temperatures may be more limiting (figure 2.20C). On these more shaded sites, melting snow keeps soil temperatures cold and research has shown that root growth is restricted below 50 °F (10 °C). Therefore, a reasonable target plant for these sites would be grown in a relatively short container to take advantage of warmer surface soils (Landis 1999).

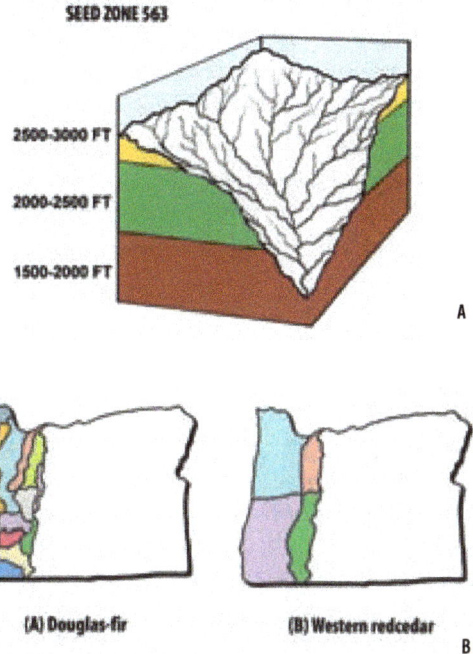

Restoration sites pose interesting challenges when evaluating limiting factors. For example, after a wildland fire, soil conditions are often severely altered and mining sites have extreme soil pH levels. Riparian restoration projects require bioengineering structures to stabilize streambanks and retard soil erosion before the site can be planted with native species (Hoag and Landis 2001). In desert restoration, low soil moisture, hot temperatures, high winds with sand blast, and heavy grazing pressure have been listed as limiting factors (Bainbridge and others 1992). Where populations of deer or other browsing animals are high, animals may be the most limiting factor, and nursery plants may have to be protected with netting or fencing.

One limiting factor deserves special mention—mycorrhizal fungi. These symbiotic organisms provide their host seedling with many benefits, including better water and mineral nutrient uptake, in exchange for food produced by the host plant. Reforestation sites typically have an adequate complement of mycorrhizal fungi that quickly infest outplanted seedlings,

A

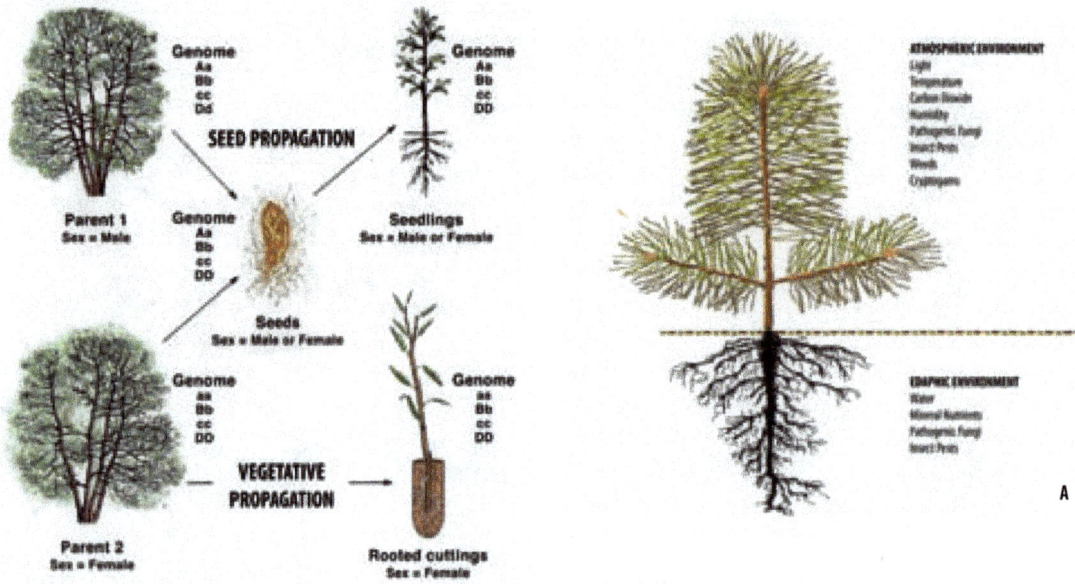

Figure 2.19—*When collecting cuttings from dioecious plants, such as willows and cottonwoods, the sex of the parent plant must be considered to ensure a mixture of both males and females.* From Landis and others (2003).

whereas many restoration sites do not. For example, severe forest fires or surface mining eliminate all soil microorganisms including mycorrhizal fungi. Therefore, plants destined for severely altered sites should be inoculated with the appropriate fungal symbiont before outplanting. See Chapter 14, *Beneficial Microorganisms*, for a complete discussion on this topic.

Step 5 — What Is the Best Season for Outplanting? The "Outplanting Window"

Conditions on most native plant project sites are harsh and nursery stock often suffers severe "transplant shock" (figure 2.21A). Each site has an ideal time when chances for native plant survival and growth are greatest; this is known as the "outplanting window." This time period is usually defined by the limiting factors discussed in the previous section, and soil moisture and temperature are the usual constraints. In most of the continental United States, nursery stock is outplanted during the rains of winter or early spring when soil moisture is high and evapotranspirational losses are low (figure 2.21B).

One real advantage of container plants is that they can be started at different dates and then cultured to

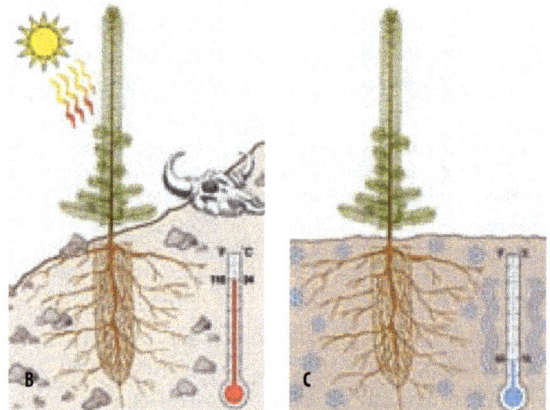

Figure 2.20—*(A) Outplanting sites should be characterized by which environmental factors are most limiting to plant growth. (B) On a sandy hillside with a southern exposure, water is most limiting. (C) At high elevations and latitudes, cold soil temperature restricts root growth.* Illustrations by Jim Marin.

be physiologically conditioned for outplanting during various times of the year. For the traditional midwinter outplanting window, nursery stock can be harvested and stored until the outplanting site is accessible. In recent years, a renewed interest in autumn outplanting has developed because of the availability of specially conditioned container stock. Summer outplanting is a relatively new practice that has been developed in the boreal regions of Canada (Revel and others 1990) and has since found some application at high-elevation

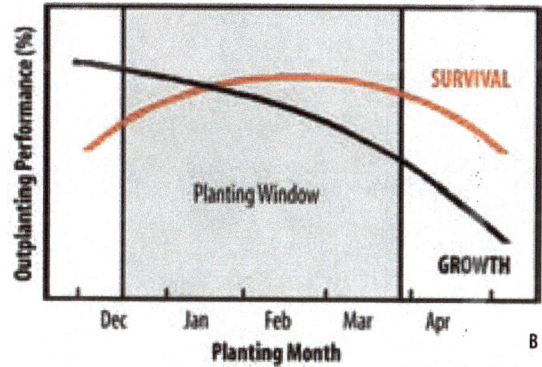

Figure 2.21—*(A) Transplant shock. (B) One way to minimize the chances of transplant shock is to identify the outplanting window, the period of time in which site conditions are least stressful. In much of the United States, the outplanting window occurs during the rainy period of midwinter.* Photo A by Thomas D. Landis, B modified from South and Mexal (1984).

sites in the Rocky Mountains (Page-Dumroese and others, in press). Target plant characteristics are similar for both summer and autumn outplanting: hardened container stock with minimal handling and storage.

Step 6 — What Are the Best Outplanting Tools and Techniques for the Project Site?

Each outplanting site is different in terms of climate and soil type, so tools and outplanting techniques must be considered in the Target Plant Concept. All too often, foresters or restoration specialists develop a preference for a particular implement because it has worked well in the past. No single tool or technique, however, will work well under all site conditions. Hand tools such as shovels and planting hoes ("hoedads" in figure 2.22A) have been very popular for outplanting native plants on a variety of sites. In large planting projects or projects planted by inexperienced planters, it may be best to have one person locate the planting spot and excavate a hole with a power auger (figure 2.22B). Be careful when selecting planting tools and ask other project managers for their opinions about planting tools if you do not have hands-on experience. The dibble was developed as an easy and quick way to outplant container seedlings. Experience has shown that dibbles work reasonably well on sandy soils, but, in silt or clay soils, dibbles compact soil around the outside of the hole, which inhibits root egress. Dibbles are ideal

for outplanting wetland plants because they work well in standing or running water. Kloetzel (2004) does an excellent job of discussing the various outplanting tools that are available and the site conditions that are best for each tool.

Different types of native plant materials may require specialized outplanting equipment. Nonrooted cuttings—even poles—can be successfully planted with specialized equipment such as the waterjet stinger (Hoag and others 2001). The "tall pots" used in many restoration projects require specialized outplanting equipment. Again, nursery managers must work closely with reforestation or restoration project managers to make certain that their target plants can be properly outplanted in the soil conditions on the project site.

The type of outplanting tool must be given special consideration when working with volunteers or other inexperienced planters. Many of these people lack the skills or strength to properly plant native plants on wildland sites. One option is to have a professional excavate planting holes with a machine auger (figure 2.22B) and let volunteers insert plants and tamp them into place. This technique has several benefits: the professional chooses the proper planting spot, creates the desired pattern, and makes certain the planting holes are large and deep enough so that the seedlings can be planted properly.

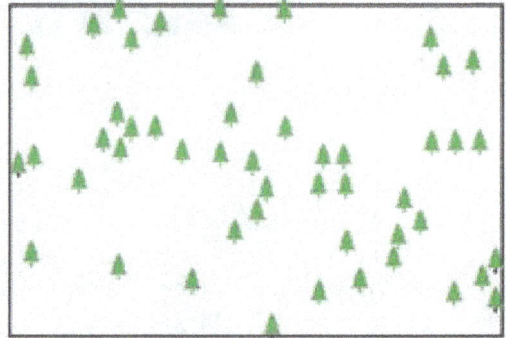

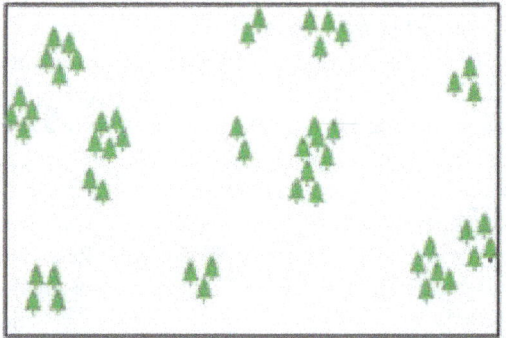

Figure 2.22—*(A) The type of outplanting tool and planting design are critical to the Target Plant Concept. Traditional tools such as the hoedad are designed for steep slopes or rocky soils. (B) Power augers work well on level sites and when working with groups. (C) Most restoration projects do not want the "cornfield" look. (D) To prevent this look, plants are spaced in more random patterns, (E) or in clumped patterns that mimic natural conditions.*
Photos by Thomas D. Landis, illustrations by Jim Marin.

The pattern and spacing of outplanted nursery stock should reflect project objectives. In commercial forestry projects where timber production is the primary objective, plants are outplanted in a regularly spaced pattern (figure 2.22C). A similar pattern is used for Christmas tree plantations, where tree growth and form are main concerns. Where ecological restoration is the objective, however, outplanting plants randomly (figure 2.22D) or in random groups (figure 2.22E) will result in more natural vegetation patterns. Again, project objectives must be considered. Regular spacing is required when quick crown closure is desired to shade out invasive species or when machinery access is needed for controlling weeds after outplanting (Davenport 2006).

For planning purposes, nursery managers must know in advance which planting tools will be used so they can develop proper plant material specifications, especially root length and volume or cutting length and diameter.

SUMMARY AND RECOMMENDATIONS

The Target Plant Concept is a way to think through the entire nursery and outplanting process and to encourage communication between clients and nursery managers. Describing an ideal plant for a particular restoration project and following a series of sequential steps will also be a useful exercise for native plant nurseries and plant users. Instead of the traditional linear process that begins in the nursery and ends on the outplanting site, the Target Plant Concept is a circular feedback system in which information from the project site is used to define and refine the best type of plant material.

The Target Plant Concept is not static, it must be continually updated and improved. At the start of the project, the supervisor and nursery manager must agree on certain specifications. These prototype target plant materials are then verified by outplanting trials in which survival and growth are monitored for up to 5 years. The first few months are critical because plant materials that die immediately after outplanting indicate a problem with stock quality. Plants that survive initially but gradually lose vigor indicate poor planting or drought conditions. Therefore, plots must be monitored during and at the end of the first year for initial survival. Subsequent checks after 3 or 5 years will give a good indication of growth potential. This performance information is then used to give valuable feedback to the nursery manager, who can fine-tune target specifications for the next crop.

LITERATURE CITED

Bainbridge, D.A.; Sorensen, N.; Virginia, R.A. 1992. Revegetating desert plant communities. In: Landis, T.D., tech. coord. Proceedings, Western Forest Nursery Association. General Technical Report RM-221. Ft. Collins, CO: U.S. Department of Agriculture, Forest Service, Rocky Mountain Forest and Range Experiment Station: 21-26.

Bean, T.M.; Smith, S.E.; Karpiscak, M.M. 2004. Intensive revegetation in Arizona's hot desert: the advantages of container stock. Native Plants Journal 5: 173-180.

Benedict, L.; David, R. 2003. Propagation protocol for black ash (*Fraxinus nigra* Marsh.). Native Plants Journal 4: 100-103.

Davenport, R. 2006. Knotweed control and experiences restoring native plants in the Pacific Northwest. Native Plants Journal 7: 20-26.

Dumroese, R.K.; Wenny, D.L.; Morrison, S.L. 2003. A technique for using small cuttings to grow poplars and willows in containers. Native Plants Journal 4: 137-139.

Guinon, M. 1993. Promoting gene conservation through seed and plant procurement. In: Landis, T.D., tech. coord. Proceedings, Western Forest Nursery Association. General Technical Report. RM-221. Ft. Collins, CO: U.S. Department of Agriculture, Forest Service, Rocky Mountain Forest and Range Experiment Station: 38-46.

Hoag, J.C.; Landis, T.D. 2001. Riparian zone restoration: field requirements and nursery opportunities. Native Plants Journal 2: 30-35.

Hoag, J.C.; Simonson, B.; Cornforth, B.; St. John, L. 2001. Waterjet stinger: a tool for planting dormant nonrooted cuttings. Native Plants Journal 2: 84-89.

Kloetzel, S. 2004. Revegetation and restoration planting tools: an in-the-field perspective. Native Plants Journal 5: 34-42.

Landis, T.D. 1999. Seedling stock types for outplanting in Alaska. In: Stocking standards and reforestation methods for Alaska. Fairbanks, AK: University of Alaska Fairbanks, Agricultural and Forestry Experiment Station, Misc. Pub. 99-8: 78-84.

Landis, T.D.; Dreesen, D.R.; Dumroese, R.K. 2003. Sex and the single *Salix*: considerations for riparian restoration. Native Plants Journal 4: 110-117.

Landis, T.D.; Lippitt, L.A.; Evans, J.M. 1993. Biodiversity and ecosystem management: the role of forest and conservation nurseries. In: Landis, T.D., ed. Proceedings, Western Forest Nursery Association. General Technical Report RM-221. Fort. Collins, CO: U.S. Department of Agriculture, Forest Service, Rocky Mountain Forest and Range Experiment Station: 1-17.

Luna, T. 2000. Propagation protocol for wild rice (*Zizania palustris*). Native Plants Journal 1: 104-105.

Page-Dumroese, D.S.; Dumroese, R.K.; Jurgensen, M.F.; Abbott, A.; Hensiek, J. 2008. Effect of nursery storage and site preparation techniques on field performance of high-elevation *Pinus contorta* seedlings. Forest Ecology and Management 256: 2065-2072.

Pequignot, S.A. 1993. Illinois—an example of how public nurseries can help meet the need for non-traditional plant materials. In: Landis, T.D., ed. Proceedings, Western Forest Nursery Association. General Technical Report RM-221. Fort. Collins, CO: U.S. Department of Agriculture, Forest Service, Rocky Mountain Forest and Range Experiment Station: 72-77.

Revel, J.; Lavender, D.P.; Charleson, L. 1990. Summer planting of white spruce and lodgepole pine seedlings. FRDA Report 145. Vancouver, BC: British Columbia Ministry of Forests; Forestry Canada. 14 p.

South, D.B.; Mexal, J.G. 1984. Growing the "best" seedling for reforestation success. Forestry Department Series 12. Auburn, AL: Auburn University. 11 p.

St. Clair, B.; Johnson, R. 2003. The structure of genetic variation and implications for the management of seed and planting stock. In: Riley, L.E.; Dumroese, R.K.; Landis, T.D., tech. coords. National proceedings: forest and conservation nursery associations—2003. Proceedings RMRS-P-33. Ogden, UT: U.S. Department of Agriculture, Forest Service, Rocky Mountain Research Station: 64-71.

APPENDIX 2.A. PLANTS MENTIONED IN THIS CHAPTER

bitterbrush, *Purshia* species

bitterroot, *Lewisia rediviva*

black ash, *Fraxinus nigra*

camas, *Camassia quamash*

cosmopolitan bulrush, *Schoenoplectus maritimus*

cottonwood, *Populus* species

dogbane, *Apocynum* species

Douglas-fir, *Pseudotsuga menziesii*

eastern white pine, *Pinus strobus*

oaks, *Quercus* species

ponderosa pine, *Pinus ponderosa*

red pine, *Pinus resinosa*

redosier dogwood, *Cornus sericea*

Russian-olive, *Elaeagnus angustifolia*

saltcedar, *Tamarix ramosissima*

sweetgrass, *Hierochloe* species

tule, *Schoenoplectus acutus*

turkey oak, *Quercus laevis*

western redcedar, *Thuja plicata*

white oaks, *Quercus* species

white-rooted sedge, *Carex barbarae*

wild rice, *Zizania palustris*

willow, *Salix* species

wormwood, *Artemisia* species

- Put PIPO, PICO, AMAL
  CAQU into STRAT
5- CLEAN GH, CHECK
  BOOMS
6- START FILLING
  160/90s FOR FEB 1
  SOW DATE - NEED 200
7- SAFETY MEETING 8 AM

  ORDER NEW GH PLASTIC

  CHECK BOKES OF LAOC
    FOR STORAGE MOLD
8- SEND CREW TO CUT
  ANOTHER 2500 SAEX
  CUTTINGS - PREPARE
  WORK PLAN FOR NEXT
  WEEK. LOOK AT NEXT
  MONTH'S PLAN

Planning Crops and Developing Propagation Protocols

Douglass F. Jacobs and Kim M. Wilkinson

3

Crop planning is an important but often neglected aspect of successful nursery management. Crop planning enables proper scheduling of the necessary time, materials, labor, and space to produce crops. Many painstaking details, such as the careful design of nursery facilities; working with clients; collecting and propagating seeds and cuttings; and making improvements in media, irrigation, fertilization, handling, and storage, go into good nursery operations. All the benefits associated with improvements in these areas, however, will not be realized without excellent crop planning. It is essential to plan crops so that high-quality plants can be delivered to clients at the agreed-on time (figure 3.1).

Native plant nurseries vary in the amount of organization necessary to plan crop production. At a minimum, the crop production process is visualized so that the crop's needs can be anticipated and met. Keeping a daily log or journal to track crop development and nursery conditions is a practice embraced by the best nursery managers. Even if written records are not used, it is valuable to consider the level of detail that can be used to plan crops. As the nursery grows in size and complexity, the value of written records correspondingly increases.

The process of crop planning usually includes the following components:

— Identify the seed dormancy of each species and apply treatments to overcome dormancy so that a reasonably uniform crop develops within a target timeframe.

Reviewing the crop schedule by R. Kasten Dumroese.

- Understand the three growth phases crops go through (establishment, rapid growth, and hardening) and the distinct requirements for each phase.
- Develop growing schedules for crop production from propagule procurement through outplanting and detail changes as the growing cycle progresses.
- List space, labor, equipment, and supplies required to support the crop during the three growth stages.
- Keep written records, including a daily log and plant development record.
- Develop and record accurate propagation protocols so that success can be replicated next time.

Exact recordkeeping is an important part of effective nursery management. A common limitation to nursery productivity is lack of species-specific and site-specific knowledge about seed treatments, germination requirements, plant development, and special crop needs. One of the greatest potential benefits of good recordkeeping is the development of specific, successful propagation protocols. A propagation protocol is a document that details all the steps necessary to propagate a plant, from the collection of seeds or cuttings all the way through shipping the plants to the field. An example propagation protocol that describes the typical development of serviceberry in a nursery in Montana is included in this chapter. Creating a propagation protocol for each species grown has these benefits:

- Invaluable resource for crop planning and scheduling.
- Beneficial for improving nursery productivity and seedling quality over time.
- Useful for teaching and sharing information about the plants to clients, the public, or nursery staff.
- A way to preserve and perpetuate propagation information.

The most important record to keep is a daily log that tracks what happens with each crop. Eventually, protocols can be developed from these logs and tailored to the unique growing conditions of a specific nursery to allow nursery managers to more readily repeat success from year to year.

This chapter will show how a propagation protocol is used to create a schedule and to plan facilities to produce a crop of a given species. Planning the schedule, management practices, and facilities for each crop

Figure 3.1—*(A) The diversity of species grown in native plant nurseries calls for detailed crop planning and scheduling to ensure that (B) high-quality plants are delivered to clients when they need them.* Photos by Tara Luna.

through each phase of growth will help maximize seedling growth and quality. It is recognized that crops rarely conform to the exact specifications of the protocol, but protocols and planning are essential guides to keep plants on target and to preclude potential problems at each development phase.

KEY PLANNING COMPONENTS

During crop planning, it is important to keep the process of plant production in mind. For native plant nurseries, six key crop planning components can be used to:

1. Determine available growing space.
2. Plan crop layout in the nursery based on the number of plants required.
3. Schedule propagule collection and processing.
4. Schedule propagule treatment.
5. Schedule propagule establishment.
6. Determine a growing schedule to meet a target date of delivery for "finished plants."

1. Available Space

A nursery is only so big—the number of plants that can be produced within it will depend on the propagule type, species, and container size. A nursery will hold more small plants than large plants. It is extremely useful to know how many plants growing in a particular container can be placed on a greenhouse bench and/or within the total nursery area.

2. Crop Layout Based on Number of Plants

Determining the crop layout is "planning what crops and stock types go where" in the greenhouse or nursery. This layout is planned to effectively provide similar growing requirements (temperatures, irrigation frequency, rates of fertilization) and other cultural requirements (frequency of shoot pruning or other treatments) for all the species and stock types grown each year.

The layout is necessary so that the total number of plants required can be accommodated by the space available. Depending on numbers, species, and container sizes grown, the layout of the crop in the greenhouse will change yearly.

The crop layout is also useful for taking advantage of microenvironments within the greenhouse or nursery. Most important, species with similar growth rates and irrigation requirements need to be grouped together.

Fast-growing species with similar growing and cultural requirements can be grouped together in one area, and moderate and slower growing species can be grouped together in another area of the greenhouse. This grouping method allows for species with similar requirements and growth rates to be treated effectively and efficiently. Ideally, species requiring cooler growing temperatures can be planned for the north and east sections of the greenhouse, and species requiring warmer temperatures can be planned for the southwest section. Likewise, the flow of plants out of a greenhouse to an outdoor nursery should be taken into consideration so that plants that finish in the same timeframe can be moved out and a second crop, if scheduled, can be planned for the available empty space.

3. Propagule Collection and Processing Schedule

This schedule needs to be closely coordinated, especially when seeds or cuttings are not on hand for all species or the supply in storage is insufficient. Seeds of species that need to be treated or sown immediately after collection should be delivered to the greenhouse as soon as possible, and seeds collected and cleaned in late autumn should reach the greenhouse by a predetermined target date so they can be treated as needed for the spring crop. To develop the seed treatment schedule, collectors of seeds or cuttings need to regularly communicate with staff at the nursery regarding when delivery to the greenhouse will occur. If seeds are provided by the client, they must be received in time to undergo treatment.

4. Propagule Treatment Schedule

Having a schedule of when to treat propagules is important for planning a target sowing or cutting establishment date for the entire crop. The propagule treatment schedule is one of the most basic and necessary crop planning tasks for native plant nurseries that grow a wide variety of species. Native species vary widely in their seed dormancy, so the seed treatments need to be scheduled properly. It is important to remember that seed dormancy requirements may vary among seedlots of the same species and slightly longer or shorter seed treatment durations should be adjusted accordingly. Similarly, cuttings vary in their collection, treatment, and storage needs. Consideration of these factors allows plants of multiple seedlots and species to establish on time so that a relatively uniform crop develops through the growing season. Making proper adjustments will mostly come with experience, although the propagation protocols in volume 2 of this agriculture handbook and the protocols available at the Native Plant Network (http://www.nativeplantnetwork.org) can be referenced to develop some guidelines.

The propagule treatment schedule is essentially a calendar in which the grower determines a target plant establishment date for the spring and schedules backward through the calendar months to organize all necessary treatments. For a target sowing date in March, for example, serviceberry and Woods' rose have a long stratification requirement that would need to start in November, whereas rushes and sedges generally have a 60-day stratification requirement and would go into stratification in January. If two crops per year are to be grown in the greenhouse, then a second treatment schedule must be made for a later date. If two crops are to be grown per season, the treatment schedules will overlap and the grower must pay careful attention to the scheduled treatment tasks as they appear on the calendar.

The propagule treatment schedule will need to be adjusted depending on the number of species produced, quantities of each species grown, growth rates (time required to grow the crop) of each species, and different container sizes.

For example, very fast-growing species such as western larch and many wetland species can be sown at a later date than the rest of the crop so that they do not become overgrown by the end of the season or the delivery date. Scheduling based on species may vary, however, based on cultural factors such as container size. For instance, western larch grown in larger containers will generally require longer growing periods than the same species grown in smaller containers.

5. Propagule Establishment Schedule

This schedule refers to the target seed sowing or cutting establishment date for the crop. This schedule includes, for example, instructing the sowing crew about which species need to be covered with mulch and which species require light to germinate and should be surface sown. This schedule is of critical importance because mistakes can prevent or delay emergence. The establishment schedule also includes the sowing method for each species. For most species, direct seeding is used; however, a few species may need to be sown as germinants as they break dormancy during stratification. Establishment schedules for cuttings may include the application of rooting hormones, keeping in mind that different species may require varying types, levels, and timing of rooting hormones to effectively induce root proliferation.

The establishment schedule is also planned by considering the growth rates of species (time to grow per container size) and the dates on which plants are to be delivered or outplanted. Dates can be adjusted as needed each season. Slower growing, woody species should be sown first, and faster growing species in small containers can be sown later in the season.

6. The Growing Schedule

This schedule is the most complex component because the nursery manager must estimate the growth rate of the crop and determine when to start the crop to meet target plant characteristics. This schedule is refined as the grower gains experience and plant development records and daily logs are reviewed. Plant specifications for woody plants are usually expressed in terms of target height, root mass, and root-collar diameter (caliper), and these specifications need to be considered and may even be specified under contracts with clients. Specifications for herbaceous plants are often much different, but usually include the need for plants to have several true leaves and a plantable, healthy root mass.

The growing schedule should include the current plant inventory. The inventory needs to be updated, especially during the establishment phase of the crops, so that, if needed, extra containers can be planted to ensure that the target number of plants is produced.

The plant development records and plant inventory should include the following information:

— Species.
— Seedlot number.
— Date propagules were collected.
— Planting date.
— Project or client number.
— Target date of delivery.

CROP GROWTH PHASES

Understanding the growth phases that crops go through is essential to crop planning. A tiny germinant has very different needs and requirements than a large plant that is almost ready for outplanting. The development of most crops can be divided into three phases: establishment, rapid growth, and hardening. Plants in each of these phases have distinct requirements for light, water, nursery space, and the types of attention and labor necessary to keep them healthy. The nursery manager's objectives for the crop are also different at each phase in order to keep production on track to produce target plants. Table 3.1 summarizes some typical aspects of each of the three phases. Please note that these aspects are generalized and will not apply to all species.

Establishment

The establishment phase is one of the most critical for successful nursery operations. For plants grown from seeds, the establishment phase is defined as the phase from the sowing of the seeds through the germination, emergence, and development of the first true leaves or primary needles (figures 3.2 and 3.5). For plants grown from cuttings, the establishment phase extends from placing cuttings into containers through the development of roots and shoots. Depending on the species, the establishment phase typically lasts 6 to 12 weeks. The goal of this phase is to maximize the amount of growing space filled with healthy plants, thereby minimizing losses.

Rapid Growth

During this phase, plants, particularly their shoots, increase dramatically in size (figures 3.3 and 3.5). Often the terminal shoot begins to approach target size. Plants are still at least somewhat protected during this phase. Rapid (but not excessive) shoot growth is encouraged.

Hardening

During the hardening phase, energy is diverted from shoot growth to root growth (figures 3.4 and 3.5). Root-collar diameter and roots reach target specifications, and shoot growth is discouraged or even stopped. Plants are "hardened"—conditioned to endure the stresses of harvesting, shipping, and outplanting. They are also fortified so that they have the energy reserves

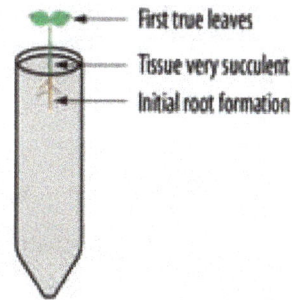

Figure 3.2—*The establishment phase.* Illustration by Jim Marin.

First true leaves
Tissue very succulent
Initial root formation

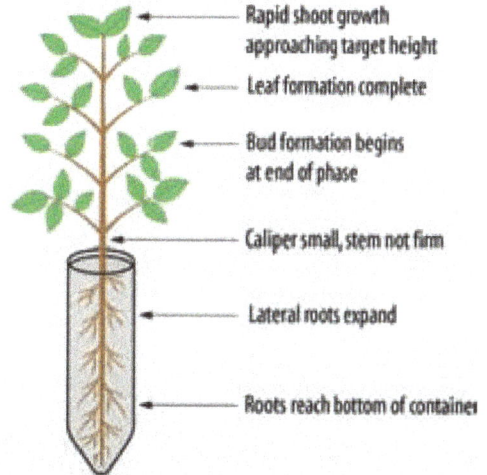

Figure 3.3—*The rapid growth phase.* Illustration by Jim Marin.

Rapid shoot growth approaching target height
Leaf formation complete
Bud formation begins at end of phase
Caliper small, stem not firm
Lateral roots expand
Roots reach bottom of container

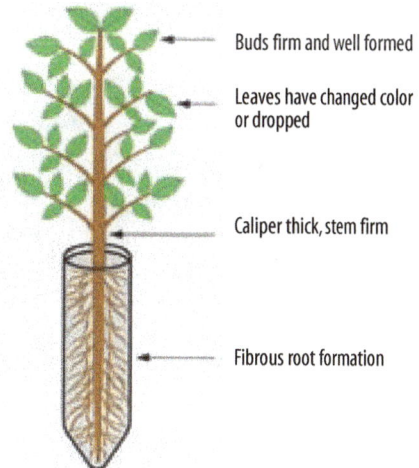

Figure 3.4—*The hardening phase.* Illustration by Jim Marin.

Buds firm and well formed
Leaves have changed color or dropped
Caliper thick, stem firm
Fibrous root formation

Table 3.1—The three phases of crop development for seedlings. After the three phases of crop development are understood for a species, the growing schedule can be developed to meet crop needs during each phase (after Landis and others 1998)

Phase	Establishment	Rapid Growth	Hardening
Definition	From germination through emergence and formation of true leaves	From emergence of true leaves to when seedling approaches target height; rapid increase in size, particularly in terminal shoot	Energy diverted from shoot to root growth; seedling reaches target height and root-collar diameter; lateral buds are set, seedling is conditioned to endure stress
Duration	Typically 14 to 21 days for germination; 4 to 8 weeks for early growth	Varies widely, typically about 10 to 20 weeks	Varies widely by species, from 1 to 4 months
Objectives	—Maximize uniform germination —Fill containers efficiently —Maximize survival —Minimize damping off	—Minimize stress —Encourage shoot growth —Maintain environmental factors near optimum levels —Monitor as seedling approaches target height and roots fully occupy container	—Stop shoot growth —Encourage root and stem diameter growth —Bring seedling into dormancy —Acclimate to natural environment —Condition to endure stress —Fortify for survival after outplanting
Special needs	—Protect from weather —Keep temps optimal —Irrigate to keep "moist, but not wet" —No or low fertilizer	—Protect from stress —Optimize temperatures —Irrigate regularly —Fertilize properly	—Induce moderate moisture stress —Decrease temperatures —Reduce photoperiod —Expose to ambient temperatures and humidity —Reduce fertilization rates and change mineral nutrient ratios
Labor	—Scout for pests and diseases —Monitor germination —Introduce beneficial microorganisms —Thin —Resow and/or transplant if necessary	—Scout for pests and diseases —Monitor environment —Modify density of crops to encourage good development —Adjust culture to avoid excessive shoot height	—Scout for pests and diseases —Monitor crops and environment carefully; see chapters 15 and 17 for details —Deliver crop to client in timely fashion to avoid problems with holdover stock.

to survive and grow after outplanting. Hardening is a crucial phase. It is a common mistake to rush hardening, resulting in plants poorly prepared for conditions on the outplanting site. When plants are not properly hardened, they may have the correct physical characteristics but survival after outplanting will be low because of an inadequate physiological condition. The goal of the hardening phase is to get plants conditioned for stress, prepared for outplanting, and ready to be delivered to the client in a timely fashion to avoid problems with holdover stock. See Chapter 12, *Hardening*, for more discussion on this topic.

Problems with Holdover Stock, Delayed Shipping, and Improper Scheduling

It is important to schedule and plan nursery production to make sure a crop goes through these three phases of development and is sent out from the nursery healthy and ready for outplanting. Although it is sometimes relatively easy to grow a seedling to target size, the tricky part is the hardening phase: slowing growth before plants get too large and conditioning them so they have energy reserves and can withstand stress. After plants are in this state, prompt outplanting is essential to ensure they can take full advantage of their hardened condition. If the stock is held over, problems quickly become apparent.

Many factors can disrupt the ability to follow a time schedule, but a common problem is the failure of clients to pick up plants on schedule. This problem can be avoided by good scheduling practices and communicating often with clients, especially periodic updates to advise them when seedlings will be ready. Clients tend to enjoy being kept abreast of the development of their crop, and updates about crop progress can become more frequent as the shipping date approaches. In some cases, having penalties, such as storage fees, in the contract for late pickups may also encourage clients to pick up their plants in a timely fashion. When communicating with clients, emphasize up front that prompt outplanting is in everyone's best interest, not only for the nursery and the health of the plants but also for the success of the client's project. See Chapter 16, *Nursery Management*, for more information about communicating with clients.

If the schedule to outplant after hardening is not met, however, a myriad of problems can develop. After

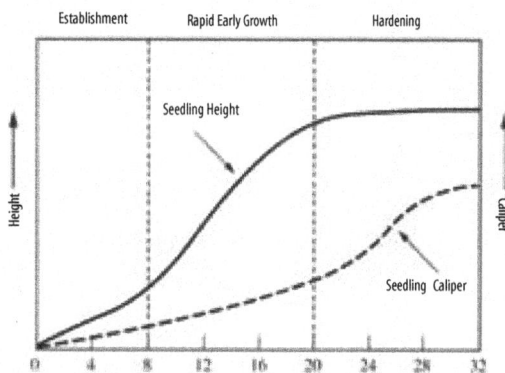

Figure 3.5—*Changes in seedling morphology during the three growth phases. Growth is relatively slow during the establishment phase. Most height growth occurs during the rapid growth phase, which ends when target height specifications are met. During hardening, roots continue to grow so long as soil temperatures are favorable, resulting in an increase in seedling root-collar diameter (caliper).* Modified from Wood (1994).

chilling requirements are met, a plant may begin to come out of dormancy, shoot growth begins, and it loses its resistance to stress as described in Chapter 12, *Hardening*. New vegetative growth after hardening must not happen until after the stresses of lifting and outplanting have occurred; otherwise, it may expose the plant to severe stress from which recovery may be difficult.

Problems with holdover stock include:

— loss of stress resistance.
— loss of cold hardiness.
— the swelling of buds, resulting in lost dormancy.
— the compacting or spiraling of roots, which reduces plant quality.

When plants are held too long in the nursery, the root system becomes woody and loses its ability to take up water and nutrients (figure 3.6). Structural problems may occur, too; roots may spiral (figure 3.7) and, instead of expanding outward and downward into the soil after outplanting, will strangle the plant or cause it to fall over in a high wind.

A solid understanding of the three phases of growth and how a particular species will develop over time in your nursery conditions is essential for good scheduling (figure 3.8). The process of developing specific information about each species and its timing and management is described next.

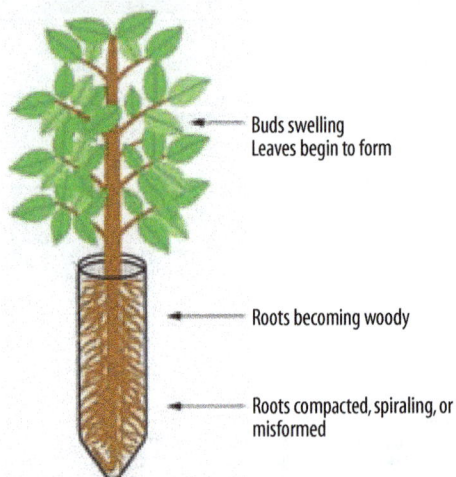

- Buds swelling
  Leaves begin to form

- Roots becoming woody

- Roots compacted, spiraling, or
  misformed

Figure 3.6—*Crops that are held too long (holdover stock) may resume shoot growth and will not be properly conditioned to endure the stresses of transportation and outplanting. They may develop severe root deformities that reduce outplanting success. Illustration by Jim Marin.*

Figure 3.7—*The spiraling roots of 2-year-old container stock. Photo by Tara Luna.*

KEEPING RECORDS MAKES PLANNING EASIER

Site-specific and species-specific information about managing a crop of plants throughout its three phases of growth can be developed by keeping some simple records. The following three kinds of records are crucial for success:

A **daily log** is a journal that notes nursery conditions and management practices on a daily basis.

Plant development records are kept for each crop of plants and record the development and management of that particular crop. These records are usually updated at least on a weekly basis as the crop develops.

Propagation protocols are created for each species and are designed to be a comprehensive guide describing how to grow that species in your nursery from propagule collection through outplanting. Propagation protocols are usually revised on a seasonal or annual basis.

These three records interrelate and support each other (figure 3.9). For species not grown before, the first step is doing some preliminary research to develop a draft propagation protocol, which is simply a best guess on how to propagate and manage that species. Literature, interviews with colleagues at other nurseries, a search for protocols written by other nurseries posted on the Native Plant Network (http://www.native plantnetwork.org), and personal experiences will inform this draft protocol. After the plants start growing, direct personal experiences (as recorded in the daily log and plant development records) will be used to refine and update the protocol regularly to improve production. The daily log is used to fill in any gaps or to track issues that come up in the plant development records, and the plant development records enable you to compare actual crop development with the protocol.

Daily Log

It is easy to get caught up in the day-to-day details of running a nursery and lose sight of how important it is to write down what is happening. Nothing compares, however, with that sinking feeling that occurs after shipping out a successful crop of beautiful plants and suddenly realizing that nobody knows how to replicate that successful crop. How long did it take to produce the crop? What materials were purchased? How was the crop fertilized, watered, and managed during each growth phase? Keeping records of plant development and general nursery activities is an essential part of good nursery management.

A daily log or journal is simply a record of what was done and what happened in the nursery each day (figure 3.10). Appendix 3.C includes a blank daily log form and a daily log example that a small nursery with a staff of just one or two people might use. Make it a habit to at least jot down something each day, even if only a minute is spent on it. Large nurseries may keep more complicated daily records and may have separate logs for irrigation, fertilization, and the like. Tailor the daily log to suit the nursery. The important thing is just to do it. What is recorded in the daily log about management practices, environmental conditions, and general crop performance will become a priceless resource for many years to come.

Some growers choose to record a large amount of detail in their general daily log and then go back to the

Figure 3.8—*Serviceberry at different growth phases: (A) germination and establishment, (B) establishment phase 4 weeks after germination, (C) early rapid growth phase in early summer, (D) late rapid growth phase in midsummer, and (E) hardening phase in outdoor nursery in late autumn.* Photos by Tara Luna.

daily log at slower times of the year to summarize specific information about each crop into a plant development record. Many growers, however, find it easier to keep a separate plant development record for each crop, as described in the following paragraphs.

Plant Development Record

Keeping a simple plant development record (or register) for each crop is a great way to build a foundation for accurate, site- and species-specific protocols. A plant development record notes what is happening with a crop of plants from crop initiation through delivery. The plant development record helps you track and remember exactly what you did to produce a crop. These notes about management practices and timing for each phase of growth are invaluable records. At a minimum, you can simply put a couple of fresh sheets of paper in a notebook or three-ring binder and jot down notes on a regular basis (at least weekly) as the crop progresses. One way to make it easier to keep track of this valuable informa-

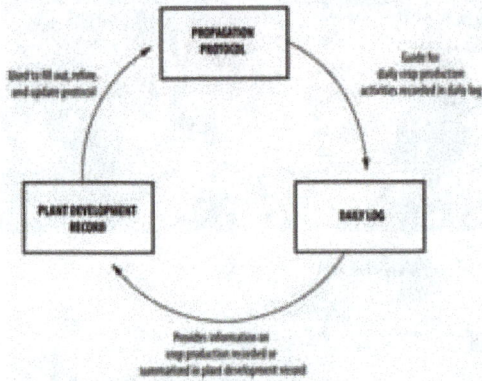

Figure 3.9—*The three basic records and how they relate. Use a protocol (originally drafted from research, literature, and experience) to plan and schedule daily activities. Daily activities are recorded in the daily log. Keeping a log helps with collection of specific information about the development of each crop. This information is recorded in the plant development record for each crop. The protocol is then refined and expanded based on this new information, which will improve scheduling and production practices the next time the crop is grown.*

Figure 3.10—*Jot down a few notes every day about what was done and what happened with the crop. These records become invaluable resources when adjusting protocols and fine-tuning crop schedules.* Photo by Tara Luna.

tion is to create a form such as the one provided in Appendix 3.D. You can photocopy this form, put it in a three-ring binder or clipboard, and fill it out as each crop progresses. You can also refine the form to match the conditions and the crops grown in your nursery. Use a blank form for each crop grown, even if it is just a small trial of a few plants or even if the species has been grown before. It is sometimes easier to remember to record these notes if the form with a pen attached to it is conveniently accessible a short distance from the crop. Filling out the form can also be done on a computer. Every time you work with the crop or make an observation about it, jot a note in the record. These records quickly become great storehouses of information for developing and updating protocols.

Propagation Protocols

A propagation protocol describes all the steps necessary to grow a species in a specific nursery and is meant to be a guide to producing and scheduling a crop of that species. A protocol is ideally comprehensive, systematic, and detailed like a cookbook recipe, although new nurseries may begin with relatively skeletal protocols that become more detailed as experience accumulates. A protocol contains significantly more detail than the plant development record, usually providing information on propagule collection and some background information about ecology and distribution. The more information a protocol contains, the easier it will be to plan and schedule crops.

The example protocol in Appendix 3.B shows the type of information usually included. Many completed protocols for native plants are available online at the Native Plant Network (http://www.nativeplantnetwork.org), an excellent place to start research as you develop protocols for species in your nursery. Remember, however, that a protocol used by another nursery cannot be explicitly applied to any other nursery, but it can be used as a guide to help develop a site-specific protocol.

A protocol typically describes the following aspects and characteristics:

— Species name and ecotype.
— Time to grow.
— Plant target specifications (for example, height, root system, root-collar diameter).
— Propagule sources and collection.

- Propagule characteristics and processing.
- Preplanting propagule treatments.
- Growing area preparation.
- Management for, and length of, establishment phase.
- Management for, and length of, rapid growth phase.
- Management for, and length of, hardening phase.
- Harvesting, storing, and shipping practices.
- Duration of storage.
- Outplanting notes.

Gathering Information

Ideally, each nursery has a protocol that provides detailed information for each species it grows. Nurseries that work with native or culturally important plants may need to develop protocols from scratch. Even when a familiar and widely propagated species is grown, the exact schedule and performance of that species will vary greatly depending on the unique conditions of the nursery and on other variables such as seed sources and weather patterns. In other words, no one else's propagation "cookbook" can be applied exactly to your nursery; you need to develop your own unique "recipe" to produce the best plants. It can be very helpful to create and share protocols with clients so that they understand the nuances and timing factors necessary to provide them with high-quality plants.

The development of the protocol is informed by both firsthand experiences and outside sources of information. Keeping a good daily log and plant development record during crop production can be used as the foundation for building protocols. Outside information sources should also be explored, including published literature, personal observations of the plants in the field, and information from other growers (Landis and others 1999).

PUBLISHED LITERATURE

A systematic search can reveal information on how to propagate the species. Trade journals, native plant societies, and botanical gardens may be able to help. An excellent source of propagation protocols, continually updated by growers and free of charge, is online at the Native Plant Network (http://www.nativeplantnetwork.org). If specific information on the species to be grown is not currently available, try to find a similar species grown in similar climatic zones to see if any information may be applicable.

PERSONAL OBSERVATIONS OF THE PLANT IN THE FIELD

Clues can be gained from studying how the plant grows in nature. This information may be gained firsthand from observation and also from published literature and/or community members who are familiar with the plant. Although collectors of plant materials for cultural uses may have never grown the species, they are likely to be knowledgeable about the species life cycles.

INFORMATION FROM OTHER PROFESSIONALS

Private nurseries may be disinclined to share their propagation methods, but government nurseries or botanical gardens are often excellent sources of information about growing the same or similar species. Again, information generously shared by other growers in the Native Plant Network is invaluable.

Writing the First Draft

Based on information gathered and firsthand experiences, a draft protocol is created. This first draft is the "best informed guess" of what will be required and how the species may perform in your nursery. This new protocol will serve as a guide as you work through the first crop cycle. The plants themselves will prove the protocol right or wrong as they grow. The daily log and plant development record will enable a comparison between projected development and actual growth. Information will be expanded and improved as the nursery gains more site-specific experience. Revise the protocol regularly according to how the species and seedlot actually behaved under local conditions. Ultimately, the nursery will have a very accurate guide for how to grow that species.

Testing and Adjusting Protocols

Refine and update the protocol with site-specific information from your nursery after the production of each crop. Do not be discouraged if a protocol drafted from background research or another nursery's experience does not produce the same results; the goal is to adjust the protocol to reflect local conditions. Remember that year-to-year variations in weather or unforeseen operational changes may prevent crops from growing exactly as projected. Allow some room for flexibility and make adjustments based on observed factors. Sometimes the protocol will need to be adjusted to more accurately reflect how crops actually develop;

sometimes management practices are adjusted to enable crop development to match the specifications of the protocol. The protocol is continuously updated and revised as plants and experiences grow, making it more accurate year by year.

Protocols are not only a guide for producing current crops but also a foundation for the improvement of crop production. Along with plant development records, they provide baseline information that enables nursery managers to determine if intended improvements (such as the introduction of new beneficial microorganisms or a different type of fertilizer) actually have a positive impact on plant health and growth compared with what was normally done. New nurseries especially benefit from updating protocols, as refinements are made and dramatic improvements are seen in crop production efficiency and effectiveness from season to season. Propagation protocols serve as an essential guide for planning and scheduling each future crop.

Planning Crop Production: Time and Space

After a draft protocol has been developed for a species, it is time to plan the crop. Two main factors in crop planning are time and space. For illustration purposes, the example protocol in Appendix 3.B will be used to create a schedule and facilities plan to produce a crop of serviceberry. Crop planning covers all phases of crop production, from the procurement of plant materials through outplanting. It also provides an overview of the schedule for bringing crops through the three growth phases (establishment, rapid growth, and hardening) and of storage. It is also essential to plan for the space, materials, and other facilities that the crops will require: containers and benches in the nursery, equipment, supplies, and materials and labor needed at each phase to produce a successful crop.

Crop Growing Schedules

Crop growing schedules show all phases of nursery production from the procurement of plant materials through outplanting. It is important to account for the time needed to obtain seeds or cuttings and the time required for seed processing, testing, and treatments, if necessary. This schedule creates a visual illustration of each step in the protocol and the time necessary to complete each step. When the timing for nursery crops is understood, appropriate dates for sowing seeds or sticking cuttings can be calculated by counting *backward* from the desired date of outplanting. The accurate calculation of field-ready dates is essential for successful client relations. For example, serviceberry for spring outplanting requires 15 months to grow, including time for stratification prior to sowing. So, if plants are needed for outplanting in March, as shown on the schedule, the stratification must begin in January of the previous year (figure 3.11). (Note that if seeds of this species are not in storage, they will need to be collected in late summer.) The total time required for the production of each crop (a variety is shown in figure 3.11) will vary widely by species, season, and nursery environment. Genetics and the variability of seedlots will also cause variations in crop scheduling.

A detailed schedule of the three different phases of crop development should be created based on the protocol. This schedule will show how to accommodate the changing needs of the crops as they develop. Table 3.2 provides examples of the necessary steps in each phase (transplanting emergents, thinning, moving from the propagation structure to the rapid growth structure, changing fertilizer and watering regimes, and so forth) and the time required to complete each step. This schedule should be posted in the nursery so that staff can track the crop's development and understand what cultural practices are required. The schedule should clearly answer questions such as, "What water and fertilizer requirements does the crop have today? What aspects of crop development should receive special attention?" If appropriate, the schedule can also be shown to clients so they fully understand the time required to produce their crop.

For more complicated crop production, schedules can be created easily with computer spreadsheets and divided into two separate calendars: (1) a year-long, monthly calendar with week-by-week general steps and (2) a second, more detailed schedule displaying day-by-day activities for the production of each crop.

Because the three distinct phases of crop development usually involve moving crops from one nursery structure to another, discussion about matching growth phases with nursery structures is continued in the next section.

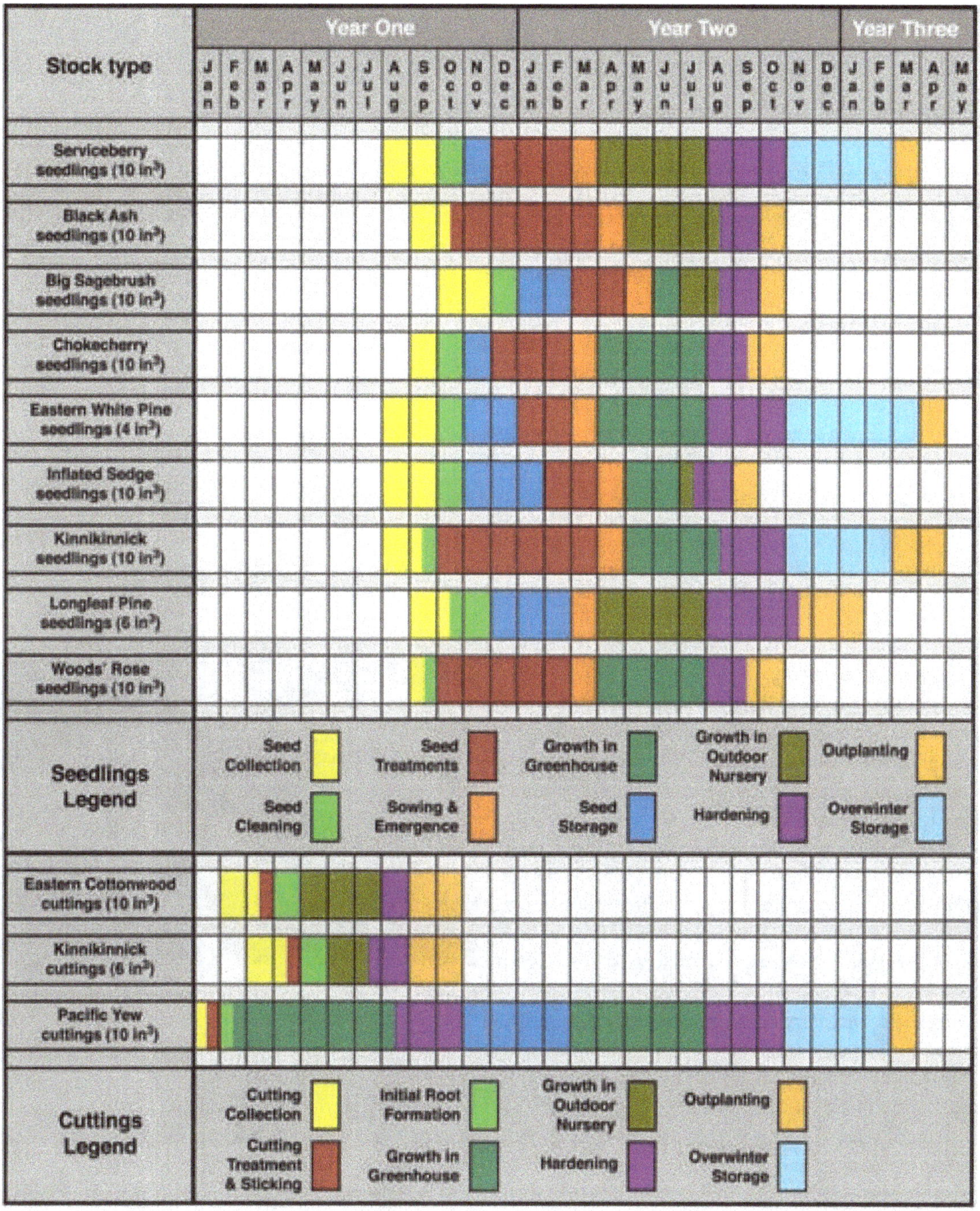

Year One **Year Two** **Year Three**

Stock type	Jan	Feb	Mar	Apr	May	Jun	Jul	Aug	Sep	Oct	Nov	Dec	Jan	Feb	Mar	Apr	May	Jun	Jul	Aug	Sep	Oct	Nov	Dec	Jan	Feb	Mar	Apr	May

Serviceberry seedlings (10 in³)

Black Ash seedlings (10 in³)

Big Sagebrush seedlings (10 in³)

Chokecherry seedlings (10 in³)

Eastern White Pine seedlings (4 in³)

Inflated Sedge seedlings (10 in³)

Kinnikinnick seedlings (10 in³)

Longleaf Pine seedlings (6 in³)

Woods' Rose seedlings (10 in³)

Seedlings Legend

- Seed Collection
- Seed Cleaning
- Seed Treatments
- Sowing & Emergence
- Growth in Greenhouse
- Seed Storage
- Growth in Outdoor Nursery
- Hardening
- Outplanting
- Overwinter Storage

Eastern Cottonwood cuttings (10 in³)

Kinnikinnick cuttings (6 in³)

Pacific Yew cuttings (10 in³)

Cuttings Legend

- Cutting Collection
- Cutting Treatment & Sticking
- Initial Root Formation
- Growth in Greenhouse
- Growth in Outdoor Nursery
- Hardening
- Outplanting
- Overwinter Storage

Figure 3.11—*The production of a crop of serviceberry plants, the example used throughout this chapter, would require about 19 months according to this hypothetical crop growing schedule. Other stock types and species may require more or less time depending on many factors, including the availability of seeds or cuttings, seed dormancy issues, propagation environments, and the scheduled time of outplanting.* Illustration by Jim Marin.

Table 3.2—An example of a crop development schedule

Activity	Seed Treatment	Establishment Phase	Rapid Growth Phase	Hardening Phase	Storage
Length of phase	17 weeks	4 weeks	16 weeks	4 weeks	16 to 20 weeks
Date	Jan 12 – May 4	May 5 – Jun 1	Jun 2 – Sep 21	Sep 22 – Oct 12	Oct 13 – Mar 1
Temperature	37 °F (3 °C)	60 to 77 °F (16 to 25 °C)	60 to 77 °F (16 to 25 °C) (ambient outdoors)	40 to 68 °F (5 to 20 °C) (ambient outdoors)	28 to 50 °F (–2 to 10 °C)
Propagation environment	Refrigerator	Germinant house	Main outdoor growth area	Main outdoor growth area	Outdoor compound under insulation
Fertilization	None	1 g Osmocote® per 172 ml Cone-tainer®	13N:13P$_2$O$_5$:13K$_2$O at 100 ppm weekly	10N:20P$_2$O$_5$:20K$_2$O at 100 ppm every other week	N/A
Irrigation	N/A (keep moist)	Daily handwatering to saturation	Daily auto overhead to saturation	Gradual reduction	N/A
Target size at end of phase	N/A	N/A	10 cm (4 in) height, 4 mm caliper	10 cm (4 in) height, 7 mm caliper	Same as hardening phase
Activity	Cold storage stratification	Thinning at 2 weeks	Weed and pest management	Protect from early frost	Monitoring temperature and insulation

N/A = not applicable

Space and Facilities Planning

The space requirements in each facility for each crop during the different stages of propagation must be planned. These plans also include labor, equipment, and supplies needed to support crop development through each growth phase.

As plants develop through the growing cycle, their needs will change. Except for nurseries with elaborate climate control systems, crops are often moved from one structure to another as they progress through the three development phases (see Chapter 4, *Propagation Environments*, for additional information on this topic). Using the serviceberry example, crops are protected in a special germination area during the establishment phase and then moved to an outdoor growth area for rapid growth and hardening (table 3.3). Likewise, the amount of space the crop will require varies by growth phase: germinants may take up very little room if they are concentrated in trays, but plants take up much more space after they have been transplanted into larger containers or spaced more widely as they grow bigger. Although the example in table 3.3 does not go into such detail, the facilities schedule should calculate how much space each crop will use, how many

Figure 3.12—*Plants are often moved from one area of the nursery to another as they go through the three growth phases. Good planning and scheduling ensures that the space will be available to meet the needs of the crops, as shown here at the Santa Ana Pueblo Nursery in New Mexico.* Photo by Tara Luna.

hours of labor will be needed, and the quantities of materials (for example, growing media) required during crop production.

It is especially important to consider the material and labor needs at each phase of crop production. The facilities schedule should answer questions such as, "What materials will be needed in the next weeks or months? What action do we need to take: for example, pot media, clean containers, and clear off benches to make room for the new crop? What labor and attention does the crop require this week and this month? Will this effort involve fertilizing, watering, and/or moving the crop to a new structure?"

Planning how each facility will be used is indispensable in determining how resources within a nursery can be best distributed to maximize production and minimize conflicts associated with overlapping needs (figure 3.12). The facilities schedule (table 3.3) may be combined with or posted side by side with the crop schedule, and the staff should have easy reference to it.

Nursery Inventory

The recordkeeping and scheduling efforts described previously make it very simple to keep track of nursery inventory. An inventory should include a listing of all plants in the nursery by bench or structure number, the current developmental stage of the crop, and details of delivery (site, name of owner, anticipated delivery date).

Keeping the inventory and growing schedules posted in a central place in the nursery helps all the staff under- stand the current needs of the existing crops; equipment and supplies needed, as dictated by schedules; and the necessary practices to keep the crop on schedule.

SUMMARY

Crop planning is an important process to help schedule time and facilities and commit to delivery dates for plants. A daily log or journal is a key aspect of nursery management, because it records a way to track what happens with each crop and provides a history of crop development. This information helps create accurate, site- and species-specific protocols for growing plants. For new species, the first protocol may be drafted from outside sources of information and experience and then revised based on actual crop performance in your nursery. The protocol is used to make a good schedule for the crops and bring them through the three growth phases so that the plants are healthy and conditioned for outplanting. The daily log and the records of plant development should be used to continuously refine the protocol on a seasonal or annual basis, resulting in increasingly

Table 3.3—An example of a facilities schedule

Activity	Seed Treatment	Establishment Phase	Rapid Growth Phase	Hardening Phase	Storage
Length	17 weeks	4 weeks	16 weeks	4 weeks	20 weeks
Date	Jan 12 – May 4	May 5 – Jun 1	Jun 2 – Sep 21	Sep 22 – Oct 12	Oct 13 – Mar 1
Labor	—Clean seeds in hydrogen peroxide: soak (20 minutes) followed by 48-hour rinse; —put in mesh bags or into flats; —keep moist; —rinse seeds weekly	—Make growing media; —fill containers; —plant seeds; —hand-water daily; —monitor germination; —thin or consolidate as necessary; —monitor greenhouse temperature	—Move to outdoor growth area after danger of frost is past; —use liquid fertilizer weekly; —monitor overhead automatic irrigation; —monitor outdoor temperature; —monitor growth; —manage weeds/pests	—Monitor growth; —monitor and gradually reduce irrigation and fertilization	—Move to storage area; —cover with insulating foam —monitor temperatures and insulation —pack and distribute when planting time comes
Facility/ space needed	Refrigerator	Benches in germinant house	Benches in main outdoor growth area	Benches in main outdoor growth area	Space in outdoor storage area
Materials needed	—Seeds; —3% hydrogen peroxide; —water; —stratification media (*Sphagnum* peat moss); —net bags; —flats or trays	—Stratified seeds; —containers and trays; —potting media (peat, perlite, vermiculite, Osmocote®)	$13N:13P_2O_5:13K_2O$ liquid fertilizer	$10N:20P_2O_5:20K_2O$ liquid fertilizer	Insulating foam; boxes or containers for shipping out seedlings

accurate information to support successful crop production. The amount of detail incorporated into schedules and plans for nursery crops is a personal decision and is influenced by how complicated the production system is. It may be unrealistic for smaller nurseries that grow small quantities of many different species to consistently maintain all the records described here. If so, start with a daily log and expand to more written records as time and resources permit.

LITERATURE CITED

Landis, T.D.; Tinus, R.W.; Barnett, J.P. 1999. The container tree nursery manual: volume 6, seedling propagation. Agriculture Handbook 674. Washington, DC: U.S. Department of Agriculture, Forest Service. 167 p.

Luna, T.; Hosokawa, J.; Wick, D.; Evans, J. 2001. Propagation protocol for production of container *Amelanchier alnifolia* Nutt. plants (172 ml conetainer); Glacier National Park, West Glacier, Montana. In: Native Plant Network. http://www.nativeplantnetwork.org (30 Jun 2004). Moscow, ID: University of Idaho, College of Natural Resources, Forest Research Nursery.

Wood, B. 1994. Conifer seedling grower guide. Smoky Lake, AB: Environmental Protection. 73 p.

ADDITIONAL READINGS

Landis, T.D.; Dumroese, R.K. 2000. Propagation protocols on the Native Plant Network. Native Plants Journal 1: 112-114.

Landis, T.D.; Dumroese, R.K. 2002. The Native Plant Network: an on-line source of propagation information. The International Plant Propagators' Society, Combined Proceedings 51: 261–264.

APPENDIX 3.A. PLANTS MENTIONED IN THIS CHAPTER

big sagebrush, *Artemisia tridentata*

black ash, *Fraxinus nigra*

chokecherry, *Prunus virginiana*

eastern cottonwood, *Populus deltoides*

eastern white pine, *Pinus strobus*

huckleberry, *Vaccinium* species

inflated sedge, *Carex utriculata*

kinnikinnick, *Arctostaphylos uva-ursi*

longleaf pine, *Pinus palustris*

Pacific yew, *Taxus brevifolia*

red alder, *Alnus rubra*

rushes, *Juncus* species

sedges, *Carex* species

serviceberrry, *Amelanchier alnifolia*

western larch, *Larix occidentalis*

Woods' rose, *Rosa woodsii*

Native Plant Nursery
Glacier National Park
West Glacier, Montana 59936

Common Name: Serviceberry
Family Scientific Name: Rosaceae
Family Common Name: Rose family
Scientific Name: *Amelanchier alnifolia* Nutt.
Ecotype: Forest margin, Saint Mary, 1616 meter (5300 foot) elevation, Glacier National Park, Glacier Co., MT
Propagation Goal: Plants
Propagation Method: Propagated from seeds
Product Type: Container (plug), 172 ml Cone-tainer® (10 in^3 Ray Leach Super Cell)
Time to Grow: 11 months (for fall planting); 15 months (if overwintered and planted in spring)
Target Specifications:
Height: 10 cm (4 in)
Caliper: 7 mm
Root system: Firm plug in 172 ml Cone-tainer®
Propagule Collection: Seeds are hand collected in late summer when fruit turns dark purple. Seeds are tan at maturity. Fruits are collected in plastic bags and kept under refrigeration prior to cleaning.
Propagule Processing: Seeds are cleaned by maceration using a Dyb-vig seed cleaner followed by washing and screening. Seed longevity is 5 to 7 years at 3 to 5 °C (37 to 41 °F) in sealed containers. Seed dormancy is classified as physiological dormancy.
Seeds/kg: 180,400 (82,000/lb)
Purity: 100%
Germination: 15% to 100%
Pre-Planting Treatments: 3:1 water:hydrogen peroxide (3 parts water to 1 part 3% hydrogen peroxide), soak for 20 minutes followed by a 48 hour water rinse. Seeds are placed into a 120 day cold moist stratification after pretreatment. Seeds are placed in fine mesh bags and buried in stratification media in ventilated containers under refrigeration (3 °C [37 °F]). It is very important to wash stratified seeds weekly; remove net bags from artificial stratification and rinse well to remove mucilaginous material. Lower germination percentages were noted with seedlots that did not receive the hydrogen peroxide and water rinse prior to stratification. This pretreatment of seed appears to significantly improve germination percentages.
Growing Area: Greenhouse and outdoor nursery growing facility.
Sowing Method: Direct Seeding. Seeds are lightly covered with medium.
Growing medium: 6:1:1 milled sphagnum peat, perlite, and vermiculite with Osmocote controlled release fertilizer (13N:13P$_2$O$_5$:13K$_2$O; 8 to 9 month release rate at 21°C [70 °F]) at the rate of 1 gram of fertilizer per 172 ml Cone-tainer®.
Environment/Water: Greenhouse temperatures are maintained at 21 to 25°C (70 to 77 °F) during the day and 16 to 18 °C (60 to 65 °F) at night. Seedlings are hand watered and remain in greenhouse until mid May, after establishment.

Seedlings are then moved to outdoor nursery for the remainder of the growing season. In the outdoor area, seedlings are irrigated with Rainbird automatic irrigation system in early morning to saturation (until water drips out the bottom). Average growing season of nursery is from late April after snowmelt until October 15th.
Establishment Phase: Germination is uniform and is usually complete in 3 weeks. True leaves appear 2 weeks after germination. Seedlings are thinned at this stage.
Length of Establishment Phase: 4 weeks
Rapid Growth Phase: Seedlings grow at a rapid rate after establishment. Plants are fertilized with soluble 13N:13P$_2$O$_5$:13K$_2$O at 50 to 75 ppm during the growing season. Plants average 10 cm (4 in) in height and 4 mm caliper in 4 months.
Length of Rapid Growth Phase: 16 weeks
Hardening Phase: Plants are fertilized with soluble 10N:20P$_2$O$_5$:20K$_2$O at 100 ppm during August and September. Irrigation is gradually reduced in September and October. Plants are given one final irrigation prior to winterization.
Length of Hardening Phase: 4 weeks
Harvesting, Storage, and Shipping: Total time to harvest: 11 months for fall planting, 15 months if overwintered and outplanted in spring.
Harvest Date: Fall or spring
Storage Conditions: Overwinter in outdoor nursery under insulating foam cover and snow.
Length of Storage: 4 to 5 months.
Outplanting performance on typical sites: Outplanting site: Saint Mary, Glacier National Park, MT. Outplanting date: spring or fall. Outplanting survival at 4 years: 86%.
Other cultivation comments: Seedlings in 3-liter (1-gallon) containers average 65 cm (25 in) in height with 10 mm caliper 16 months following germination.
General comments: Distribution: *A. alnifolia* occurs from southern Alaska to California, east across Canada to western Ontario, south through the Rocky Mountains to New Mexico, and east to the Dakotas and Nebraska, in open forests, canyons, and hillsides from near sea level to the subalpine zone. *A. alnifolia* is a long-lived seral species that is widely used in restoration projects in many habitats. Deer, moose, and elk browse the foliage and twigs, and berries are an important food source to birds and mammals. There are 3 botanical varieties; var. *pumila*, var. *humptulipensis*, and var. *alnifolia*.
References:

Flora of the Pacific Northwest, Hitchcock and Cronquist, University of Washington Press, 7th printing, 1973.

Seeds of the Woody Plants in the United States, Agriculture Handbook No. 450, USDA Forest Service, Washington D.C., 1974.

Seeds of Woody Plants in North America, Young and Young, Dioscorides Press, 1992.

Seed Germination Theory and Practice, 2nd Edition, Deno, N., published June, 1993.

Glacier Park Native Plant Nursery Propagation Records, unpublished.

1998 Revegetation Monitoring Report, Glacier National Park, Asebrook, J. and Kimball, S., unpublished.

Citation: Luna, Tara; Hosokawa, Joy; Wick, Dale; Evans, Jeff. 2001. Propagation protocol for production of container *Amelanchier alnifolia* Nutt. plants (172 ml cone-tainer); Glacier National Park, West Glacier, Montana. In: Native Plant Network. URL: http://www.nativeplantnetwork.org (accessed 30 June 2008). Moscow (ID): University of Idaho, College of Natural Resources, Forest Research Nursery.

APPENDIX 3.C. DAILY LOG FORM

Date:

Environmental conditions in growing areas (light, temperature, humidity):

_____

_____

Sunrise/Sunset times:

Moon phase:

Other weather notes (cloud cover, and so on):

What water did seedlings receive? (irrigation type and frequency, or precipitation):

_____

_____

Today's activities (note how many person-hours per activity) (fertilization, pest management, transplanting, packing and shipping, making potting media, moving crops from one structure to another, treating or sowing seeds, and so on):

_____

_____

_____

Growth phase status (make notes when a crop moves from one phase to another):

_____

_____

Purchases (what supplies or equipment were purchased and their cost):

_____

_____

Orders (what plant materials were delivered and payments made):

_____

_____

General crop/nursery observations:

_____

_____

Questions or concerns:

_____

_____

_____

_____

_____

_____

_____

DAILY LOG SAMPLE

Date: April 2, 2008

Environmental conditions in growing areas (light, temperature, humidity):

Min/Max Temp: Outdoor area: 60°F min; 75°F max, Indoor area: 68°F min; 80°F max.

No artificial lighting used, just sunlight.

Sunrise/Sunset times: Sunrise: 6:45 am Sunset: 7:00 pm

Moon phase: Full moon tonight!

Other weather notes (cloud cover, and so on): Partly cloudy, no rain today.

What water did seedlings receive? (irrigation type and frequency, or precipitation):

Hand watered germinant area with fine headed sprayer in a.m.. First thing in the morning, automatic
overhead watering for 1 hour on all seedlings in main greenhouse and benches one through six in outdoor area.

Today's activities (note how many person-hours per activity) (fertilization, pest management, transplanting, packing and shipping, making potting media, moving crops from one structure to another, treating or sowing seeds, and so on):

Mixed potting media and filled trays for the new order for 500 serviceberry seedlings, 3 hours total.

Hand watered the germinant area, 1 hour. Fixed the leak (noticed yesterday) in the main water line, 1 hour.

Answered e mail correspondence, 1 hour. Fertilized the 8 week old serviceberry and huckleberry seedlings in greenhouse area
with 200 ppm Peter's 20 20 20, 1 hour. Moved new alder seedlings from establishement area to main greenhouse, 1 hour.

Growth phase status (make notes when a crop moves from one phase to another):

The red alder seedlings sown earlier this month have entered the rapid growth phase   today I moved them
from the germinant area to the main greenhouse. They are about 3 inches tall now.

Purchases (what supplies or equipment were purchased and their cost):

Bought a new coupling for fixing the irrigation line, plus an extra one to have on hand in case there is another leak: $10.87.

Bought potting media materials for serviceberry order, $28.45.

Orders (what plant materials were delivered and payments made):

No plants going out until September. Order confirmed for 500 serviceberry.

General crop/nursery observations:

Things look good in general. The huckleberry were starting to look a little yellow, which is why I switched
from the other fertilizer to the 20 20 20 today. I'm pleased with how the aldr look...it seems the inoculation
I did 3 weeks ago might be kicking in. I don't see any nodules yet, but they look green despite no nitrogen fertilizer,
and roots have that ammonia like smell I was reading about.

Questions or concerns:

I'd like to check the media pH for the huckleberry...they like it acidic and maybe I need to adjust more. Everything seems fine so far
but I'd like to stay vigilant so I can nip any problems in the bud. There seem to be lots of slugs out...because of all the rain last week?
I'd like to explore some organic slug control options...maybe I'll call around and see if anyone I know has had success with that copper
barrier stuff, and what the cost might be.

APPENDIX 3.D. PLANT DEVELOPMENT RECORD FORM

Species name: _____

Seedlot/Seed source: _____

Date of seed collection: _____

Establishment

Type and length of propagule treatment (scarified, stratified, and so on):

_____

Date of propagule establishment:

_____

Potting media and tray or container type used:

_____

Germination notes (date begins and ends, % germination, and so on):

_____

Date transplanted (if not direct sown):

_____

Container type and potting media for transplanting:

_____

Microorganisms used?

_____

Irrigation type and frequency: daily, every other day, and so on

_____

Fertilization (type, rate, and frequency, if any):

_____

Environmental conditions for crop (light, temperature, humidity):

_____

Horticultural treatments (cultivation practices, and so on):

_____

Date establishment phase completed:

_____

Notes (resowing or thinning activities, problems, or challenges):

_____

_____

_____

Rapid Growth

Time after sowing to enter rapid growth phase:

_____

Plant size at start of phase (height):

_____

Container type and potting media:

_____

Irrigation type and frequency: daily, every other day, and so on

_____

Fertilization type, rate, and frequency:

_____

Environmental conditions for crop (light, temperature, humidity):

_____

Horticultural treatments (spacing, cultivation practices, and so on):

_____

Date rapid growth phase completed:

_____

Notes (development, vigor and health, challenges, or problems):

_____

_____

_____

Hardening:

Plant size at start of phase (height and root-collar diameter):

_____

Irrigation type and frequency: daily, every other day, and so on

_____

Fertilization type, rate, and frequency:

_____

Environmental conditions for crop (light, temperature, humidity):

_____

Horticultural treatments (spacing, cultivation practices, and so on):

_____

Plant size at end of phase (height and root-collar diameter):

_____

Date hardening phase completed:

_____

Date plants delivered:

_____

Notes (vigor and health, challenges, or problems):

_____

_____

_____

_____

_____

Other notes

Notes on performance of crop after outplanting

_____

_____

_____

_____

_____

_____

_____

Propagation Environments

Douglass F. Jacobs, Thomas D. Landis, and Tara Luna

4

An understanding of all factors influencing plant growth in a nursery environment is needed for the successful growth and production of high-quality container plants. Propagation structures modify the atmospheric conditions of temperature, light, and relative humidity. Native plant nurseries are different from typical horticultural nurseries because plants must be conditioned for outplanting on stressful sites where little or no aftercare is provided. This set of circumstances makes conditioning and hardening (see Chapter 12, *Hardening*) especially important, and these horticultural treatments require changing or modifying propagation structures.

Two essential processes in plants are photosynthesis and transpiration. Photosynthesis is the process in which light energy from the sun is converted into chemical energy in the presence of chlorophyll, the green pigment in leaves. During photosynthesis, sugars are produced from carbon dioxide from the air and water from the soil while oxygen is released back into the air (figure 4.1). Photosynthesis is a leaky process because, to allow the intake of carbon dioxide, water vapor is lost through pores, or stomata, on the leaf surfaces. This process is called transpiration. To maximize the photosynthesis necessary for plant growth, growers must manage any limiting atmospheric factors in the propagation environment.

> A propagation environment is any area that has been modified to grow plants. It may or may not involve a structure like a greenhouse.

Greenhouse operated by the Confederated Salish and Kootenai Tribes in Montana by Tara Luna.

Figure 4.1—*Two important processes occur in the leaves of green plants. In photosynthesis, sunlight triggers a chemical reaction in which water from the soil and carbon dioxide from the air are converted to sugars and oxygen, which are released back to the atmosphere. During the process, water vapor is lost from the leaves in a process known as transpiration.* Illustration by Jim Marin.

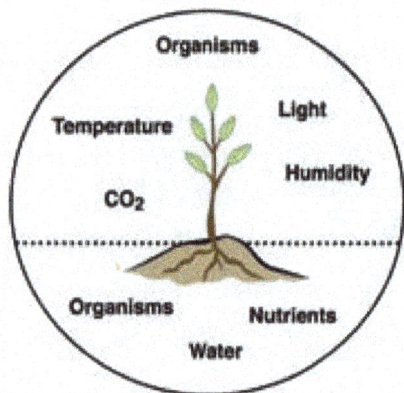

Figure 4.2—*It is useful to think of the nursery environment in terms of factors that might be limiting to plant growth. Limiting factors in the soil include water and mineral nutrients whereas, temperature, light, carbon dioxide, and humidity can be limiting factors in the atmosphere. Other organisms can either be beneficial or detrimental in both places.* Illustration by Jim Marin.

LIMITING FACTORS

Managing all the various factors that can be limiting to plant growth is the key to successful nursery management. To do this, the best possible propagation environment must be designed for a specific nursery site (Landis 1994). It is helpful to separate these limiting factors of the environment into those in the atmosphere and those in the growing medium (figure 4.2).

Atmospheric

Atmospheric limiting factors include light, temperature, humidity, carbon dioxide, and organisms that are determined by the climate at the nursery if plants are grown outside. As discussed in this chapter, propagation structures can be built to modify the local climate so that plants will grow more rapidly. For example, a greenhouse will modify light, temperature, and wind compared to the outside environment, which affects not only temperature but also humidity and carbon dioxide levels inside the greenhouse. The greenhouse also affects the organisms that interact with the crop. For example, although a greenhouse structure can exclude insect pests, it also creates a more humid environment for new pests such as algae and moss. In this chapter we discuss modifying light, temperature, carbon dioxide, and humidity with propagation structures.

Growing Medium

Growing medium limiting factors include water and mineral nutrients. The type of propagation environment can certainly affect water use; the details are discussed in Chapter 10, *Water Quality and Irrigation*. Mineral nutrients are supplied through fertilization (see Chapter 11, *Fertilization*); both water and mineral nutrients are held for plant uptake in the growing medium (see Chapter 5, *Growing Media*).

Biotic

Organisms can be limiting in either the atmosphere or the growing medium. Animal pests, including insects, can be excluded from a nursery through proper design, and beneficial microbes, such as mycorrhizal fungi, can be promoted. Beneficial organisms are covered in Chapter 14, *Beneficial Microorganisms*, and pests are discussed in Chapter 15, *Holistic Pest Management*.

Matching Propagation Environments to the Site

Whether building a new nursery or modifying an existing one, it is critical to analyze the limiting factors on the site. For example, the Hopi Reservation in northeastern Arizona is at high elevation with sunny, cloudless days for most of the year. Winters can be very cold with snow, and high winds are very common year-round. Here, the most limiting site factors are the intense sunlight, freezing temperatures, and high winds; therefore, a strong greenhouse to withstand snow and wind loads, with an aluminized shadecloth to minimize heat buildup, is desired.

A completely different propagation structure would be required on the Yurok Reservation in the coastal rainforest of the northern California coast, where fog is common and it rarely gets below freezing. Here, the limiting factors are low sunlight with heavy winter rains, so a structure with a very clear covering to maximize light transmission while protecting the crop from heavy rains would be ideal.

As expected, the costs of nursery development increase with greater control of the propagation environment. A nursery that is well matched to its environment, however, will be much less expensive to operate than a poorly designed one.

The Challenge of Growing Many Species and Stock Types

Native plant nurseries differ from other types of horticulture in which high quantities of a few crops are grown in large greenhouses. Most tribal nurseries grow small numbers of a wide variety of plants in one location. Often, these crops must be started on various dates, so, at any one time, a nursery might have everything from germinating seeds to large plants. Although some species are grown from seeds, others in the same nursery might have to be grown from rooted cuttings. So, a good native plant nursery will have to be designed with many relatively small propagation environments in which plants of similar requirements can be grown. When starting a new nursery, designing a variety of propagation environments is a luxury; unfortunately, most existing nurseries have to modify existing propagation structures.

At a given nursery, different propagation environments are needed for different growth stages at different times of the year. For example, a greenhouse is an ideal environment for germinating seeds and establishing young seedlings in their containers. Greenhouses are expensive to operate, however, because of high energy requirements. After young seedlings are established, they could be moved to a shadehouse or open compound to continue their growth. During hardening, the crops must be acclimated to the ambient environment and this is usually done in the same shadehouse or open compound. In cold climates, the crops might need to be moved to an overwintering structure to protect their root systems from freezing temperatures and the shoots from winter desiccation. Many potential problems can be averted by careful nursery design and planning.

A good nursery manager knows how to "think like a plant," and create a propagation environment that modifies all physical and biological factors that may be limiting to plant growth.

TYPES OF PROPAGATION ENVIRONMENTS

When most people think of container nurseries, they think of greenhouses; however, many other propagation environments are available. For our purposes, we define "propagation environment" as any area that is modified to encourage the growth of nursery stock. This environment could be as simple as a shady area under a tree or it could be a greenhouse with full environmental controls. It is important to realize that you do not need a greenhouse to grow native plants. Many simpler and inexpensive propagation structures can be designed to create the type of growing environments that crops require. Understanding different types of propagation structures and how they work is critical whether designing a new nursery facility or modifying an existing one.

Container nursery facilities can be distinguished by their relative amount of environmental control: minimally controlled, semicontrolled, or fully controlled. These facilities differ not only in their complexity but also in their biological and economical aspects (table 4.1).

Minimally Controlled Environments

A minimally controlled environment is the simplest and least expensive of all types of propagation environments. The most common type is an open growing compound. It consists of an area where plants are exposed to full sunlight and usually nothing more than an irrigation system and a surrounding fence.

Open Growing Compounds

Nurseries use open compounds as hardening areas to expose crops grown inside structures to ambient conditions. Some nurseries, such as the Temecula Band of the Luiseno Indians in southern California, use an open compound that incorporates trees for natural shade (figure 4.3A). Plants can be grown directly on the ground using landscape fabric to control weeds over a layer of gravel to provide drainage. The Banksavers Nursery of the Stillaguamish Tribe in coastal Washington State grows a variety of riparian and wetland plants in an open compound (figure 4.3B). Some nurseries prefer to grow their stock on pallets or benches to improve air pruning of the roots. If the nursery soil is heavy and poorly drained, then drainage tiles should be installed. Irrigation is provided by way of sprinklers for smaller containers or driplines for larger ones; plants obtain nutrients from controlled-release fertilizers that are incorporated into the growing media. The compound should be fenced to minimize animal damage, and, in windy areas, a shelterbelt of trees can improve the coverage of the irrigation system. Although open compounds are an inexpensive way to grow plants, they have the highest risk of freezing injury. Frost protection with irrigation is possible, however, the excess water can cause serious disease problems. For this reason, open growing compounds are more popular in milder climates; for example, in Louisiana, where the Clifton-Choctaw Nursery grows longleaf pine (figure 4.3C).

Wetland Ponds

Artificial ponds are another type of minimally controlled environment. They are used for growing riparian and wetland plants and are especially good for propagating sedges and rushes. Wetland ponds can be aboveground tanks, such as wading pools or cattle troughs, or they can be constructed with pond liners either in an excavated area or at ground level using a raised perimeter (figure 4.4A). These simple propaga-

Figure 4.3—*(A) The simplest nurseries are open compounds that use natural shade but have irrigation and are fenced. (B) Open compounds, like this one used by Roy Tyler of the Clifton-Choctaw Tribe in Louisiana, are most appropriate in areas with milder climates, where the risk of cold injury is minimal. (C) Even in colder climates, open compounds are often used for hardening crops grown in greenhouses or other structures.* Photo A by Tara Luna, B by Charles Mathern, and C by Thomas D. Landis.

Table 4.1—Operational considerations for selecting a propagation environment

Factors	Minimally Controlled	Semi Controlled	Fully Controlled
BIOLOGICAL			
Ambient climate	Mild	Moderate	Any
Growing season	Summer	Spring to fall	Year-round
Cropping time	6 to 24 months	3 to 12 months	3 to 9 months
Risk of crop loss	High	Low	Low
ECONOMIC			
Construction costs	Low	Medium	High
Maintenance costs	Low	Medium	High
Energy use	Low	Low to medium	High

tion environments use growing media amended with controlled-release fertilizer and require only periodic flood irrigation. For example, the Shoshone-Bannock Tribes in southeastern Idaho grow a variety of wetland and riparian plants in their nursery (figure 4.4B).

Semicontrolled Environments

This next category of propagation environments is called "semicontrolled" because only a few of the limiting factors in the ambient environment are modified. Semicontrolled environments consist of a wide variety of growing structures ranging from simple cold frames to shadehouses.

Cold Frames

Cold frames are low-to-the-ground structures consisting of a wood or metal frame with a transparent covering (figure 4.5). As their name suggests, they have no heating source except the sun. Cold frames are the most inexpensive propagation structure and are easy to build and maintain. Because temperatures inside can rise quickly, cold frames can be used to extend the growing season in spring. Seeds can be germinated and cuttings can be rooted weeks before they could be in an open compound. Cold frames are also used in late summer or autumn for hardening plants moved out from greenhouses and can be used for overwintering crops. Cold frames are labor intensive, however, because they need to be opened and closed daily to control temperature and humidity levels (figure 4.5).

The ideal location for a cold frame is a southern exposure with a slight slope to ensure good drainage and maximum solar absorption. A sheltered spot against the wall of a building or the greenhouse pro-

Figure 4.4—*(A) Wetland ponds can be constructed in the outdoor nursery for growing wetland species or (B) by using plastic tubs inside a greenhouse, such as this one operated by the Shoshone-Bannock Tribes of Idaho.* Photo A by Thomas D. Landis, B by J. Chris Hoag.

Figure 4.5—*Cold frames are an inexpensive alternative to a greenhouse. (A) Cold frames should be placed in a sheltered location for additional protection. (B) Coverings may be removed or (C) held open to manage humidity and heat levels.* Photos by Tara Luna.

vides additional protection. Some nurseries sink the frame 6 to 12 in (15 to 30 cm) into the ground to use the earth for insulation. Other nurseries make their cold frames lightweight enough to be portable so they can move them from one section of the nursery to another.

It is relatively easy to build a cold frame. Frames are usually made of wood such as redwood or cedar that will resist decay; the new recycled plastic lumber also works well. Never use creosote-treated wood or wood treated with pentachlorophenol because these substances are toxic to plants. The cold frame should be built so that it is weathertight and the top lid should be constructed so that it can be propped open at different heights to allow for various levels of ventilation, watering, and the easy removal of plants. The cover must be able, however, to be attached securely to the frame to resist wind gusts. Old storm windows make excellent covers for cold frames. Heavy plastic film is an inexpensive covering but usually lasts only a single season. Hard plastic or polycarbonate panels are more durable and will last for several years. Cold frame kits may also be purchased and are easily assembled; some kits even contain automatic ventilation equipment.

Cold frames require careful management of temperature and humidity levels. A thermometer that can be conveniently read without opening the cover is mandatory. In a cold frame, cool-season plants grow best at 55 to 65 °F (13 to 18 °C), while warm-season plants grow well at 65 to 75 °F (18 to 24 °C). If air temperature goes above 85 °F (29 °C), the top must be opened to allow ventilation. Monitor the temperature to ensure that the cold frame does not cool down too much, but close it early enough before the solar heat is lost.

Hot Frames or Hot Beds

Structurally similar to cold frames, hot frames are heated with electric heating cables and are primarily used as an inexpensive way to root cuttings. They are also ideal for overwintering nonhardy seedlings or newly rooted cuttings. Cold frames can easily be converted into hot beds. Start by removing the soil to a depth of 8 or 9 in (20 cm). Lay thermostatically controlled heating cables horizontally in loops on top of 2 in (5 cm) of sand (figure 4.6). Be sure that the cable loops are evenly spaced and do not cross each other. Cover the cable with 2 in (5 cm) of sand and cover the

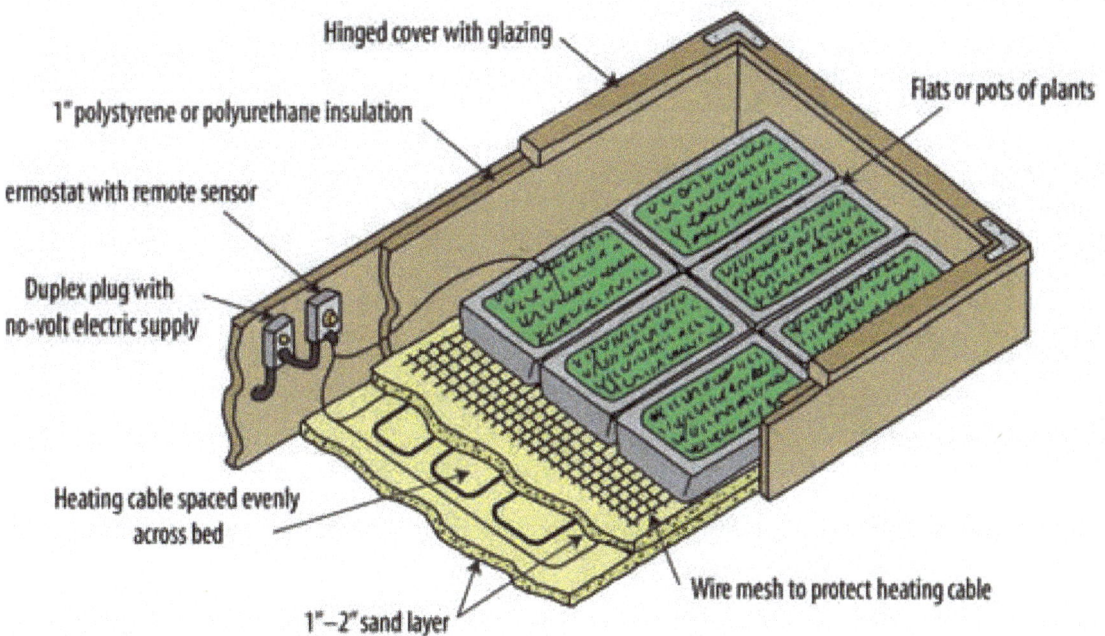

Hinged cover with glazing

1" polystyrene or polyurethane insulation

ermostat with remote sensor

Duplex plug with no-volt electric supply

Heating cable spaced evenly across bed

1"–2" sand layer

Flats or pots of plants

Wire mesh to protect heating cable

Figure 4.6—*A hot bed is structurally similar to a cold frame but is heated by electric heating cables.* Illustration courtesy of John W. Bartok, Jr.

Figure 4.7—*Hoop houses, like this one at the Colorado River Indian Tribes Nursery in Arizona, can be used for a variety of propagation environments by changing or removing the coverings.* Photo by Tara Luna.

sand with a piece of wire mesh (hardware cloth). Trays of cuttings or seedlings can be placed directly on top of the wire mesh.

Hoop Houses and Polyethylene Tunnels

Hoop houses and polyethylene ("poly") tunnels are very versatile, inexpensive propagation environments. The semicircular frames of polyvinyl chloride (PVC) or metal pipe are covered with a single layer of heavy poly and are typically quite long (figure 4.7). The end walls are made of solid material such as water-resistant plywood. The cover on hoop houses is changed during the growing season to provide a different growing environment, eliminating the need to move the crop from one structure to another. Generally, in early spring, a clear plastic cover is used during seed germination and seedling establishment. As the days become longer and warmer, the plastic cover can be pulled back on sunny days to provide ventilation. After the danger of frost is past, the plastic cover is removed and replaced with shadecloth. Sometimes, a series of shadecloths, each with a lesser amount of shade, are used to gradually expose crops to full sun. During hardening, the shadecloth is completely removed to expose the plants to the ambient environment. When covered with white plastic sheeting, hoop houses can be used for overwintering. See Chapter 13, *Harvesting, Storing, and Shipping*, for further discussion on this topic.

Shadehouses

Shadehouses are the most permanent of semicontrolled propagation environments and serve several uses. Shadehouses are used for hardening plants that have just been removed from the greenhouse and also serve as overwintering structures. In the Southwestern United States, many tribal nurseries have shadehouses similar to that of the White Mountain Apache Tribe in Arizona (figure 4.8A). Shadehouses, however, are also being used to propagate plants, especially in locations where sunlight is intense; for example, at the Colorado River Indian Tribes Nursery in Arizona (figure 4.8B). One popular shadehouse design consists of upright poles with a framed roof; snow fencing is installed over the roof and sides to provide about 50 percent shade (figure 4.8C). Other shadehouses consist of a metal pipe frame covered with shadecloth. When used for growing, they are equipped with sprinkler irrigation and fertilizer injectors (figure 4.8D). When the shade is installed on the sides of the structure, shadehouses are very effective at protecting crops from wind and so reduce transpiration. In areas with heavy snow, removable shadecloths are used for the roofing and sides so that they can be removed to avoid damage during winter months. This can be done after plants are put into storage.

Fully Controlled Environments

Fully controlled environments are propagation structures in which all or most of the limiting environmental factors are controlled. Examples include growth chambers and greenhouses. Fully controlled environments are often used because they have the advantage of year-round production in almost any climate. In addition, most crops can be grown much faster than in other types of nurseries. These benefits must be weighed against the higher costs of construction and operation. Murphy's law of "Anything that can go wrong will go wrong" certainly applies to nurseries, so the more complicated a structure is, the more problems that can develop. This concept is particularly true in the remote locations of many tribal lands, where electrical power outages are more common and it is difficult, time consuming, and expensive to obtain specialized repair services (see table 4.1). The following is a brief description of common greenhouse designs. Some good references that provide much more detail

Figure 4.8—*(A) Shadehouses are semicontrolled environments that are used for hardening and overwintering plants and (B) are also used for propagation. (C) They are constructed of wood frames with snow fencing or (D) metal frames with shadecloth and are equipped with irrigation systems that can also apply liquid fertilizer.* Photos by Thomas D. Landis.

include Aldrich and Bartok (1989), Landis and others (1994), and Bartok (2000).

Growth Chambers

Growth chambers are high-cost options that are used almost exclusively for research. The high cost of the equipment and the expense to operate them make them unsuitable for normal nursery production.

Greenhouses

Tribes with large forestry and reforestation programs, such as the Mescalero Apache Tribe in New Mexico, the Red Lake Band of the Chippewa Indians in Minnesota, and the Confederated Salish and Kootenai Tribes in Montana, use greenhouses. In addition to growing native trees, tribes have grown many other kinds of native plants in different types of greenhouses, but all the greenhouses are transparent structures that allow natural sunlight to be converted into heat (figure 4.9A). On the other hand, greenhouses are poorly insulated and require specialized heating and cooling systems. Most people think that keeping a greenhouse warm during cold weather is the most critical consideration, but it is not that difficult to accomplish with modern heating systems. Actually, keeping a greenhouse cool during sunny days in spring and summer is more of a challenge, and so ventilation systems must be carefully engineered. In climates with periods

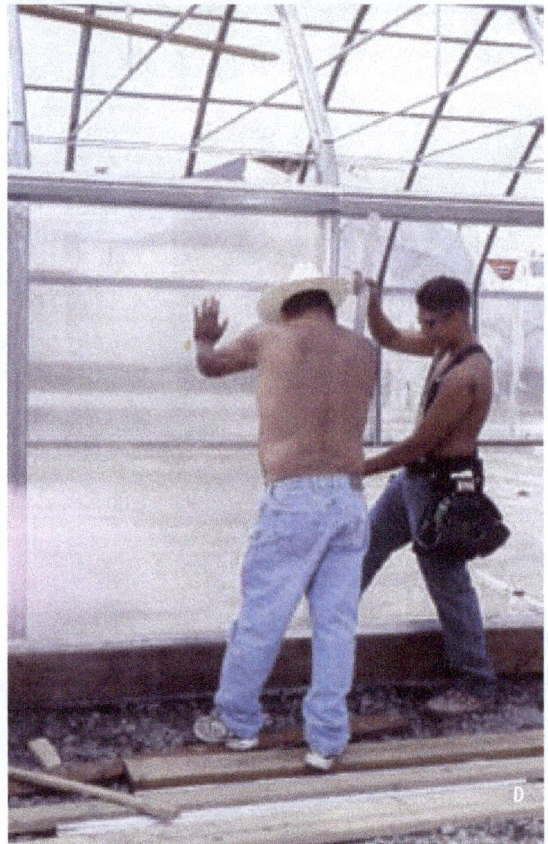

Figure 4.9—(A) Greenhouses, like this one on the Hopi Reservation in Arizona, are the most sophisticated propagation environments. (B) Retractable roof greenhouses allow crops to be exposed to the outside environment during hardening. (C) Workers with specialized skills are needed from the initial surveying to (D) the final construction. Photo A by Thomas D. Landis, B and C by Tara Luna, D by Joseph F. Myers.

of clear winter weather, greenhouses can heat up considerably during the day so, for this reason, it is usually not a good idea to use them for overwintering plants.

Retractable-roof greenhouses, the newest type of propagation structure, have become popular in temperate regions. Their major advantage is that the roof can be opened to expose plants to the outside environment when cooling is required (figure 4.9B). They are particularly useful during the hardening phase because the crop does not have to be moved to another structure such as a shadehouse. Some nursery managers without retractable-roof greenhouses remove the plastic from their greenhouses to help harden their crops.

ENGINEERING CONSIDERATIONS

It is important to understand that greenhouse construction is a very specialized business. Even a licensed contractor who is skilled at general construction will be challenged by the specialized demands of building a greenhouse (figures 4.9C and D). Here are a few general terms that anyone building a greenhouse or shadehouse should be familiar with.

Design Loads. *Dead loads* are the weight of the structure, whereas *live loads* are caused by building use. Live loads include people working on the structure and the weight of equipment, such as irrigation systems, heaters, lighting systems, and even hanging plants. Weather-related loads (snow and wind) must also be taken into consideration. In developed areas, be aware that greenhouses may be regulated by municipal building codes and zoning; this is another good reason to work with a professional firm before buying a greenhouse.

Foundations, Floors, and Drainage. The foundation connects the greenhouse to the ground and counteracts the design load forces. Inexpensive floors can consist of gravel covered with ground cloth, but the ground beneath the floor must drain water freely. Nursery crops require frequent irrigation and in most systems much of this water will end up on the floor. Drain tiles might be needed to make sure that the nursery floor will not turn into a bog. It may be necessary to design the greenhouse so that all wastewater drains into a pond or constructed wetland to prevent contamination of water sources. This water can sometimes be used for other purposes on the site, such as growing plants in wetland ponds as described previously in this chapter.

If wheeled equipment will be used to move plants, concrete walkways between the benches are necessary. Note that black asphalt heats up rapidly and becomes soft, so concrete is a better but more expensive option. Full concrete floors will eliminate many pest problems, especially algae, moss, and liverworts that thrive in the humid nursery environment. Make sure that floors are engineered to drain freely to prevent standing water, which is a safety hazard. Full floors can be engineered with drains so that all water and fertilizer runoff can be contained on site; runoff containment is a legal requirement in some parts of the country.

Framing Materials. Ideal framing supports the covering with minimal shading and heat loss while allowing ease of access and handling. Framing materials include galvanized steel, aluminum (lightweight but expensive), and treated wood.

Greenhouse Kits. The heating and cooling systems of fully controlled greenhouses must be carefully engineered to match both the size of the structure and the ambient environment. Be aware that inexpensive greenhouse kits often have vents or fans that are too small for the size of the greenhouse. Kit greenhouses were designed for some "average" environment and will probably have to be modified to handle the limiting environmental factors on your site. Before purchasing a greenhouse kit, it is a good idea to hire an experienced consultant, speak with a knowledgeable company representative, and discuss designs with other growers or professionals.

GREENHOUSE COVERINGS

A wide variety of greenhouse coverings are available and the selection of a particular type is usually based on cost, type of structure, and the environmental conditions at the nursery site.

Poly tarps are relatively cheap but require replacement every 2 to 4 years depending on the grade of plastic. Double layers of poly sheeting that are inflated with a fan are stronger and provide better insulation longer than a single layer (figure 4.10A). The two layers are attached to the framing with wooden furring strips or specially designed fasteners. This process is relatively simple so many growers change their own coverings. Because they are so well insulated and airtight, poly greenhouses require good ventilation to prevent condensation.

Figure 4.10—*(A) Greenhouses are covered with transparent coverings such as plastic sheeting or (B) hard plastic panels to maximize the amount of sunlight reaching the crop.* Photo A by Tara Luna, B by Thomas D. Landis.

Polycarbonate ("polycarb") panels, the most popular permanent greenhouse covering, have about 90 percent of the light transmission properties of glass. Polycarb is strong and durable but is one of the more expensive coverings (figure 4.10B).

These are the most common greenhouse coverings, and a more detailed description with costs, engineering and operational considerations can be found in Landis and others (1994) Some good references that provide much more detail on environmental controls include Aldrich and Bartok (1989), Landis and others (1994), and Bartok (2000).

SPECIALIZED PROPAGATION ENVIRONMENTS

This section discusses smaller propagation environments that have very specific functions. Often, they are constructed inside greenhouses or shadehouses.

Rooting Chambers

The most common type of vegetative propagation is the rooting of cuttings. Often, this form of propagation requires a specialized environment known as a rooting chamber that creates specific conditions to stimulate root initiation and development. Because cuttings do not have a root system (figure 4.11A), rooting chambers must provide frequent misting to maintain high humidity to minimize transpiration. Root formation is stimulated by warm temperatures and moderate light levels; these conditions maintain a high level of photosynthesis. Therefore, many rooting chambers are enclosed with poly coverings that, in addition to maintaining high hu-

midity, keep the area warm. If the chambers are outside, the covering further protects cuttings from drying winds and rain. Usually, bottom-heating cables are placed below the flats of cuttings or rooting medium to warm the medium, which speeds root development. Experience has shown that cuttings of many native plants root better when the growing medium is kept warmer than the shoots. See Chapter 9, *Vegetative Propagation*, for more information on rooting cuttings.

The two most common rooting chambers in native plant nurseries are enclosed chambers without irrigation and chambers with intermittent-mist systems.

Enclosed Rooting Chambers

Because it is easy to construct and very economical, a simple enclosed rooting chamber is essentially the same as the hot frame discussed earlier. Because they rely on manual operation, enclosed systems require diligent daily inspection to regulate humidity and air temperatures, and the rooting medium must be watered as needed. They are typically covered with shadecloth to moderate temperatures, but, if heat or humidity becomes excessive, enclosed chambers need to be opened for ventilation. One design is known as a "poly propagator" because it is covered with plastic sheeting and is a simple and inexpensive way to root cuttings (figure 4.11B).

Intermittent-Mist Rooting Systems

These propagation environments are the most common way to root cuttings and are either enclosed (figure

4.12A) or open (figure 4.12B). Rooting cuttings is much easier in these environments because intermittent-mist rooting chambers have a high degree of environmental control (figure 4.12E). Clock timers (figure 4.12D) control the timing and duration of mistings from specialized nozzles (figure 4.12E). These frequent mistings maintain very high humidity that reduces water loss from the cuttings, while evaporation of the small droplets moderates air and leaf temperature (figure 4.12F). Bottom heat is supplied to the rooting zone of the cuttings by means of insulated electrical cables at the bottom of the rooting medium (figure 4.12G) or with specially designed rubber heating mats placed under the rooting trays.

Mist systems require high water pressure that is supplied through PVC pipes and delivered through special nozzles (figure 4.12E). Very practical timing of the mistings is controlled by a series of two timeclocks that open and close a magnetic solenoid valve in the water line to the nozzles. One clock turns the system on during the day and off at night and the other controls the timing and duration of the mists. Because the aperture of the mist nozzles is so small, a cartridge filter should be installed in the water line after a gate valve. A thermostat controls the temperature of the heating cables or mat, which are protected by a wire mesh (figure 4.12C).

Because of the proximity of water and electricity, all employees should receive safety training. All wiring used for mist propagation must be grounded and must adhere to local building codes. Electrical outlets and components must be enclosed in waterproof outlets. The high humidity encourages the growth of algae and mosses, so the mist propagation system should be cleaned regularly. Mist systems require water low in dissolved salts; "hard" water can result in whitish deposits that can plug mist nozzles. See Chapter 10, *Water Quality and Irrigation*, for more information on this topic.

MANAGING THE PROPAGATION ENVIRONMENT

In nurseries, a variety of horticultural techniques can be used to modify the propagation environment. The type of propagation environment dictates the extent to which environmental conditions may be controlled. Ways of controlling the propagation environment are discussed in other chapters. The main way, and one of

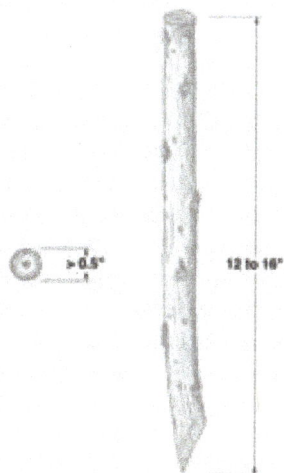

Figure 4.11—*(A) Rooting cutting requires a specialized propagation environment because cuttings lack a root system. (B) The "poly propagator" is the most simple and inexpensive rooting chamber; this one being used by Shannon Wetzel is at the Salish Kootenai College Nursery in Montana.* Illustration by Steve Morrison, photo by Tara Luna.

the most critical, in which growers control their crops is by the type of container. Container volume, plant spacing, and characteristics such as copper coating can greatly affect the size and quality of the crop (see Chapter 6, *Containers*). Different crops, as well as different stock types of the same species, may require different growing media. For example, a very porous media containing perlite or pumice is used for rooting cuttings, whereas a media with more water-holding capacity is required for germinating seedlings (see Chapter 5, *Growing Media*). A steady supply of high-quality water is one of the most critical needs of growing plants (see Chapter 10, *Water Quality and Irrigation*). It is possible to greatly accelerate the growth of native plants with fertilizer, especially for very slow-growing species (see Chapter 11, *Fertilization*). Certain organisms can be extremely important for the health and growth of some nursery crops. For example, the rooting of some difficult-to-root native plants has been shown to improve after treatment with a beneficial bacterium (Tripp and Stomp 1998) (see Chapter 14, *Beneficial Microorganisms*). Because of the high light intensity in greenhouses, controlling the light and temperature can be challenging.

Modifying Light in the Greenhouse

Light affects plants in several ways. As mentioned earlier, light is necessary for photosynthesis, which provides energy for plant growth. For light-loving species, more light equals more growth (figure 4.13A), but greenhouse light levels are often too intense to grow some native plants (table 4.2). As a result, growers apply shadecloths to lessen light intensity and the resultant heat (figure 4.13B). Shadecloths are rated by the amount of shade they produce, ranging from 30 to 80 percent. Black has been the traditional color but now shadecloth also comes in white, green, and reflective metal. Because black absorbs sunlight and converts it into heat that can be conducted into the propagation structure, black shadecloth should never be installed directly on the covering (figure 4.13C) but instead should be suspended above it to facilitate air movement. White shadecloth absorbs much less heat than black, and other colors absorb intermediate amounts of heat. New aluminized fabrics do a great job of reflecting incoming sunlight (figure 4.13D). Applying a series of shadecloths, each with a lesser amount of shade, over a period of time is a good way to gradually harden nursery stock and prepare it for outside conditions.

Supplemental Lighting

Another way that sunlight affects plants is the relative length of day and night, which is known as "photoperiod." Some native plants, especially those from high latitudes or high elevations, are very sensitive to daylength, a process controlled by the plant pigment phytochrome. When days are long, shoot growth occurs, but, when daylength drops below a critical level, shoot growth stops (figure 4.14A). Native plants

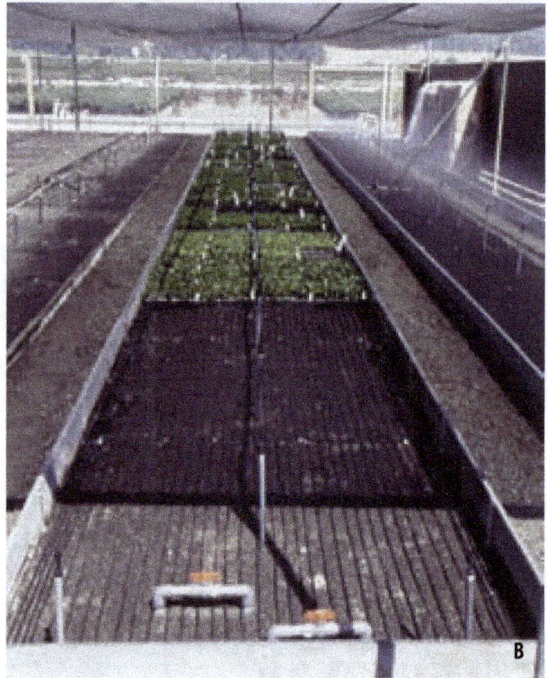

Figure 4.12—*(A) Intermittent-mist systems can be either enclosed or (B) in outdoor growing compounds. (C) Their environments can be easily controlled. (D) Programmable clock timers control the timing and duration of (E) specialized mist nozzles, which (F) keeps humidity levels high, reduces transpiration, and provides cooling. (G) Heating cables or mats under the growing medium keep rooting medium temperatures high.*
Photos A, B, D–G by Thomas D. Landis, C by Tara Luna.

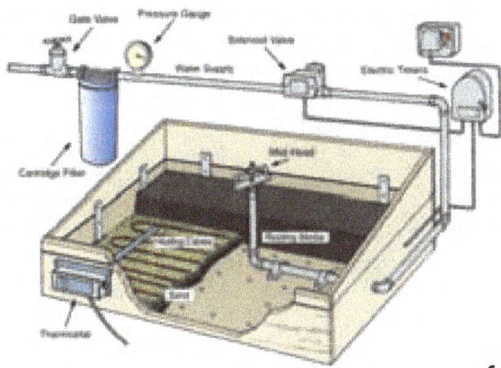

C

F

D

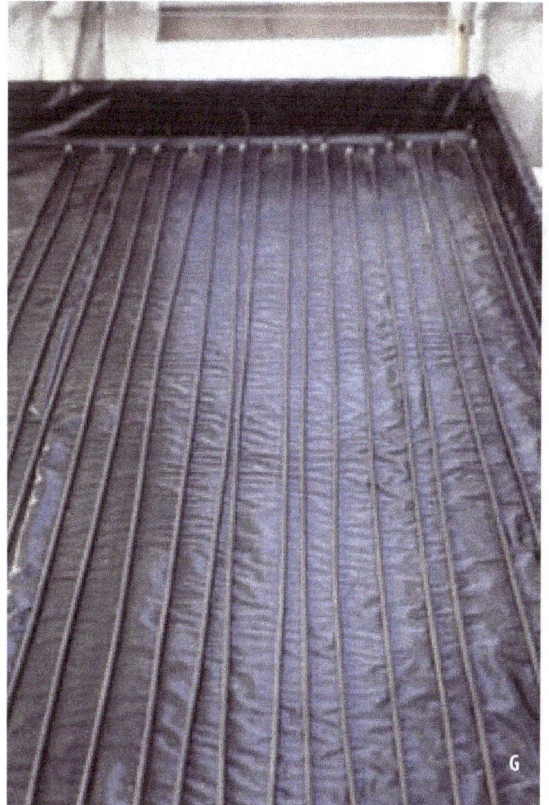

E

G

Table 4.2—Shade requirements of a variety of native plant species

Scientific Name	Common Name	Shade Requirement		
		Sun	Shade	Either
Artemisia tridentata	Big sagebrush	X		
Carex aquatilis	Water sedge	X		
Prunus virginiana	Chokecherry	X		
Dryopteris filis-mas	Male fern		X	
Chimaphila umbellata	Pipsissewa		X	
Gymnocarpium dryopteris	Oakfern		X	
Ceanothus sanguineum	Redstem ceanothus			X
Rubus parviflorus	Thimbleberry			X
Pteridium aquilinum	Bracken fern			X

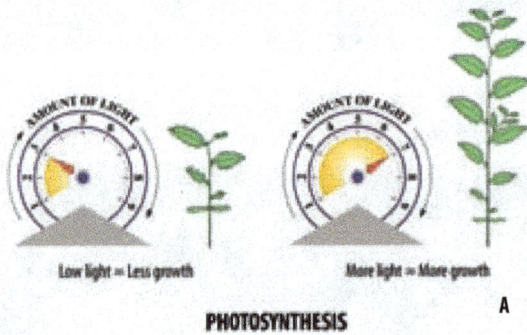

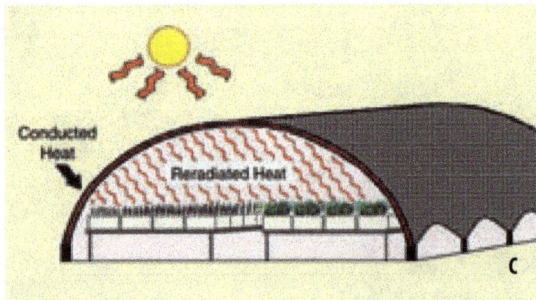

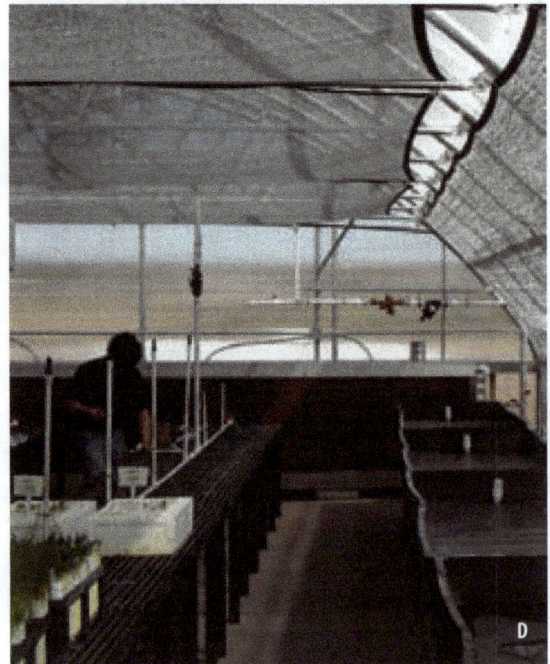

Figure 4.13—*(A) Sunlight provides the energy necessary for plant growth but is converted to heat inside propagation structures. (B) By reducing light intensity, shadecloth cools the environment. (C) Black shadecloth absorbs heat that is radiated back into the propagation environment, but (D) new aluminized shadecloth diffuses light without generating heat.* Photos by Thomas D. Landis, illustrations by Jim Marin.

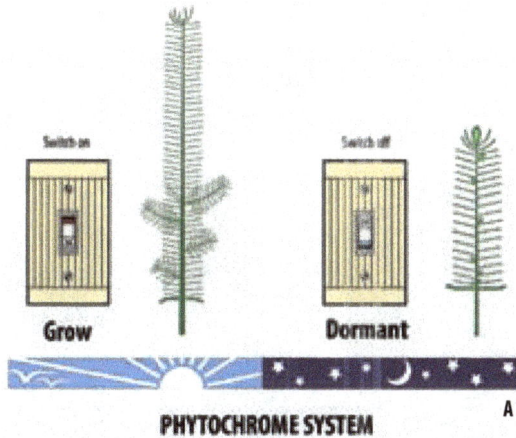

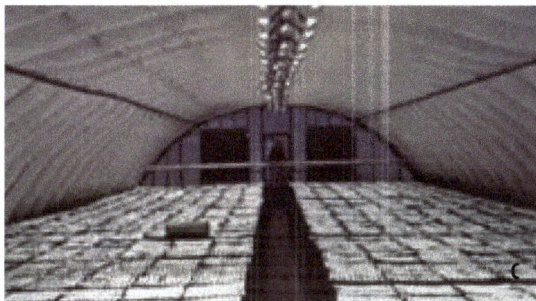

PHYTOCHROME SYSTEM

Grow

Dormant

Switch on

Switch off

A

B

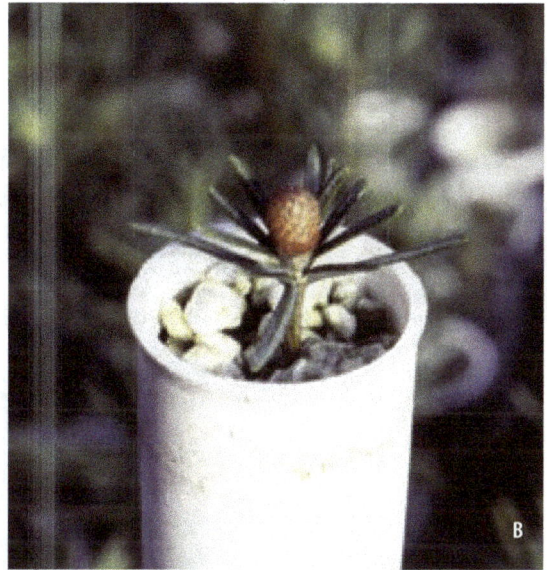

C

Figure 4.14—*(A) The relative length of day compared to night affects plant growth like a switch; long days stimulate growth, but short days cause dormancy. (B) Native plants from northern climates or high elevations are particularly sensitive to daylength and will quickly set a dormant bud under short day conditions. (C) Nurseries use photoperiod lights to artificially keep days long and their crops in an active state of shoot growth.* Photos by Thomas D. Landis, illustration by Jim Marin.

from northerly latitudes or high elevations are particularly sensitive to daylength and will "set bud" (stop shoot growth) quickly when days begin to shorten (figure 4.14B). This process is genetically controlled and protects plants against early fall frosts.

Container nurseries use photoperiod lights to extend daylength to force continued shoot growth (figure 4.14C). The lights are turned on as soon as seeds germinate and are shut off when height growth is adequate and the hardening phase begins. Several different lighting systems are used in nurseries; for more details see Landis and others (1992).

Modifying Temperature

One of the most challenging aspects of nursery management is controlling temperature in propagation structures. Temperature directly affects chemical reactions involved in plant metabolism and also affects rates of transpiration. As just discussed, sunlight is converted into heat, but this can be managed with shadecloth. Control units (figure 4.15A) automatically operate cooling and heating systems within green-

houses. Vents and fans are used to keep air moving inside the greenhouse (figure 4.15B) and exhaust heat from the structure (figure 4.15C). In dry environments, wet walls use the power of evaporation to cool incoming air (figure 4.15D). Growers can also use short bursts of their irrigation system for cooling (figure 4.15E).

Heating greenhouses is much easier than cooling them. During cold weather, heaters keep the propagation environment at the ideal temperature for growth (figure 4.16A). Rising fuel costs are becoming more of a concern and nurseries are adjusting their growing schedules and using other management strategies to reduce fuel costs. Many growers start their crops in heated greenhouses and then move them outside as soon as the danger of frost has passed (figure 4.16B)

Temperature Monitoring and Control Systems

Fortunately, temperature is very easy to measure and should be monitored at all times during the growth of the crop. Automatic sensing instruments are available that can be connected with cooling and heating equipment to trigger a cooling or heating cycle for

Figure 4.15—(A) Fully controlled greenhouses contain heating and cooling systems and sophisticated controllers. (B) Greenhouses heat up quickly during sunny weather and must be designed for cooling with interior circulation fans and (C) exhaust fans and vents. (D) In dry climates, fans pull air through wet walls, where it is cooled by evaporation. (E) A quick application of irrigation water can also be used to cool crops. Photo A by Jeremy R. Pinto, B-E by Thomas D. Landis.

Figure 4.16—(A) The heat from greenhouse heaters is circulated through the growing area. (B) Rising fuel costs are causing many nurseries to start their crops in a heated greenhouse and then move them to an open compound like this one at the Confederated Tribes of the Colville Reservation Nursery in Washington State. Photo A by Thomas D. Landis, B by R. Kasten Dumroese.

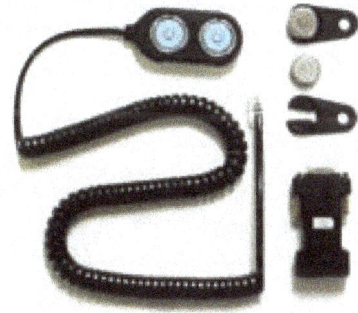

Figure 4.17—*(A) Monitoring and controlling temperature is critical to successfully growing a crop of native plants, and monitoring equipment is inexpensive. Many nurseries use thermometers that record daily maximum and minimum temperatures and (B and C) small, programmable, self-contained temperature sensors.* Photo A by Thomas D. Landis, B and C by David Steinfeld.

the greenhouse. Mechanical thermostats consist of a temperature sensor and switch that can be used to activate motorized vents, fans, and unit heaters within a greenhouse. Thermostats provide the best and most economical form of temperature control. Sophisticated control systems that can maintain a designated temperature through a series of heating and cooling stages is a very necessary and wise investment, considering current fuel costs.

Thermometers that record the maximum and minimum temperatures during the day are simple and economical instruments (figure 4.17A) that can help growers monitor subtle microclimates within the propagation environment. For example, the south side of the greenhouse is usually warmer than the north side or areas closest to the vents or cooling system. Thus, you could plan your greenhouse space by placing species that require slightly cooler temperatures for germination and growth on the north side of the greenhouse and use the south side of the greenhouse for species that prefer warmer temperatures. New devices, such as self-contained, programmable temperature sensors, are revolutionizing the ways in which temperature can be monitored in nurseries (figures 4.17B and C). Many of these sensors are small enough to be placed within a container or storage box and can record temperatures (between -40 and +185 °F [-40 and 85°C]) for more than 10 years. Because these

single-chip recording devices can be submersed in water and are resistant to dirt and impact, they can be used to monitor most temperatures encountered in the nursery. The data recorded on the sensors must be downloaded to a computer and can then be easily placed into a computer spreadsheet. The small size of the sensor can also be a drawback; they are easy to misplace. Attach a strip of colorful flagging to indicate where they are located and write any necessary information on the flagging with a permanent marker.

EQUIPMENT MAINTENANCE

Even if you purchase the best "automatic" environmental control equipment, it must be monitored and maintained. The hot and humid nursery environment is particularly hard on equipment; regular maintenance ensures longevity, reduces costly repairs, and may help avoid disasters. When selecting equipment, it is helpful to consult with other nurseries in your area that are growing similar species.

Routine maintenance of all greenhouse and nursery operation equipment should be a top priority. Someone who is mechanically inclined should be given the responsibility for equipment maintenance. Write everything down. The best place to do this is in a daily log book. See Chapter 3, *Crop Planning and Developing Propagation Protocols,* and Chapter 16, *Nursery Management,* for more details. These log books can be filed away by year and will prove invaluable when solving problems, budgeting, and developing maintenance plans. A system of "promise cards" specifies when servicing needs to be done and can be incorporated into the nursery computer system. Keep a supply of spare parts on hand, especially parts that may not be readily available or may take a long time to receive. It is a good idea to have a spare cooling fan motor on stand-by. A handy supply of hardware items such as washers, screws, and bolts is also good idea. Familiarize all employees on the operation of all equipment so that problems can be detected early. The instruction manuals for all equipment need to be kept on hand.

SUMMARY

A propagation environment must be carefully designed and constructed to modify the limiting factors on the nursery site. Each site is unique and there is no ideal type of nursery structure. Crop size, species, length of crop rotation, and the number of crops grown per year are important design considerations. The need for different propagation environments for different species and at different growth stages should also be considered. If only a few species with similar growing requirements are to be grown, then a large single growing structure is feasible. If the plan is to grow many different species with very different growth require- ments, then a variety of propagation structures will be needed. Having several smaller propagation structures provides more flexibility; these structures can be added over time. Propagation structures are modified based on the location and prevailing environmental conditions at the nursery and for the species that are being grown. In fully controlled greenhouses, temperature and humidity are usually provided by automatic heating, cooling, and ventilation equipment. Automatic control systems are a necessary investment in greenhouses. The diversity of species grown in native plant nurseries requires the provision of ambient conditions during the different phases of growth, including germination and establishment, rapid growth, and hardening. Thus, most native plant nurseries have a variety of propagation structures to meet these needs. With careful planning, these structures can be used for a variety of purposes throughout the year.

LITERATURE CITED

Aldrich, R.A.; Bartok, J.W., Jr. 1989. Greenhouse engineering. NRAES-33. Ithaca, NY: Cornell University, Northeast Regional Agricultural Engineering Service. 203 p.

Bartok, J.W., Jr. 2000. Greenhouse for homeowners and gardeners. NRAES-137. Ithaca, NY: Cornell University, Northeast Regional Agricultural Engineering Service. 200 p.

Landis, T.D. 1994. Using "limiting factors" to design and manage propagation environments. The International Plant Propagators' Society, Combined Proceedings 43: 213-218.

Landis, T.D.; Tinus, R.W.; McDonald, S.E.; Barnett, J.P. 1992. The container tree nursery manual: volume 3, atmospheric environment. Agriculture Handbook 674. Washington, DC: U.S. Department of Agriculture, Forest Service. 145 p.

Landis, T.D.; Tinus, R.W.; McDonald, S.E.; Barnett, J.P. 1994. The container tree nursery manual: volume 1, nursery planning, development, and management. Agriculture Handbook 674. Washington, DC: U.S. Department of Agriculture, Forest Service. 188 p.

Tripp, K.E.; Stomp, A.M. 1998. Horticultural applications of *Agrobacterium rhizogenes* ("hairy-root") enhanced rooting of difficult-to-root woody plants. The International Plant Propagators' Society, Combined Proceedings 47: 527-535.

ADDITIONAL READING

Clements, S.E.; Dominy, S.W.J. 1990. Costs of growing containerized seedlings using different schedules at Kingsclear, New Brunswick. Northern Journal of Applied Forestry 7(2): 73-76.

APPENDIX 4.A. PLANTS MENTIONED IN THIS CHAPTER

big sagebrush, *Artemisia tridentata*

bracken fern, *Pteridium aquilinum*

chokecherry, *Prunus virginiana*

longleaf pine, *Pinus palustris*

male fern, *Dryopteris filix-mas*

oakfern, *Gymnocarpium dryopteris*

pipsissewa, *Chimaphila umbellata*

redstem ceanothus, *Ceanothus sanguineus*

rushes, *Juncus* species

sedges, *Carex* species

thimbleberry, *Rubus parviflorus*

water sedge, *Carex aquatilis*

Growing Media

Douglass F. Jacobs, Thomas D. Landis, and Tara Luna

5

Selecting the proper growing medium is one of the most important considerations in nursery plant production. A growing medium can be defined as a substance through which roots grow and extract water and nutrients. In native plant nurseries, a growing medium can consist of native soil but is more commonly an "artificial soil" composed of materials such as peat moss or compost.

When people first began to grow plants in containers, they used ordinary field soil but soon found that this practice created horticultural problems. The very act of placing soil in a container produces conditions drastically different from those of unrestricted field soil. In the first place, plants growing in containers have access to a very limited amount of growing medium compared to field-grown plants. This limited rooting volume means that nursery plants can access only a small amount of water and mineral nutrients and these resources can change quickly. Second, containers create a perched water table, which means that water cannot drain freely out the bottom of the container (Swanson 1989). Third, native soils contain many microorganisms, such as bacteria and fungi, which do not exist in artificial growing media. Finally, native soils have texture (particle size) and structure (particle aggregations) that create porosity. An artificial growing medium has a texture based on the size and shape of its particles but does not have structure because the individual particles of the various components do not bind together. Therefore, the textural properties of

Potential components of growing media by Tara Luna.

growing media components must be carefully chosen and blended to produce the right mixture of porosity that will persist throughout the growing cycle.

It is important to realize that three different types of growing media are used in container nurseries:

1. Seed Propagation Media. For germinating seeds or establishing germinants, the medium must be sterile and have a finer texture to maintain high moisture around the germinating seeds.

2. Media for Rooting Cuttings. Cuttings are rooted with frequent mistings, so the growing medium must be very porous to prevent waterlogging and to allow good aeration, which is necessary for root formation.

3. Transplant Media. When smaller seedlings or rooted cuttings are transplanted into larger containers, the growing medium is typically coarser and contains compost instead of *Sphagnum* peat moss. Native plant growers often add 10 to 20 percent of soil or duff to encourage the development of mycorrhizal fungi and other beneficial microorganisms.

In this chapter, we explore the important media characteristics that can affect plant growth and discuss how nursery growers may use these basic principles to select and manage their growing media. More detailed information can be found in Bunt (1988) and Landis and others (1990).

FUNCTIONS OF GROWING MEDIA

In a native plant nursery, a growing medium serves four functions: (1) it physically supports the plant, (2) large pores promote oxygen exchange for root respiration, (3) small pores hold water, and (4) mineral nutrients are carried in the water to plant roots (figure 5.1).

1. Physical support

Although it might seem obvious, the growing medium must be porous to allow roots to grow out and provide physical support. Young plants are very fragile and must remain upright so that they can photosynthesize and grow. With larger nursery stock in individual containers, a growing medium must be heavy enough to hold the plant upright against the wind. Bulk density is the responsible factor and will be discussed in a later section.

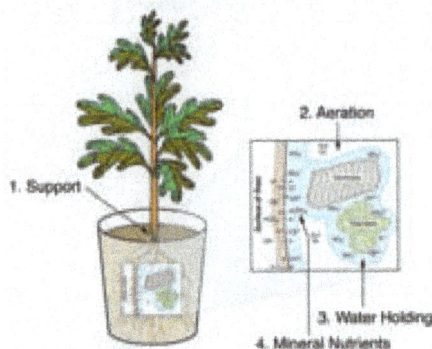

Figure 5.1—*Primary functions of growing media include the capacity to hold water and nutrients for root uptake, providing adequate root aeration, and ensuring structural support to the plant.* Illustration by Jim Marin.

2. Aeration

Plant roots need oxygen to convert the photosynthate from the leaves into energy so that the roots can grow and take up water and mineral nutrients. This process is called "aerobic respiration" and requires a steady supply of oxygen. The by-product of this respiration is carbon dioxide that must be dispersed into the atmosphere to prevent the buildup of toxic concentrations within the root zone. This gas exchange occurs in the large pores (macropores) in the growing medium. Because nursery plants are growing rapidly, they need a medium with good porosity—a characteristic termed "aeration" that will be discussed in more detail in the next section.

3. Water Supply

Nursery plants use a tremendous amount of water for growth and development, and this water supply must be provided by the growing medium. Artificial growing media are formulated so they can hold water in the small pores (micropores) between their particles. Many growing media contain a high percentage of organic matter such as peat moss and compost because these materials have internal spaces that can hold water like a sponge. Therefore, growing media must have adequate microporosity to absorb and store the large amounts of water needed by the growing plants.

4. Supply of Mineral Nutrients

Most of the essential mineral nutrients that nursery plants need for rapid growth must be obtained through the roots from the growing medium. When they are taken up by plants, most mineral nutrients are electrically charged ions. Positively charged ions (cations) include ammonium nitrogen (NH_4^+), potassium (K^+), calcium (Ca^{+2}), and magnesium (Mg^{+2}). These cations are attracted to negatively charged sites on growing medium particles until they can be extracted by roots (figure 5.1). The capacity of a growing medium to adsorb these cations is referred to as cation exchange capacity (CEC), and this important characteristic is discussed in the next section. Different types of media components vary considerably in their CEC, but peat moss, vermiculite, and compost have high CEC values, which explains their popularity in artificial growing media.

CHARACTERISTICS OF AN IDEAL GROWING MEDIA

Because no single material can meet all of the above criteria, artificial growing media often consist of at least two components. Therefore, growers must be familiar with the positive and negative characteristics of the various components and how they will affect plant growth in order to select a commercial medium or make their own. For our discussion, these characteristics can be divided into physical and chemical properties.

Physical Properties

Water-Holding Capacity

Micropores absorb water and hold it against the pull of gravity until plants can use it (figure 5.2). The water-holding capacity of a medium is defined as the percentage of total pore space that remains filled with water after gravity drainage. A good growing medium will have a high water-holding capacity but will also contain enough macropores to allow excess water to drain away and prevent waterlogging. Water-holding capacity is determined by the types and sizes of the growing medium components. For example, a peat moss particle will hold much more water than a piece of pumice. The degree of compaction is also extremely important. If growing medium particles are damaged during mixing or compacted when the containers are filled, the percentage of macropores is severely reduced. Overmixed or compacted media will hold too much water and roots will suffocate. Finally, the height of the growth container affects the water-holding capacity. Because nursery stock must be supported to allow air pruning of the roots, a certain amount of water will always remain in the bottom of the container. When filled with the same medium, short containers will have a higher percentage of waterlogging than taller ones (see figure 5.2).

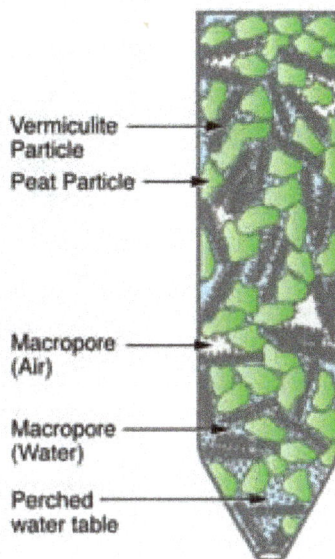

Figure 5.2—*A good growing medium contains micropores, which hold water, and macropores, which allow for air exchange. All containers also have a perched water table in the bottom.* Illustration by Jim Marin.

Aeration

The percentage of total pore space that remains filled with air after excess water has drained away is known as "aeration porosity." As we have already discussed, oxygen for good healthy roots is supplied through the larger macropores (figure 5.2), which also allow carbon dioxide from respiration to dissipate. A good growing medium, especially for rooting cuttings, contains a high percentage of macropores.

Porosity

The total porosity of a growing medium is the sum of the space in the macropores and micropores; as we have discussed, plants need both. A growing medium composed primarily of large particles will have more aeration and less water-holding capacity than a medium of smaller particles (figure 5.3). Either of these media would restrict plant growth. Plants growing in a medium with all large particles would dry out too quickly, and those growing in a medium with all small particles would suffer from waterlogging. For a single-component medium, the ideal particle range to promote both water-holding capacity and aeration is about 0.8 to 6 mm. In actual practice, however, a good growing medium will contain a mixture of components with different particle sizes and characteristics, for example, peat moss and vermiculite.

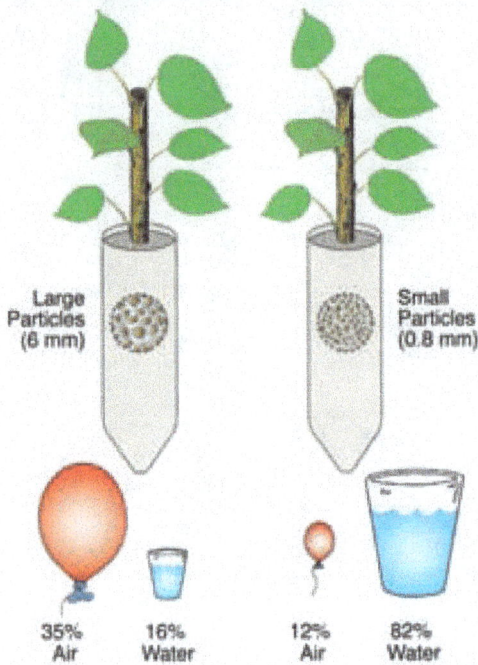

Figure 5.3—*These two containers each contain growing media with different particle sizes, either large (left) or small (right). The corresponding ballons show relative amounts of air being held, and the glasses show relative amounts of water being held.* Illustration by Jim Marin.

In the figure labels: Large Particles (6 mm), Small Particles (0.8 mm), 35% Air, 16% Water, 12% Air, 82% Water.

Shrinking and Swelling

Some soil-based media, especially those containing clays, shrink when drying or swell when wet. Shrinking and swelling is not a problem with the artificial growing media typically used in native plant nurseries.

Bulk Density

Bulk density means weight per volume. For any container type, weight per volume varies with the inherent bulk density of the growing medium components and how much they are compressed. For small-volume containers, an ideal growing medium will be light-weight to facilitate handling and shipping. For larger, free-standing containers, however, a good growing medium must have enough weight to provide physical support.

For a given container type and growing medium, excessive bulk density is a measure of compaction. Bulk density and porosity are inversely related; when bulk density increases, porosity decreases. Even a very porous growing medium can be ruined if it is compressed when the containers are filled.

Chemical Properties

Fertility

Because proper nutrition is so important for growing healthy nursery stock, fertility is the most important chemical property. Rapidly growing young plants use up the stored nutrients in their seeds soon after emergence. For the rest of the season, plants must rely on the growing medium to meet their increasing demands for mineral nutrients. As described in Chapter 11, *Fertilization*, many container nurseries prefer media with inherently low fertility (for example, peat-vermiculite) and add soluble fertilizers to media throughout the growing season. If fertilizers are difficult to obtain or cost prohibitive, organic amendments such as mature compost can be included in the growing media. Some native plants just grow better under low fertilization; in addition, beneficial microorganisms, such as mycorrhizal fungi, sometimes require low fertility to become established on plant roots. See Chapter 14, *Beneficial Microorganisms*, for more discussion on this topic.

pH

Another important chemical property is pH of growing medium, which is a measure of its relative acidity or alkalinity. pH values range from 0 to 14; those below 7 are acidic and those above 7 are alkaline. Most native plants tend to grow best at pH levels between 5.5 and 6.5, although some species are more pH tolerant. The main effect of pH on plant growth is its control on nutrient availability (figure 5.4). For example, phosphorus availability drops at extreme pH values because it binds with iron and aluminum at low pH levels and with calcium at high pH levels. The availability of micronutrients, such as iron, is even more affected by pH. Iron chlorosis, caused by high pH, is one of the most common nutrient deficiencies around the world. Exceptionally high or low pH levels also affect the abundance of pathogens and beneficial microorganisms. For example, low pH can predispose young plants to damping-off fungi.

CEC

CEC refers to the ability of a growing medium to chemically hold positively charged ions. Because most artificial growing media are inherently infertile, CEC is a very important consideration. In the growing medium, plant roots exchange excess charged ions for charged nutrient ions (see figure 5.1), and then these nutrients are transported to the foliage, where they are used for growth and development. Because the CEC of a growing medium reflects its nutrient storage capacity, it provides an indication of how often fertilization will be required. Because substantial leaching occurs with high irrigation rates, container nurseries prefer a growing medium with a very high CEC. One reason why native soils are not recommended for growing media is that clays adsorb cations so strongly that the cations may become unavailable for plant uptake, while the very low CEC of sandy soils causes most nutrients to be lost by leaching.

Biological Properties

Artificial growing media are preferred in nurseries because they are generally pest free. Although peat moss is not technically sterile, it should not contain pathogens or weed seeds when obtained from reliable sources. Vermiculite and perlite are rendered completely sterile during manufacturing, when they are exposed to temperatures as high as 1,832 °F (1,000 °C). In comparison, one of the most serious problems with soil-based growing media is that native soil can contain a variety of pests, such as pathogenic fungi, insects, nematodes, and weed seeds. For this reason, soil needs to be pasteurized with heat or sterilized with chemicals before it is used in growing media.

Well-prepared composts, however, are generally pest free because high temperatures during composting kill all pathogens. Another benefit of composting is that beneficial microorganisms increase in the final stages of the process. Composted pine bark, for example, contains microbes that suppress common fungal pathogens and nematodes. These suppressive effects depend on the parent material and composting time (Castillo 2004). Some commercial mixes advertise that they contain products that are antagonistic to pathogenic fungi.

COMPONENTS OF ARTIFICIAL MEDIA

Most native plant nurseries prefer artificial growing media and either mix it themselves or purchase premixed commercial brands. Although pure peat moss is used in some northern container tree nurseries, most growing media consist of two or more components

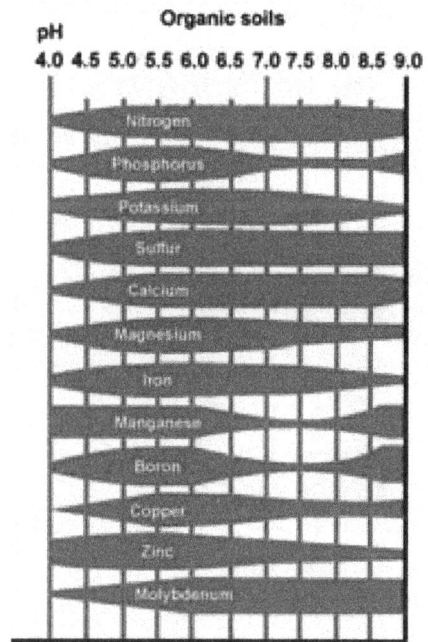

Figure 5.4—*The availability of all mineral nutrients is affected by the pH of the growing medium. In growing media such as organic soils, maximum availability occurs between 5.5 and 6.5.*

that are selected to provide certain physical, chemical, or biological properties. Other amendments, such as fertilizer or wetting agents, are sometimes added during the mixing process. By definition, a growing medium component usually constitutes a large percentage (more than 10 percent) of the mixture, whereas an amendment is defined as a supplemental material that contributes less than 10 percent.

A typical growing medium is a composite of two or three components. Mixtures of organic and inorganic components are popular because these materials have opposite, yet complementary, physical and chemical properties (table 5.1). Common organic components include peat moss, bark, compost, rice hulls, coconut coir, and sawdust. These materials are generally low in weight and have high water-holding capacity. In addition, organic components generally have a high resistance to compaction, a high CEC, and may contain significant quantities of nutrients. Inorganic components of artificial media include gravel, sand, vermiculite, perlite, pumice, and polystyrene beads. Inorganic components improve media properties by increasing aeration pore space, adding bulk density, and enhancing drainage.

Table 5.1—Different chemical and physical properties of some common materials used to create an artificial growing medium

Component	Bulk Density	Porosity		pH	Cation Exchange Capacity
		Water	Air		
Peat Moss					
Sphagnum	Very low	Very high	High	3 - 4	Very high
Hypnum	Low	Very high	Moderate	5 - 7	Very high
Vermiculite	Very low	Very high	High	6 - 8	High
Perlite	Very low	Low	High	6 - 8	Very low
Bark	Low	Moderate	Very high	3 - 6	Moderate
Sand	Very high	Low	Moderate	Variable	Low

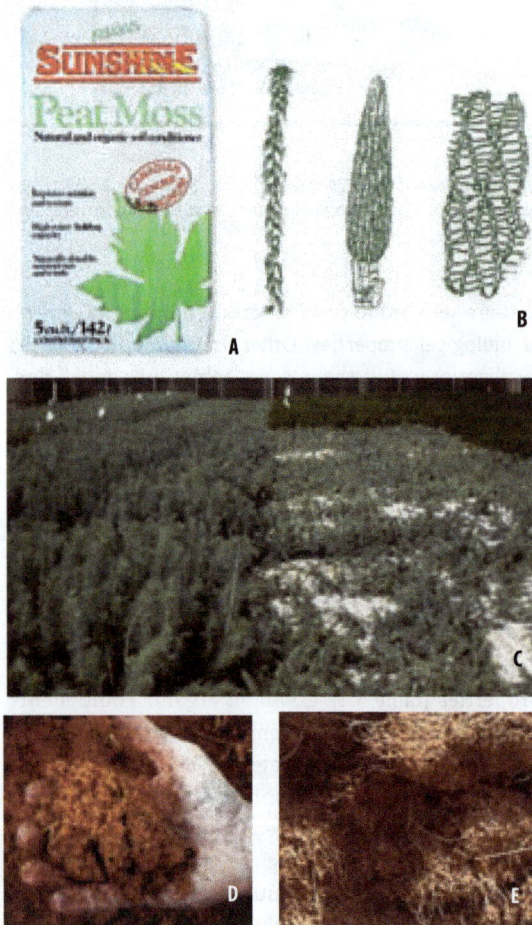

Figure 5.5—*Common organic components of growing media. (A)* Sphagnum *peat moss is the most popular because (B) its leaves contain open pores that create a high waterholding capacity. (C) Growers should avoid cheaper types of peat moss (*Sphagnum *on left;* Hypnum *on right) that can severely reduce plant growth. (D) Sawdust and (E) coconut coir have recently been used as a peat moss substitute.* Photos A–D by Thomas D. Landis, E by Tara Luna.

Organic Components

Peat moss is the most common component of artificial growing media (figure 5.5A). Peats can be composed of several species of plant, including mosses, sedges, and grasses. The species of plant, its degree of decomposition, variation in the local climate, and water quality all contribute to differences in peat moss quality and determine its value as a growing media component (Mastalerz 1977). Because *Sphagnum* moss leaves have open pores like a sponge (figure 5.5B), *Sphagnum* peat moss has a very good water-holding capacity, high CEC, low nutrient levels, and a comparatively low pH, often ranging from 3 to 4.5. In addition, the spongy texture of *Sphagnum* peat moss helps resist compaction and maintain porosity. Although types of peat moss may appear similar, they can have very different physical and chemical properties. *Sphagnum* peat moss must contain 75 percent mosses of the genus *Sphagnum*, and other peat products should never be considered (figure 5.5C). The ideal horticultural properties of *Sphagnum* peat moss (table 5.1) make it the only choice for growing native plants. In this handbook, when we use the generic term "peat moss," we mean *Sphagnum* peat moss.

Although peat moss is most popular, other organic materials such as bark, sawdust, compost, and coconut coir (figures 5.5D and E) have potential as growing media, especially in warmer climates where the cost of peat moss can be prohibitive. Tree bark is probably the most promising of alternative organic materials, and, when prepared properly, both pine and hardwood bark have found wide acceptance, especially for larger volume containers. One reason is that particle size can be controlled by the hammermilling and screening process (Gordon 2004). Composted pine bark, which has natural fungicide properties, can be produced on a small scale and has reduced pesticide use in nurseries (Castillo 2004). Raw Douglas-fir sawdust, combined with peat moss and vermiculite, has recently been used as a component in growing media for conifer seedling nurseries.

Growing media containing composts made from yard waste or pine bark were shown to be a viable alternative for peat moss when growing a variety of native plants in Florida (Wilson and others 2004). The chopped fiber from coconut husks is known as "coir" and has proven to be an excellent organic component for container growing media. Mixes containing coir are commercially available but are relatively more expensive than those with *Sphagnum* peat.

Inorganic Components

Inorganic materials are added to growing media to produce and maintain a structural system of macropores that improves aeration and drainage and decreases water-holding ability (Mastalerz 1977). Many inorganic components have a very low CEC and provide a chemically inert base for the growing medium. Inorganic materials with high bulk densities provide stability to large, freestanding containers.

Several materials are routinely being used as inorganic components in growing media in container tree nurseries in the United States and Canada. Vermiculite is the most common component (figure 5.6A) and is a hydrated aluminum-iron-magnesium silicate material that has an accordion-like structure (figure 5.6B). Vermiculite has a very low bulk density and an extremely high water-holding capacity, approximately five times its weight. This material also has a neutral pH, a high CEC (table 5.1), and contains small amounts of potassium and magnesium. Vermiculite is produced in four grades. The grades are based on particle size, which determines the relative proportion of aeration and water-holding porosity. Grades 2 and 3 are most commonly used in growing media; grade 2 is preferred when more aeration porosity is desired, whereas grade 3 produces more water-holding porosity. A mixture of 50 percent *Sphagnum* peat moss and 50 percent coarse vermiculite is considered a standard artificial growing media.

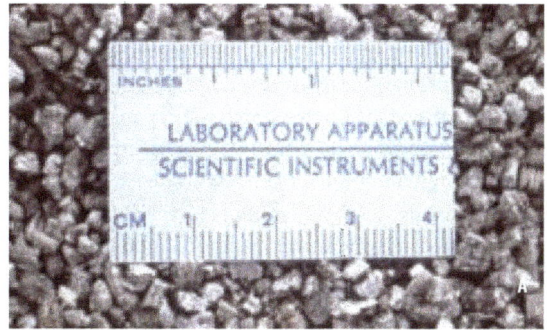

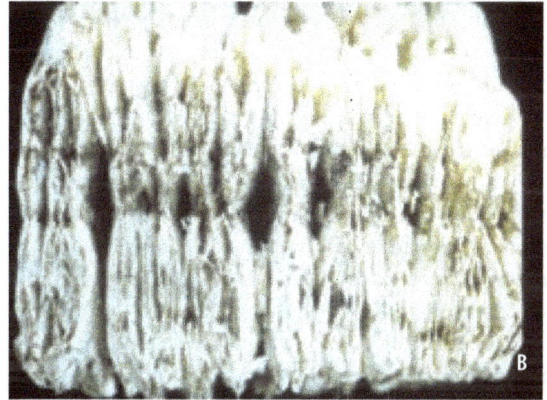

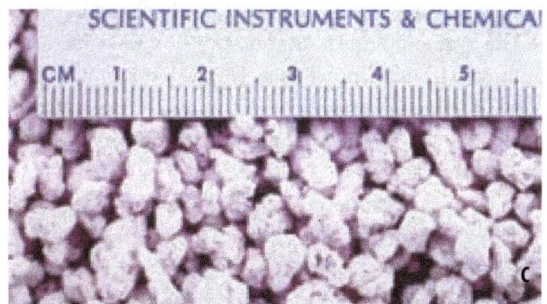

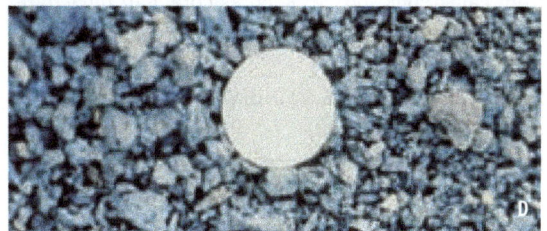

Figure 5.6—*Common inorganic components of growing media. (A) Horticultural vermiculite particles (B) look like accordions because of their expanded structure of parallel plates, which allow vermiculite to absorb water and mineral nutrients like a sponge. (C) Perlite particles have a closed-cell structure that prevents water absorption and improves aeration and drainage. (D) The particles of pumice also improve aeration but do absorb some water.* Photos by Thomas D. Landis.

Perlite is the second most popular inorganic component (figure 5.6C) and is a siliceous material of volcanic origin. Perlite particles have a unique closed-cell structure so that water adheres only to their surface; they do not absorb water as peat moss or vermiculite do. Therefore, growing media containing perlite are well drained and lightweight. Perlite is also rigid and does not compress easily, which promotes good porosity. Because of the high temperatures at which it is processed, perlite is completely sterile. It is essentially infertile, has a minimal CEC, and has a pH around neutral (table 5.1). Perlite is typically included to increase aeration, and commercial mixes contain no more than 10 to 30 percent of perlite. Perlite grades are not standardized, but grades 6, 8, or "propagation grade" are normally used in growing media. Perlite grades often contain a range of particle sizes, depending on the sieve sizes used during manufacturing. One safety concern is that perlite can contain considerable amounts of very fine particles that cause eye and lung irritation during mixing.

Pumice (figure 5.6D) is a type of volcanic rock consisting of mostly silicon dioxide and aluminum oxide with small amounts of iron, calcium, magnesium, and sodium. Readily available and relatively cheap in most of the Western United States, the porous nature of pumice particles improves aeration porosity but also retains some water within the pores. Pumice is the most durable of the inorganic components and so resists compaction.

Sand was a traditional component in the first artificial growing media but is almost never used now because of its weight (Gordon 2004). If you must use sand, choose siliceous sands because those derived from calcareous sources such as coral can have dangerously high pH values.

SELECTING A GROWING MEDIUM

A wide variety of commercial mixes are available that feature combinations of the components mentioned above (figure 5.7A). Although most media, such as peat-vermiculite, contain only two to three components, the exact composition of a brand may vary by location (figure 5.7B). Always read the label before purchasing a commercial mix. To appeal to a broader market, many brands contain a wide variety of additional amendments including fertilizers, wetting agents, hydrophilic

A

B

C

Figure 5.7—*(A) Many commercial growing media are available (B) so always check the label to determine the exact composition and whether amendments such as pine bark, coconut coir, or sand, have been added. (C) Many native plant growers prefer to mix their own species-specific media using a variety of components.* Photos A and B by Thomas D. Landis, C by Tara Luna.

gels, and even beneficial microorganisms. Again, always check the label to be sure of exactly what is being purchased. More details on amendments are provided in a later section.

Many native plant growers prefer to purchase components separately and mix their own custom growing media (figure 5.7C). In addition to saving money, custom mixing is particularly useful in small native plant nurseries, where separate mixes are needed to meet propagation requirements of different species. A very porous and well-drained medium, for example, might be needed for plants from very dry habitats.

When considering a new growing medium, first test it on a small scale with several species. In this way new media can be evaluated and plant quality compared before making a major change with the whole crop. Because of the diverse characteristics of the various growing media components, a grower can formulate a growing medium with almost any desired property. Be aware, however, that the physical, chemical, and biological properties of each growing medium are strongly affected by cultural practices, particularly irrigation, fertilization, and the type of container. Because the growing medium controls water and nutrient availability, it is easiest and most efficient to design custom mixes when several species are grown in the same irrigation zone. For the same reason, it is not a fair test to place a few containers of a new medium on a bench under existing irrigation and fertilization. See Chapter 17, *Discovering Ways to Improve Crop Production and Plant Quality*, for proper ways to install tests in the nursery.

CREATING A HOMEMADE GROWING MEDIUM

Although standard commercial mixes, such as peat-vermiculite, are generally superior for growing crops, some native plant nurseries prefer to formulate their own homemade medium. Reasons include poor availability of commercial media, price, shipping costs, lack of adequate storage, or simple preference. Many native plant nurseries are located in remote areas, where shipping costs for media components or commercial mixes may exceed their actual price.

Use of Field Soil

Most container nurseries prefer artificial growing media, but owners of some native plant nurseries think that soil-based media are more natural or organic. When considering native soil, several things should be kept in mind. Soils are naturally variable, so it is difficult to maintain the same quality from container to container or crop to crop. Ecological sustainability should also be considered. Harvesting topsoil is actually a mining operation that uses up a limited resource that took thousands of years to develop. If the decision is made to use native soil, we still recommend a sterile, uniform artificial media for germinating seeds, rooting cuttings, and any plants growing in smaller containers. The safest use of native soils is to incorporate a small amount (10 to 20 percent) into the mix when transplanting into larger containers. Adding a small amount of topsoil introduces desirable microorganisms into the medium and adds weight for greater stability. Be aware that most topsoil contains weed seeds that will germinate quickly in the ideal growing environment of a nursery.

When selecting soil, use dark topsoil that has a high percentage of organic matter; and lighter sandy loams are better than heavy clays. Harvesting soil from beneath healthy plants of the same species being grown ensures that the proper microorganisms will be present. For example, to grow normally in nurseries, the roots of *Ceanothus* plants must be inoculated with a bacterium called *Frankia*. Tests revealed that neither *Frankia* grown in artificial cultures or crushed *Frankia* root nodules were effective. However, 75 percent of the plants inoculated with soil that was collected under native *Ceanothus* stands became inoculated and exhibited superior vigor and growth (Lu 2005). After collection, sift the soil through a 0.5 in (12 mm) screen to remove debris and large objects such as rocks. When using native soils, heat pasteurization (described in the following paragraphs) will eliminate fungal pathogens, insect pests, nematodes, and weeds.

Making Compost

Because of the risks of using soils, many native plant nurseries prefer organic compost as a "green" alternative to peat moss. Composts are an excellent sustainable organic component for any growing medium and significantly enhance the medium's physical and chemical characteristics by improving water retention, aeration porosity, and fertility. Some compost has also been found to suppress seedborne and soilborne pathogens.

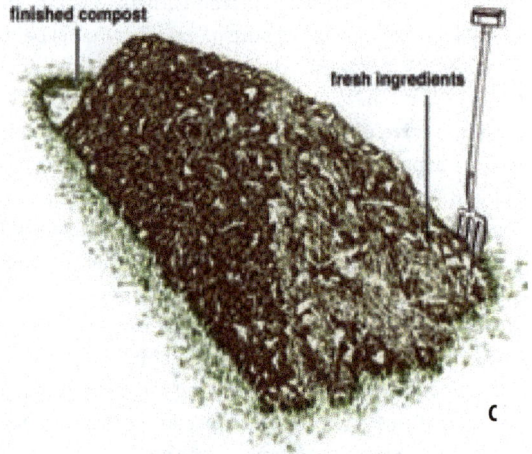

finished compost

fresh ingredients

C

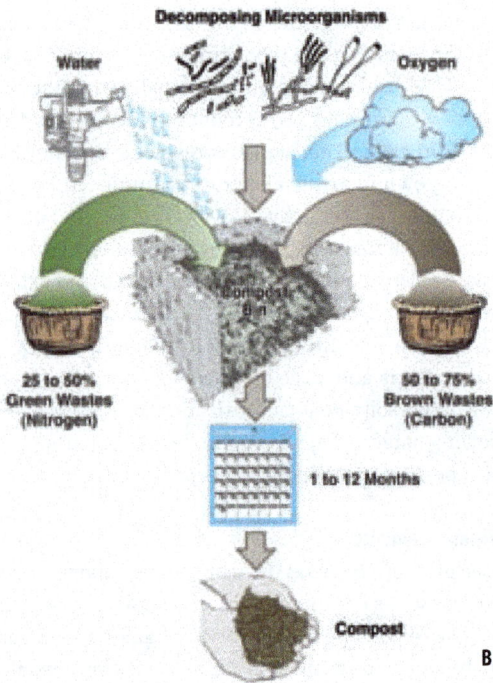

A

Decomposing Microorganisms

Water

Oxygen

Compost Bin

25 to 50%
Green Wastes
(Nitrogen)

50 to 75%
Brown Wastes
(Carbon)

1 to 12 Months

Compost

B

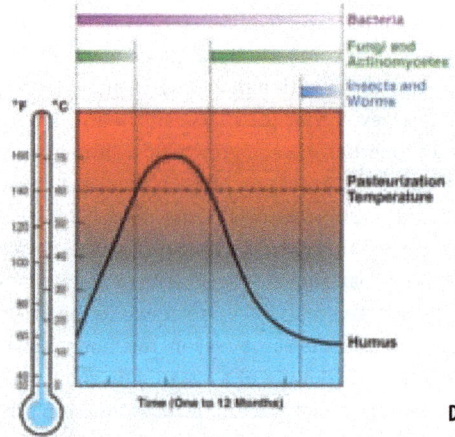

Bacteria

Fungi and
Actinomycetes

Insects and
Worms

Pasteurization
Temperature

Humus

Time (One to 12 Months)

D

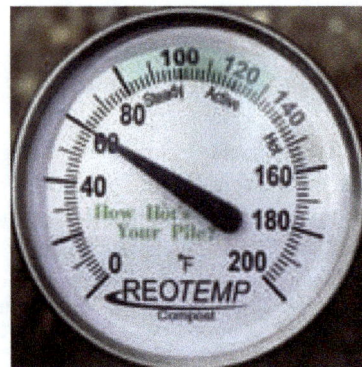

E

Figure 5.8—*(A) Many commercial composts are available. (B) Creating good compost takes several months and requires the proper mix of organic materials and creating the ideal environment for the microorganisms that decompose the materials. (C) Frequent mixing to foster good aeration is critical. (D) Compost goes through a typical temperature curve due to a succession of different microorganisms, (E) so the process should be monitored with a long-stemmed thermometer.* Photos by Thomas D. Landis, illustrations by Jim Marin.

Many brands of commercial compost are available (figure 5.8A). Compost can also be made on-site. Locating suitable organic materials for composting will vary considerably depending on the region where the nursery is located. Examples include grass, leaves, fruit wastes, coffee shells, rice hulls, wood waste such as sawdust or bark, sugar cane, manure, and even chicken feathers. Another benefit is that organic nursery wastes, such as used growing media or cull seedlings, can be composted, which reduces the costs and hassle of their disposal.

Making good compost is a rather technical process and takes some practice to learn. Here, we present some of the basic principles for creating compost for use in container production of native plants. To present these principles, we have synthesized information on composting from several excellent publications, including Martin and Gershuny (1992), Wightman (1999), and Castillo (2004). The Internet also has many sites devoted to composting. This brief description is meant only to introduce basic concepts and principles, not to serve as a step-by-step guide.

Composting is a natural process in which a succession of insects, fungi, and bacteria decompose organic matter and change its composition. The purpose of composting is to accelerate and control this process by modifying environmental conditions, especially moisture and temperature. Other factors that can be controlled include carbon-to-nitrogen ratio (C:N), aeration, and particle size (figure 5.8B).

Finished composts should have a C:N of about 30:1. Materials such as sawdust have much higher C:N that slows the composting process unless nitrogen fertilizer is added. When choosing materials for composting, maintain a mixture of 25 to 50 percent green organic matter and 50 to 75 percent brown organic matter (figure 5.8B). Green organic materials (fresh grass and fruit wastes) have a higher proportion of nitrogen compared with brown materials (tree leaves or sawdust), which contain more carbon. The particle size of your organics is very important. Particles that are too large reduce the surface area available for microbial attack whereas particles that are too small become compacted and create anaerobic conditions. A mixture of particles in the 0.5-to-2 in (1.2-to-5 cm) range works best. In well-aerated compost piles, however, particles can generally be at the smaller end of this range.

Maintaining adequate aeration is an important, yet often overlooked, factor. Microorganisms need an adequate and continual supply of oxygen, so it is important to turn over your compost pile once or twice a week. Poor aeration can delay or even stop the composting process. One good procedure to make certain the components are well mixed and all parts of the pile get proper aeration is to create an elongated windrow by turning over the pile in one direction (figure 5.8C). Moisture is another critical factor; the moisture content must be high enough to promote microbial activity but not so high as to reduce aeration and decrease temperature. Compost with a moisture content of approximately 50 to 60 percent has the feel of a damp sponge, which is ideal.

Several independent microbial decomposition phases occur during the composting process, creating a distinctive temperature sequence (figure 5.8D). Within the first few days, temperatures rise steadily to 100 to 120 °F (38 to 49 °C) as the smaller and easily biodegradable materials decompose. In the next step, temperatures rise to 130 to 150 °F (54 to 65 °C) as materials that are more resistant decay. A peak temperature of about 160 °F (71 °C) should be maintained for several days to kill weed seeds and fungal pathogens. Finally, temperatures fall to around 105 °F (40 °C) and lower during the "curing" stage. Growers can check the progress and detect problems in their compost heaps by monitoring with long-stemmed thermometers (figure 5.8E).

Determining When Compost Is Mature

Mature compost should be dark in color and have a rich, earthy smell (figure 5.9A). The texture should be friable and crumbly; the original organic materials should not be recognizable. Earthworms and soil insects invade mature composts and are an excellent sign that the compost is complete and ready to use (figure 5.9B).

Several tests help determine if your compost is mature and safe to use. These procedures, described in the following list, should also be used when purchasing commercial composts:

— **Sniff-and-Feel Test.** Place a small amount of compost into a plastic bag and seal it. Let this mixture sit for a day. If it feels hot or smells like manure or ammonia, it has not yet finished composting and should not be used.

Figure 5.9—*(A) The maturity of commercial or homemade compost should be checked before use in growing media. (B) In a mature compost, original components are no longer visible, earthworms and other soil insects are often visible, and it feels crumbly.* Photos by Tara Luna.

- **Germination Test.** Collect a sample of compost and put it into a small container. Sow seeds of a rapidly growing plant such as radish or lettuce, and place the sample in a window or in the greenhouse. If the compost is mature, the seeds should germinate and grow normally within a week or so.
- **Compost Maturity Tests.** Commercial test kits, such as the Solvita®, use a colorimetic process to measure the carbon dioxide and ammonium levels in a sample of compost. The level of these two factors correlates well with compost maturity.

After making the first batch of compost, it is a good idea to have a soils laboratory test it so that any nutritional deficiencies can be detected and corrected. Finally, before using composts, they should be sifted through a screen to remove large particles.

MIXING GROWING MEDIA

The mixing process is critical to producing custom growing media; the quality of the best components is compromised if the growing medium is improperly mixed. Whitcomb (2003) emphasized that improper mixing is one of the major causes of variation in container plant quality. The proper operating procedures are just as important as purchasing the right type of mixing equipment. Mixing should be performed by diligent, experienced workers who will faithfully monitor the quality of the growing media.

The special paddle-and-belt mixers used by commercial growing media companies do the best job of thoroughly mixing components without breaking down their structure. Most small native plant nurseries, however, cannot afford such specialized equipment and prefer to prepare small batches of growing medium by hand. Up to 5 or 6 cubic ft (0.15 cubic m or 155 L) of a medium can be mixed on any clean, hard surface by workers with hand shovels. Be sure to screen soil or compost to remove sticks and break up large clods (figure 5.10A). Pile the components on top of one another and broadcast any amendments over the pile. Then work around the edge of the pile with a large scoop shovel, taking one shovel full of material at a time and turning it over onto the top of the pile. As this material is added to the top, it tumbles down all sides of the pile and is mixed (figure 5.10B). Make sure that the center of the pile is mixed by gradually moving the location of the pile to one side during the mixing procedure. Some organic components repel water when dry, so misting the pile with water at frequent intervals during mixing makes the medium absorb water better. Continue this procedure until samples from the pile appear to be well mixed.

Nurseries that require larger quantities of custom growing media on a regular basis should purchase a mixer. A cement mixer (figure 5.10C) is often used and works well as long as care is taken to avoid excessive mixing ("overmixing"), which breaks down the size and texture of components. Fragile materials such as vermiculite and peat moss are particularly vulnerable to overmixing. Mechanized mixing can be easily overdone if the mixers are run too long or are overfilled or if the components are too wet. All too often, workers think that more mixing is better than less. On the contrary, overmixed media compacts easily during container filling and leads to reduced aeration and waterlogging.

Safety Considerations

Workers should follow certain precautions when handling growing media or its components. Dust is the most common concern, so work areas should be well ventilated. Spraying growing media and work areas with a water mist will also reduce dust. Workers handling and mixing growing media should wear protective dust masks and safety glasses (figure 5.11A). These same safety precautions should be taken when filling containers (figure 5.11B).

All growing media and components will generate dust, but perlite has been linked to silicosis, an inflammation that occurs over time when dust containing silica is inhaled into the lungs. Based on medical studies, however, no relationship exists between handling perlite and the development of silicosis (Schundler Company 2002). Still, dust is irritating and common sense dictates that masks should be worn when handling growing media or filling containers.

Asbestos contamination of vermiculite became another concern after the W.R. Grace mine in Libby, Montana, was closed in 1990. This mine produced a unique type of vermiculite that contained asbestos. Because of concerns about horticultural vermiculite, the National Institute for Occupational Safety and Health studied a broad range of commercial growing media containing vermiculite. It found that the use of commercial vermiculite hor-

Figure 5.10—(A) Nurseries that mix their own media should first screen materials and then (B) mix them thoroughly using the moving pile technique. (C) To reduce labor, a cement mixer can also be used but care must be taken to avoid overmixing and the resultant damage to particle size and structure. Photos A and B by Thomas D. Landis, C by Tara Luna.

Figure 5.11—(A) When mixing growing media (B) or filling containers, nursery workers should wear dust masks and safety glasses. Photo A by J. Chris Hoag, B by Thomas D. Landis.

ticultural products presents no significant asbestos exposure risk to commercial greenhouse or home users (Chatfield 2001).

Workers handling and mixing *Sphagnum* peat moss should not have cuts or abrasions on their hands because of the possibility of infection. A more serious concern is sporotrichosis, which is an infection caused by a fungal pathogen sometimes found in peat moss and other organic materials. The spores of this fungus can invade cuts on the hands or arms of workers or can even be inhaled (Padhye 1995). Preventative measures include:

— Storing peat moss and peat-based growing media under dry conditions.
— Ventilating work areas well.
— Wearing gloves, dust masks, and long-sleeved shirts to protect hands and arms.
— Thoroughly washing arms and other exposed parts of the body with soap and water to reduce the risk of infection.
— Treating any injury that breaks the skin with a disinfectant, such as tincture of iodine.
— Regularly sweeping and washing the work areas.

Sterilization of Growing Media

Sterilization refers to the complete elimination of all living organisms in the medium; pasteurization is less drastic. Completely sterile growing media may not be particularly desirable because many beneficial micro-

organisms, including bacteria, actinomycetes, and fungi, normally found in growing media can actually be antagonistic to pathogens. Some commercial brands of growing media are sterilized to prevent the introduction of pests, weeds, and diseases into the nursery (figure 5.12A). If concerned, growers should contact their supplier to find out if their media has been treated. For those mixing their own media, common inorganic components, such as vermiculite and perlite, are inherently sterile, but peat moss and other organic components are suspect. When using field soil or compost, growers should seriously consider pasteurization.

Although chemical fumigation is very effective, it is also expensive and hazardous and should be done only by registered pesticide personnel. Besides, the most common and effective chemical fumigant, methyl bromide, is being phased out because of concerns about ozone degradation. Heat pasteurization is the most common way of treating growing media and is traditionally done with steam. The standard recommendation is to heat the growing medium to 140 to 177 °F (60 to 80 °C) for at least 30 minutes. Commercially, media is pasteurized with large, expensive equipment but some native plant nurseries have developed their own portable pasteurization equipment (figure 5.12B). A practical technique would be to enclose small batches of media in black plastic tarps on an inclined table to expose it to maximum sunlight. Long-stemmed thermometers should be used to make sure that temperatures stay in the recommended range for the proper amount of time.

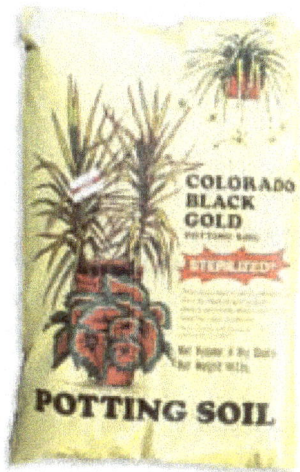

Figure 5.12—*(A) Some commercial growing media have been pasteurized to kill pathogens. (B) For nurseries making their own media, pasteurization with steam or solar heat is simple, effective, and done with portable equipment.* Photos by Thomas D. Landis.

INCORPORATION OF AMENDMENTS IN GROWING MEDIA

A variety of materials are routinely added to growing media during the mixing process; these include fertilizers, lime, surfactants, superabsorbents, and mycorrhizal inoculum. The uniform incorporation of these materials is important because plant roots have access to only a limited volume of growing media in the relatively small containers used in native plant nurseries Whitcomb (2003). When purchasing commercial media, growers should check the label and question the supplier to find out exactly what amendments have been added (see figure 5.7B).

Limestone. Called "lime" in horticulture, dolomitic limestone has traditionally been added to growing media to raise the pH and to supply calcium for plant nutrition. Better ways of supplying calcium exist, so we do not recommend limestone amendments unless you are growing plants that require a neutral or alkaline pH.

Starter Fertilizers. Some commercial media contain a small "starter dose" of soluble granular fertilizer. If fertigation (irrigation water containing liquid fertilizer) is not possible, then starter fertilizer may be a good idea to ensure that young, developing plants have quick access to mineral nutrients. Incorporating larger quantities of soluble fertilizer is never recommended because of the high potential for salt injury.

Controlled-Release Fertilizers. Several brands of growing media contain controlled-release fertilizers, and it is important to know their formulation and release rate. See Chapter 11, *Fertilization*, for more discussion about these fertilizers.

Surfactants. These chemical amendments, also known as "wetting agents," break down the surface tension of water and increase the wettability of hydrophobic organic materials such as peat moss and pine bark. Some surfactants have been shown to adversely affect the growth of pine seedlings. Because even less is known about their effects on other native plants, growers should ask other nurseries about their experiences and perform small tests before using surfactants operationally. See Chapter 17, *Discovering Ways to Improve Crop Production and Plant Quality*, for proper ways to set up tests.

Superabsorbents. Superabsorbents are cross-linked polymers that absorb many times their own weight in water. They have been proposed as additives to increase the water-holding capacity of growing media. Several brands of growing media contain superabsorbents but this is mainly an advertising gimmick. Because nursery crops are regularly irrigated, the use of superabsorbents is rarely justified. If growers want to try them, they should first test them on a limited basis.

Mycorrhizal Inoculum. One method of inoculating native plants with beneficial mycorrhizal fungi is to incorporate inoculum into the growing medium at the time of mixing. Again, this practice should be confirmed with small tests before adopting it on a large scale. See Chapter 14, *Beneficial Microorganisms*, for more discussion on this topic.

SUMMARY AND RECOMMENDATIONS

The selection of a growing medium is one of the most important decisions in the container culture of native plants. The physical, chemical, and biological characteristics of a growing medium affect not only seedling growth but also other aspects of nursery operations. Growers should carefully consider both biological and operational aspects when evaluating different types of growing media.

The decision to purchase a commercial brand of growing media or to custom mix depends on many factors, including the availability of components and mixing equipment and the size of the nursery operation. Several good commercial brands of growing media are available, but, for complete quality control, nursery managers should consider custom mixing their own media.

Whether purchasing a commercial media or custom mixing, the selection of growing medium components is critical. For most native plants, a growing medium consisting of *Sphagnum* peat moss and vermiculite is a good first choice if these materials are available and reasonable in price. The proportion of peat moss to vermiculite on a volume-to-volume basis can range from 1:1 to 3:1. Coarse-grade peat moss should be used whenever possible, and the coarser grades of vermiculite are preferred. A small proportion of perlite (10 to 30 percent) can be substituted for part of the vermiculite if better drainage is desired. Tree bark, especially composted pine bark, has shown promise, and sawdust has also been used as a peat moss substitute. Substitution of alternative organic materials for peat moss should be approached cautiously, however, and only composted organics should be considered.

When making your own growing media, make sure that mixing is complete but not so severe as to damage particle size or structure. Even though most components are considered sterile, nursery managers should consider pasteurization to eliminate any pathogens. As far as incorporating limestone, fertilizers, mycorrhizal fungal inoculum, surfactants, or superabsorbents, small-scale trials are always recommended before using the growing medium operationally.

LITERATURE CITED

Bunt, A.C. 1988. Media and mixes for container grown plants. London: Unwin Hyman, Ltd. 309 p.

Castillo, J.V. 2004. Inoculating composted pine bark with beneficial organisms to make a disease suppressive compost for container production in Mexican forest nurseries. Native Plants Journal 5(2): 181-185.

Chatfield, E.J. 2001. Review of sampling and analysis of consumer garden products that contain vermiculite. http://www.vermiculite.org/pdf/review-EPA744R00010.pdf (21 Feb 2006).

Gordon, I. 2004. Potting media constituents. The International Plant Propagators' Society, Combined Proceedings 54: 78-84.

Landis, T.D.; Tinus, R.W.; McDonald, S.E.; Barnett, J.P. 1990. The container tree nursery manual: volume 2, containers and growing media. Agriculture Handbook 674. Washington, DC: U.S. Department of Agriculture, Forest Service. 119 p.

Lu, S. 2005. Actinorhizae and *Ceanothus* growing. International Plant Propagators' Society, Combined Proceedings 54: 336-338.

Martin, D.L.; Gershuny, G. 1992. The Rodale book of composting. Emmaus, PA: Rodale Press. 278 p.

Mastalerz, J.W. 1977. The greenhouse environment. New York: John Wiley & Sons. 629 p.

Padhye, A.A. 1995. Sporotrichosis—an occupational mycosis. In: Landis, T.D.; Cregg, B., tech. coords. National Proceedings, Forest and Conservation Nursery Associations. General Technical Report PNW-GTR-365. Portland, OR: U.S. Department of Agriculture, Forest Service, Pacific Northwest Research Station: 1-7.

Schundler Company. 2002. Perlite health issues: studies and effects. http://www.schundler.com/perlitehealth.htm (20 Feb 2002).

Swanson, B.T. 1989. Critical physical properties of container media. American Nurseryman 169 (11): 59-63.

Whitcomb, C.E. 2003. Plant production in containers II. Stillwater, OK: Lacebark Publications. 1,129 p.

Wightman, K.E. 1999. Good tree nursery practices: practical guidelines for community nurseries. International Centre for Research in Agroforestry. Nairobi, Kenya: Majestic Printing Works. 93 p.

Wilson, S.B.; Mecca, L.K.; Stoffella, P.J.; Graetz, D.A. 2004. Using compost for container production of ornamental hammock species native to Florida. Native Plants Journal 4(2): 186-195.

ADDITIONAL READING

Anonymous. 2004. How to make compost: a composting guide. http://www.compostguide.com (17 Feb 2006).

APPENDIX 5.A. PLANTS MENTIONED IN THIS CHAPTER

ceanothus, *Ceanothus* species

Douglas-fir, *Pseudotsuga menziesii*

Containers

Tara Luna, Thomas D. Landis, and R. Kasten Dumroese

6

The choice of container is one of the most important considerations in developing a new nursery or growing a new species. Not only does the container control the amount of water and mineral nutrients that are available for plant growth, a container's type and dimensions also affect many operational aspects of the nursery such as bench size and type of filling and harvesting equipment. After a container is selected, it can be very expensive and time consuming to change to another type.

Many terms have been used to describe containers in nurseries and some are used interchangeably. In the ornamental trade, large individual containers are called "pots" or "cans" but they are simply called "containers" in native plant nurseries. Restoration seedlings, typically grown in small-volume containers, are often referred to as "plugs." Plug seedlings are usually grown in individual containers called "cells" or "cavities" that are aggregated into "blocks," "trays," or "racks." In general, individual cavities are permanently aggregated into blocks, and cells are independent containers that can be inserted in or removed from trays or racks.

Most nurseries will grow a wide variety of species and therefore several different containers will be required (figure 6.1A). The choice of container for a particular native plant species depends on root system morphology, target plant criteria (see Chapter 2, *The Target Plant Concept*), and economics. These factors are interrelated, which makes discussing them difficult, but, fortunately, some generalizations, hold true:

Alex Gladstone of the Blackfeet Nation in Montana by Tara Luna.

Figure 6.1—*(A) Native plant nurseries use a variety of containers to produce different species and stocktypes. (B) Some plants, including most forbs, grow best in shorter containers (C) whereas taprooted species such as oaks do better in taller ones. (D) Fleshy-rooted plants should be grown in short wide containers.* Photo by Tara Luna, illustrations by Steve Morrison.

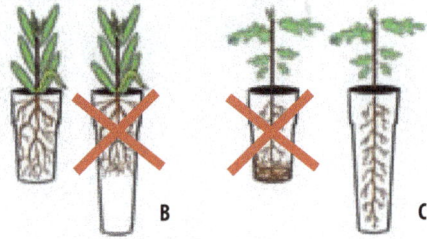

- Plants that develop shallow, fibrous root systems, as most forbs do, grow better in shorter containers (figure 6.1B).
- Plants with long taproots, such as oaks or some pines, grow better in taller containers (figure 6.1C).
- Plants with multiple, thick, fleshy roots, such as arrowleaf balsamroot and prairie turnip, and species with thick, fleshy rhizomes grow better in wide containers (figure 6.1D).

Because choosing the type of container is such a critical step, a discussion of some of the biological and operational considerations is very important.

CONSIDERATIONS IN CHOOSING CONTAINERS
Size

Container size can be described in many ways, but volume, height, diameter, and shape are most important.

Volume

The volume of a container dictates how large a plant can be grown in it. Optimum container size is related to the species, target plant size, growing density, length of the growing season, and growing medium used. For example, to grow large woody plants for an outplanting site with vegetative competition, a nursery would choose large-volume containers with low densities. These plants would be taller, with larger root-collar diameters, and have been shown to survive and grow better under these conditions.

In all nurseries, container size is an economic decision because production costs are a function of how many plants can be grown per area of bench space in a given time. Larger containers occupy more growing space and take longer to produce a firm root plug. Therefore, plants in larger containers are more expensive to produce, and they are also more expensive to store, ship to the project site, and outplant. The benefits, however, may outweigh the costs if the outplanting objectives are more successfully satisfied.

Height

Container height is important because it determines the depth of the root plug, which may be a consideration on dry outplanting sites. Many clients want the plants to have a deep root system that can stay in contact with soil moisture throughout the growing season. Height is also important because it determines the proportion of freely draining growing medium within the container. When water is applied to a container filled with growing medium, it percolates downward under the influence of gravity until it reaches the bottom. There, it stops due to its attraction for the growing medium, creating a saturated zone that always exists at the bottom of any container. Two things control the depth of this saturated layer—container height and the type of growing medium. With the same growing medium, the depth of the saturation zone is always proportionally greater in shorter containers (figure 6.2). For example, a 4-in- (10-cm-) tall container will have the same depth of saturation as a 10-in- (25-cm-) tall container, but the 4-in-tall container will have a smaller percentage of freely drained medium.

Diameter

Container diameter is important in relation to the type of species being grown. Broad-leaved trees, shrubs, and herbaceous plants need a larger container diameter so that irrigation water can penetrate the dense foliage and reach the medium.

Shape

Containers are available in a variety of shapes and most are tapered from top to bottom. Container shape is important as it relates to the type of outplanting tools used and the type of root system of the species grown.

Plant Density

In containers with multiple cells or cavities, the distance between plants is another important factor to consider. Spacing affects the amount of light, water, and nutrients that are available to individual plants (figure 6.3A). In general, plants grown at closer spacing grow taller and have smaller root-collar diameters than those grown farther apart (figure 6.3B). Plant leaf size greatly affects growing density. Broad-leaved species should be grown only at fairly low densities, whereas smaller-leaved and needle-leaved species can be grown at higher densities. Container spacing will affect height, stem straightness, root-collar diameter, and bushiness. Container spacing also affects nursery cultural practices, especially irrigation. Some of the other effects of plant growing densities are shown in table 6.1.

Root Quality

Container plant quality is greatly dependent on the plant's root system. Most native plants have very aggressive roots that quickly reach the bottom of the container and may spiral or become rootbound. Several container design features have been developed specifically to control root growth and development.

Drainage Holes

Containers must have a bottom hole large enough to promote good drainage and encourage "air pruning."

Figure 6.2—*A saturated layer of growing medium always exists in the bottom of containers. With the same growing medium, the proportion of saturated media is higher for shorter containers.* Modified from Landis and others (1989).

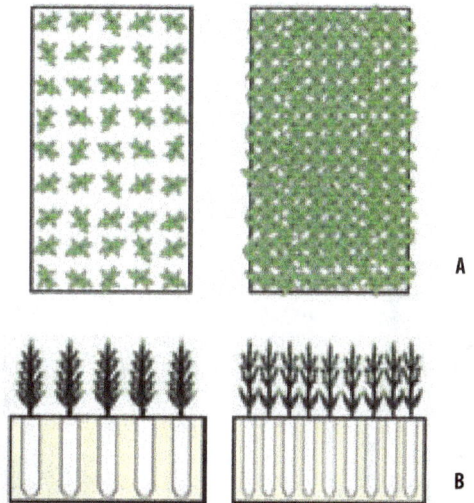

Figure 6.3—*(A) Next to volume, spacing is the most important characteristic in multicelled containers. (B) Plants grown too close together become tall and spindly and have less root-collar diameter.* Illustration by Steve Morrison.

As mentioned previously, roots quit growing when they reach an air layer under the container. Some containers feature a bottom rail to create this air layer (figure 6.4A), whereas flat-bottomed containers must be placed on specially designed benches (figures 6.4B and C). On the other hand, the drainage hole must be small enough to prevent excessive loss of growing medium during the container-filling process.

Root Pruning

Spiraling and other types of root deformation have been one of the biggest challenges for container growers, and nursery customers have concerns about potential problems with root-binding after outplanting

Table 6.1—Effects of container density on plant growth in nurseries

High Density	Low Density
Plants will be taller and have smaller root-collar diameters	Plants will be shorter and have larger root-collar diameters
Difficult to irrigate and fertilize with overhead sprinklers because water and liquid fertilizers need to penetrate dense patches of foliage	Easier to irrigate and fertilize with overhead sprinklers
Greater likelihood of foliar diseases due to poor air circulation between plants	Better air circulation between seedlings; less disease problems
Cooler medium temperature	Warmer medium temperature
Foliage in lower crown will die because of shading	Plants have fuller crowns because more light reaches lower foliage

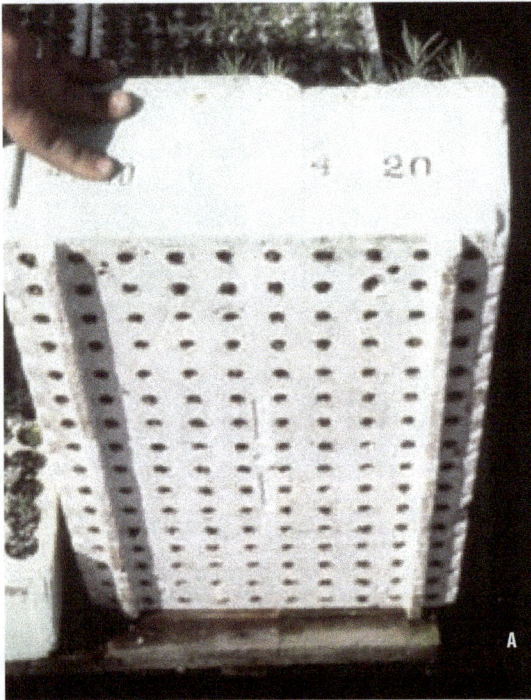

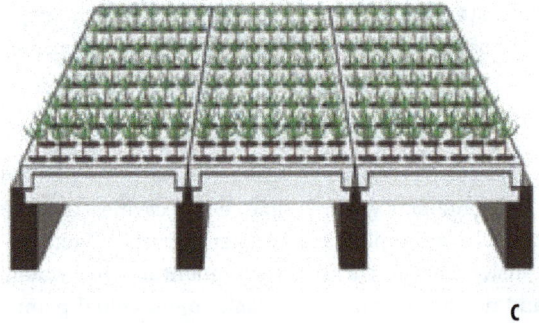

Figure 6.4—*(A) Some block containers are designed to promote air pruning. (B) Other containers must be placed on mesh-topped benches or (C) be supported to create an effective air space underneath.* Photos by Thomas D. Landis, illustration by Jim Marin.

(figure 6.5A). The research, most of which has been done with forest trees such as lodgepole pine, showed that rootbound seedlings were more likely to blow over after outplanting. The aggressive roots of native plants, however, can be culturally controlled by chemical or air pruning. Although both pruning methods have been used in forest nurseries, they are uncommon in native plant nurseries.

Chemical pruning involves coating the interior container walls with chemicals that inhibit root growth (figure 6.5B), such as cupric carbonate or copper oxychloride. Copper-coated containers are available commercially (for example, the Copperblock™) and some nurseries apply the chemical by spraying or dipping. Copper toxicity has not been shown to be a problem for most native species, and the leaching of copper into the environment has been shown to be minimal.

Several companies have developed containers that featured air slits on their sides to control spiraling and other root deformation by air pruning (figure 6.5C). The basic principle behind the "sideslit" container is simple. Just as when plant roots air prune when they reach the bottom drainage hole, they stop growing and form suberized tips when they reach the lateral slits in sideslit containers. Forest nurseries found that sideslit containers had two drawbacks: (1) roots sometimes bridged between containers, and (2) seedlings in sideslit containers dried out much faster than those in containers with solid walls.

Soft versus Hard Plugs

Better outplanting performance is usually achieved with container plants whose roots form a firm root plug, but the amount of root deformation increases with the amount of time that plants are kept in a given container. A hard or firm plug is achieved when plant roots bind the growing medium just enough to facilitate extraction from the container without the medium falling off the roots. Some customers prefer soft plugs, however, that have looser roots around the plug after extraction because they grow new roots more quickly following outplanting and better resist frost heaving or other mechanical disturbances.

Root Temperature

Color and insulating properties of the container affect medium temperature, which directly affects root growth. Black containers can quickly reach lethal temperatures in full-sun whereas white ones are more reflective and less likely to have heat buildup. In hot, sunny climates, a grower should use containers in white or other light-reflecting colors to protect against root injury (figure 6.6). Another option is to use white plastic, Styrofoam™, or other insulating material around the outside perimeter of the containers.

Economic and Operational Factors

Cost and Availability

Although the biological aspects of a specific container are important, cost and availability are often the controlling factors in container selection. Associated expenses, such as shipping and storage costs, must be considered in addition to purchase price. Many containers are produced at only one location and their shipping costs increase as a direct function of distance from the manufacturer; others, such as Styrofoam™

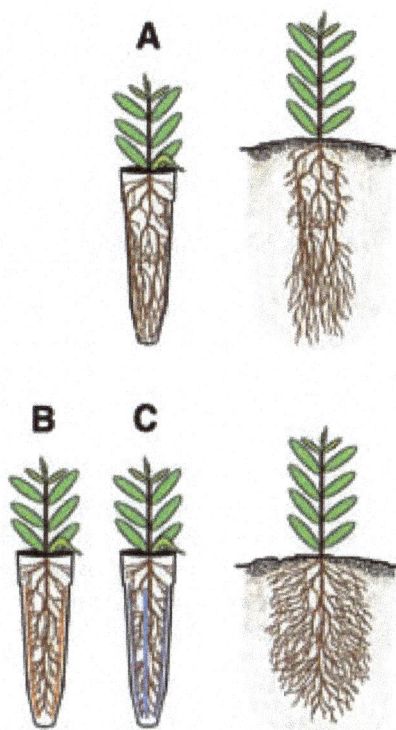

Figure 6.5—(A) Native plants with aggressive roots often exhibit spiraling and other deformities after outplanting. New roots often retain the shape of the original plug . (B) Containers coated with copper will chemically prune roots, and (C) other containers are available with lateral slits to encourage air pruning on the side of the plug. Illustration by Steve Morrison.

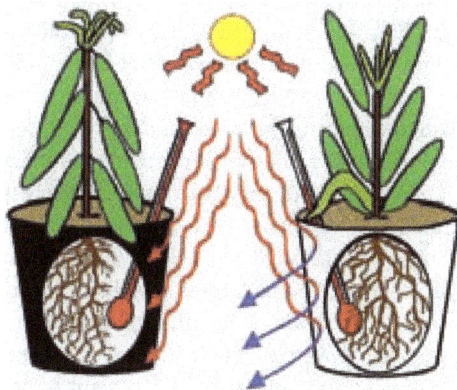

Figure 6.6—Container color is a consideration, especially when containers are exposed to direct sunlight. Roots in white containers stay cooler than those in black ones. Illustration by Steve Morrison.

blocks, are produced or distributed from various locations around the continent and are therefore widely available. Long-term availability must also be considered to ensure that ample supplies of the container can be secured in the foreseeable future.

Durability and Reusability

Containers must be durable enough to maintain structural integrity and contain root growth during the nursery period. The intense heat and ultraviolet rays in container nurseries can cause some types of plastics to become brittle, although many container plastics now contain ultraviolet (UV) inhibitors. Some containers are designed to be used only once whereas others can be reused for 10 or more crop rotations. Reusability must be considered in the container cost analysis because the cost of reusable containers can be amortized over their life span after adjusting for the cost of handling, cleaning, and sterilizing of the containers between crops (discussed later in this chapter).

Handling

Containers must be moved several times during crop production, so handling can be a major concern from logistic and safety standpoints. Collapsible or stackable containers, such as Zipset™ Plant Bands or Spencer-Lemaire Rootrainers™, have lower shipping and storage costs; they must, however, be assembled before filling and sowing and thus require additional handling. The size and filled weight of a container will affect ease of handling. Containers must be sturdy enough to withstand repeated handling.

As mentioned earlier, large containers are increasing in popularity, but they become very heavy when saturated with water and may require special pallets for handling by forklift (figure 6.7). Some block containers are easier to handle than others. Styroblock™ containers are rectangular with a smooth bottom, which makes them much easier to handle by conveyor. Containers with exchangeable cells are more difficult to handle, especially if they will be shipped to the outplanting site and must be returned. Ray Leach Cone-tainers™ are popular for growing native plants, but the plastic trays often crack after several uses and their sharp edges make conveyor handling difficult. Automated handling systems also place mechanical stress on containers. If containers will be shipped to the outplanting site, then

Figure 6.7—*Handling containers through the entire nursery cycle is a major consideration, especially when plants become large and heavy.* Photo by Thomas D. Landis.

the type of shipping and storage system must be considered during container selection. If seedlings are to remain in the container, then some sort of shipping box must be used to protect them.

Ability to Check Roots

Although it is easy to observe shoot growth and phenology, the condition of the growing medium and the degree of root activity are hidden by the container. For most containers, it is impossible to monitor root growth without disturbing the plant. Late in the growth cycle, however, the plant's roots have formed a firm root plug and can be removed from the container. Book-type containers, such as Spencer-Lemaire Rootrainers™, however, are hinged along the bottom of the containers so that they can be opened and the growing medium and roots examined during the entire growing season.

Ability to Cull and Consolidate

One advantage of tray containers with interchangeable cells, such as Ray Leach Cone-tainers™ and Deepots™, is that cells can be removed from the tray and replaced. This advantage is particularly useful during thinning, when empty cells can be replaced with cells containing a germinant, and during roguing, when diseased or otherwise undesirable plants can be replaced with cells containing healthy ones. This consolidation can save a considerable amount of growing space in the nursery. This practice is particularly valuable with seeds that germinate slowly or unevenly, and so exchangeable cells are very popular in native plant nurseries.

Holdover Stock

Some nurseries will hold onto their stock without transplanting in an effort to reduce costs and save growing space, hoping that the stock will be outplanted next year. This practice, however, is a bad idea. The shoots of holdover stock may look just fine, but, in fact, the root system is probably too rootbound to grow well after outplanting (figure 6.8). Nursery stock that has been held in containers for too long a period is much more prone to root diseases. If nursery stock must be held over, then it must be transplanted to a larger container size to keep the root systems healthy and to maintain good shoot-to-root balance.

TYPES OF CONTAINERS

Many types of containers are available and each has its advantages and disadvantages, so side-by-side comparisons are difficult. It is a good idea to try new containers for each species on a small scale before buying large quantities.

Six main types of containers are used in native plant nurseries. These containers, identified in the following list, range in volume from 0.5 cubic in (8 ml) to 25 gal (95 L) (table 6.2):

— One-time-use containers.
— Single, free-standing containers.
— Exchangeable cell containers held in a tray or rack.
— Book or sleeve containers.
— Block containers made up of many cavities or cells.

One-Time-Use Containers

The first major distinction in container types is whether they will be used once or whether they can be cleaned and used again. The idea of growing a plant in a container that can be transplanted or outplanted directly into the field is attractive and many designs have been tried. Unfortunately, most of these early attempts failed because the material broke down in the nursery or failed to decompose after outplanting. Jiffy® containers are the only one-time system in use in native plant nurseries and are discussed in a later section.

Single, Free-Standing Containers

Several types of single-cell containers are being used to grow native plants for specific conditions.

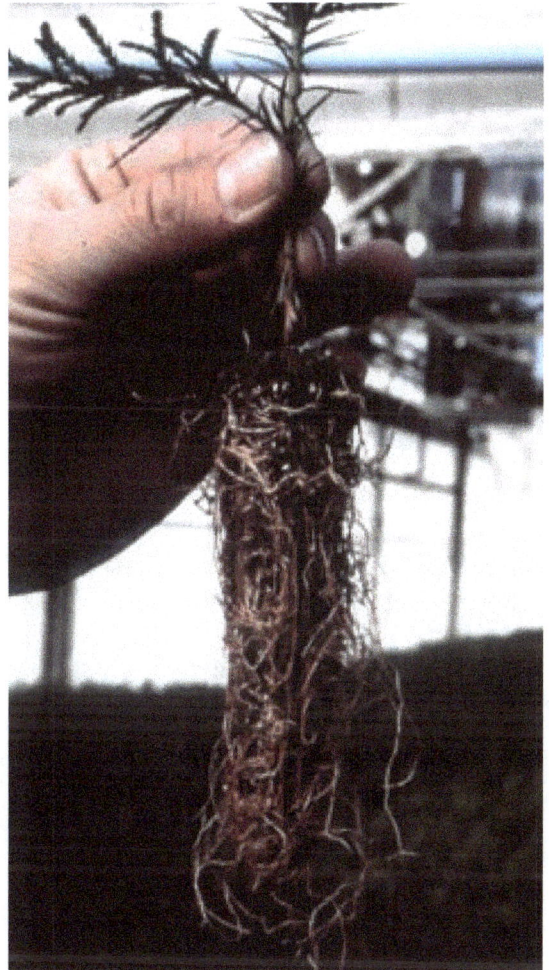

Figure 6.8—*Many native plants have aggressive roots and cannot be held over from one growing season to the next or they will become dangerously rootbound.* Photo by Thomas D. Landis.

RootMaker® Containers

These unique containers have staggered walls and a staggered bottom that prevent root circling and direct roots toward the holes in the walls and the bottom of the container. The containers were among the first to use side "air slits" to air prune plant roots (figure 6.9) and are available in many sizes of single containers that are either square or round. Smaller volume Root-Maker® cavities are joined together in blocks.

Polybags

Bags made of black polyethylene (poly) plastic sheeting are the most commonly used nursery containers in the world because they are inexpensive and

Table 6.2—Volumes and dimensions of containers used in native plant nurseries

Type	Volume (in³ [ml])	Height (in [cm])	Top Diameter (in [cm])
SINGLE FREE-STANDING CONTAINERS			
RootMaker® singles	180–930 (15,240–29,500)	6–12 (15–30)	6–10 (15–25) [a]
Polybags	90–930 (1,474–15,240)	4–8 (10–20)	6–8 (15–20)
Treepots™	101–1,848 (1,655–30,280)	9.5–24.0 (24–60)	3.8–11.0 (10–28) [a]
Round pots	90–4,500 (1,474–73,740)	6–18 (15–45)	6–14 (15–35)
CONTAINERS WITH EXCHANGEABLE CELLS HELD IN A TRAY OR RACK			
Ray Leach Cone-tainers™	3–10 (49–172)	4.75–8.25 (12–21)	1.0–1.5 (2.5–3.8)
Deepots™	16–60 (262–983)	10–14 (25–36)	10–14 (25–36)
Jiffy® pellets	0.6–21.4 (10–350)	1.2–5.9 (3–15)	0.8–2.2 (2.0–5.6)
Zipset™ Plant Bands	126–144 (2,065–2,365)	10–14 (25–36)	3.0–3.8 (7.5–10.0) [a]
BOOK CONTAINERS			
Spencer-Lemaire Rootrainers™	3.8–21.5 (62–352)	4.25–6.00 (10–15)	1.0–2.0 (2.5–5.0) [a]
BLOCKS MADE UP OF MANY CAVITIES OR CELLS			
Styroblock™ and Copperblock™	0.5–61.0 (8–1,000)	2.0–5.9 (5–15)	0.6–4.0 (1.5–10.0)
Ropak® Multi-Pots	3.5–6.0 (57–98)	3.50–4.75 (9–12)	1.2–1.5 (3.0–3.8)
IPL® Rigi-Pots™	0.3–21.3 (5–349)	1.7–5.5 (4–14)	0.6–2.3 (1.5–5.8)
Hiko™ Tray System	0.8–32.0 (13–530)	1.9–7.9 (4.9–20.0)	0.8–2.6 (2.1–6.7)
RootMaker®	11–25 (180–410)	3–4 (7.5–10.0)	1.5–3.0 (3.8–7.5) [a]
"Groove Tube" Growing System™	1.7–11.7 (28–192)	2.50–5.25 (6–13)	1.3–2.3 (3.3–5.8)

[a] Containers have square tops.

easy to ship and store (figure 6.10A). Unfortunately, polybags generally produce seedlings with poorly formed root systems that spiral around the sides and the bottoms of the smooth-walled containers. This problem becomes much worse when seedlings are not outplanted at the proper time and are held over in the containers. Copper-coated polybags are available and, compared with regular polybags, plants grown in copper polybags produce a much finer, fibrous root system that is well distributed throughout the containers (figure 6.10B).

Individual Natural Fiber Containers

Containers made of fiber or compressed peat come in a variety of sizes (figure 6.11) and are preferred by some native plant nurseries because they are ecologically friendly. The roots develop without the potential deformity problems of solid-walled containers, and peat pots can be transplanted or outplanted with minimal root disturbance or transplant shock.

Treepots™

These large-volume containers are constructed of flexible hard plastic and are good for growing trees and woody shrubs. Many sizes are available that are either square or round (figure 6.12A); square shapes increase space and irrigation efficiency in the growing area. Treepots™ feature vertical ribs on the inside wall to prevent root spiraling, are reusable, and store easily because they can be nested when empty. The depth of their root plug helps plants access soil water on dry sites and, for riparian restoration, provides stability against water erosion. Because of their large height-to-diameter ratios, Treepots™ require a support rack for growing and shipping (figure 6.12B).

Round Pots

Round black plastic pots or cans are the standard for ornamental nursery stock. They are available in many large sizes from numerous manufacturers; one encouraging feature is that some brands are recyclable (figure 6.13A). Round pots are used in some native plant nurs-

Figure 6.9—*The RootMaker®, which was the first to feature sideslit air pruning, is available as a single, free-standing container or in aggregate blocks.* Photo by Tara Luna.

Figure 6.10—*(A) Polybags are inexpensive containers that can produce good plants. (B) Root spiraling is often serious, but copper-coating has solved that problem.* Photo A by Thomas D. Landis, B by R. Kasten Dumroese.

Figure 6.11—*Fiber or peat pots are popular in some native plant nurseries because they are ecologically friendly.* Photo by Tara Luna.

eries, especially for landscaping applications (figure 6.13B). Round pots are very durable and so can be reused for many years; because they can be nested when empty, they use little storage space. Most designs have a ridged lip that makes the pots easier to move and handle when they are wet. Root deformation has been a serious problem with these containers, however, some are now available with internal ribs or copper coating to prevent root spiraling.

Pot-in-Pot System

This technique is popular for producing very large container plants and involves growing plants in one container that is placed inside another larger in-ground container (figure 6.14A). The plant container is suspended off the bottom of the in-ground pot by the shoulder lip and is supplied with trickle or drip irrigation. The pot-in-pot system is advantageous because plants are more stable and less susceptible to being blown over by the wind. Because the container is below ground, its root system is insulated against excessive cold or heat (figure 6.14B). Disadvantages include higher costs for materials and labor, and plants with aggressive root systems may grow through the drain holes of the in-ground pot into the surrounding soil.

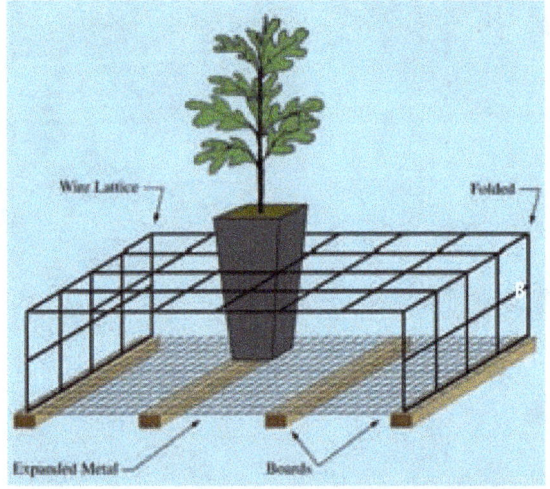

Figure 6.12—*(A) Treepots™ are popular native plant containers because of their deep root systems, but (B) they need to be held in a rack system.* Photo by Thomas D. Landis, illustration by Jim Marin.

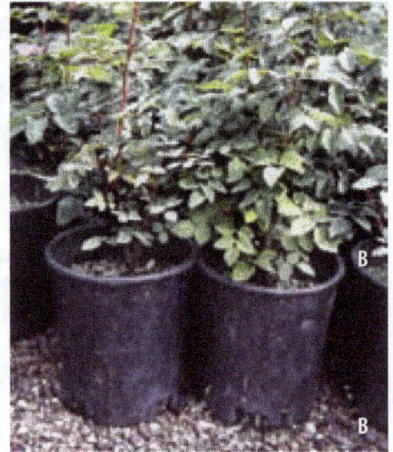

Figure 6.13—*(A) Some standard round plastic containers or "cans" are now recyclable, and (B) are sometimes used to grow native plants for ornamental landscaping.* Photo A by Thomas D. Landis; B by Tara Luna.

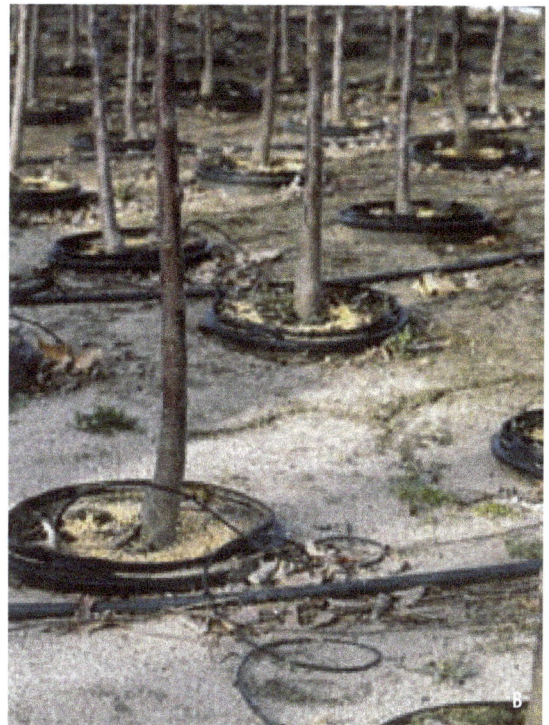

Figure 6.14—*(A) The pot-in-pot system is a popular method for growing very large container stock because it provides stability and (B) insulation against heat or freezing injury.* Photos by Thomas D. Landis.

Exchangeable Cells Held in a Tray or Rack

In this category, individual cell containers are supported in a hard plastic rack or tray. The major advantage of this container system is that the individual cells are interchangeable. Following germination, empty cells can be removed and their spaces in the rack filled with cells with plants. This process is known as consolidation and makes efficient use of nursery space. For native plants that germinate over a long period, plants of the same size can be consolidated and grown under separate irrigation or fertilizer programs. Another unique advantage is that cells can be spaced farther apart by leaving empty slots; this practice is ideal for larger-leaved plants and also for promoting good air circulation later in the season when foliar diseases can become a problem. Racks are designed to create enough air space underneath to promote good air pruning. Plastic cells can be reused for several growing seasons.

Ray Leach Cone-tainers™

This is one of the oldest container designs on the market and is still popular with native plant growers. In this system, individual soft, flexible plastic cells are supported in a durable hard plastic tray (figure 6.15). Trays are partially vented to encourage air circulation between cells, and have a life expectancy of 8 to 10 years. Cells come in three types of plastic: recycled, low density, and low density with UV stabilizers. All have antispiral ribs and a center bottom drainage hole with three or four side-drain holes on the tapered end.

Deepots™

These single cells are constructed of thick plastic and held together by hard plastic racks (figure 6.16). Available in several sizes, they have internal vertical ribs for root control and supports on the bottom of each container provide stability. Racks hold the containers together but do not create an air space underneath, so Deepots™ must be grown on wire mesh to facilitate air pruning of the roots. Due to their large volume and depth, Deepots™ are popular with native plant nurseries growing woody shrubs and trees.

Figure 6.15—*Ray Leach Cone-tainers™ are one of the most popular container types for growing native plants because they can be consolidated in the racks.* Photo by Tara Luna.

Figure 6.16—*Deepots™ come in large sizes and are popular for growing woody shrubs and trees.* Photo by Tara Luna.

Figure 6.17—*(A) Jiffy® pellets are composed of dry compressed peat surrounded by mesh and expand when watered . (B) Smaller pellets are being used to start germinants can be transplanted into larger Jiffy® containers or other containers.* Photo A by Thomas D. Landis; B by Stuewe & Sons, Inc.

Jiffy® Pellets

Jiffy® products consist of dry, compressed peat growing media inside a soft-walled, meshed bag and come in a variety of sizes (figure 6.17A). When sown and irrigated, the pellet expands into a cylindrical plug surrounded by mesh that encourages air pruning all around the plug. Pellets are supported in hard plastic trays, so individual pellets can be consolidated to ensure full occupancy. Irrigation schedules must be adjusted due to greater permeability of the container wall. Some root growth occurs between the pellets, so they must be vertically pruned prior to harvesting. Jiffy® Forestry Pellets are popular in forest nurseries in the Northeastern United States and Eastern Canada, where they are outplanted directly into the field.

Smaller Jiffy® pellets are used for starting plants that are then transplanted into larger Jiffy® sizes or other containers (figure 6.17B). This system is ideal for species that germinate very slowly or over a long period of time.

Zipset™ Plant Bands

Zipset™ Plant Bands are square, one-use containers composed of high, internal-sized, bleached cardboard that are assembled in a hard plastic tray (figure 6.18). Zipset™ Plant Bands maintain their integrity in the nursery but biodegrade after 9 to 18 months. Some native plant nurseries prefer Zipset™ Plant Bands because they protect the root plug during storage and shipping.

Figure 6.18—*Zipset™ Plant Bands are inexpensive containers that can be shipped directly to the field.* Photo by Stuewe & Sons, Inc.

Figure 6.19—*(A) Spencer-Lemaire Rootrainers™ are designed to allow easy inspection of growing media and the root plug. (B) The hinged soft plastic sheets are assembled and placed into hard plastic trays or wire "baskets" to form blocks.* Photos by Thomas D. Landis.

Book or Sleeve Containers: Spencer-Lemaire Rootrainers™

These unique "book" containers are composed of flexible plastic cells that are hinged at the bottom, allowing the growing media and root system to be examined when the books are open (figure 6.19A). The books are held together in plastic or wire trays or "baskets" to form blocks of cells (figure 6.19B). As the name implies, Rootrainers™ have an excellent internal rib system to guide plant roots to the drainage hole and to prevent spiraling. One real advantage of using the books is that they nest easily and can be shipped inexpensively; the nesting feature also makes for efficient storage. The plastic is less durable than other container types, but the books can be reused if handled properly. Rootrainers™ have been used to grow conifers and other native plants in tribal nurseries in the Southwestern United States for many years.

Blocks Made up of Many Cavities or Cells

Multicelled containers are popular for growing native plants and range in size from very small "miniplugs" to some around 1 gal (4 L) in volume (table 6.2).

Styroblock™ and Copperblock™

Styroblock™ containers are the most popular type of container used in forest nurseries in the Western United States and are available in a wide range of cavity sizes and spacings (figure 6.20A) although outside block dimensions are standard to conform to equipment handling. This container has also been used for grow-

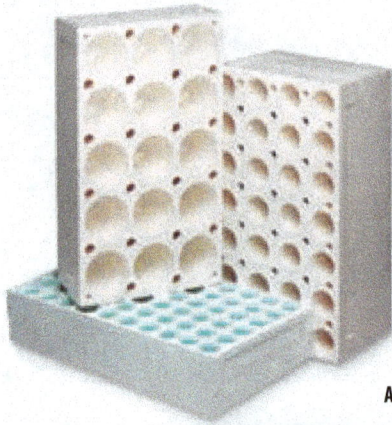

Figure 6.20—*(A) A wide assortment of Styroblock™ containers is available. (B) The cavity walls of Copperblock™ containers are coated with copper, which causes chemical root pruning of species with aggressive roots systems. Styroblock™ and Copperblock™ containers have been used to grow a variety of native plants from (C) grasses to (D) oaks.*
Photo A by Stuewe & Sons, Inc, B and D by Thomas D. Landis, C by Tara Luna.

ing native grasses (figure 6.20B), woody shrubs, and trees (figure 6.20C). The insulation value of Styrofoam™ protects tender roots from cold injury and the white color reflects sunlight, keeping the growing medium cool. Styroblock™ containers are relatively lightweight yet durable, tolerate handling, and can be reused for 3 to 5 years or more. One major drawback is that plants cannot be separated and consolidated and so empty cavities and cull seedlings reduce space use efficiency. Species with aggressive roots may penetrate the inside walls of the cavities (especially in older containers reused for several crops), making the plugs difficult to remove. The Copperblock™ container is identical to the Styroblock™ except that it is one of the few com-

mercially available containers with copper-lined cavity walls to promote root pruning (figure 6.20D).

Hardwall Plastic Blocks

Hardwall plastic blocks are available in a variety of cavity sizes and shapes and outside block dimensions (table 6.2). Extremely durable, these containers have a life expectancy of more than 10 years. The thick plastic is impervious to root growth.

Ropak® Multi-Pots are white in color, available in square and round cavity shapes, and have been used to grow a wide variety of species (figure 6.21A). Because they are so durable, they are popular in mechanized nurseries and have been used to grow herbaceous and

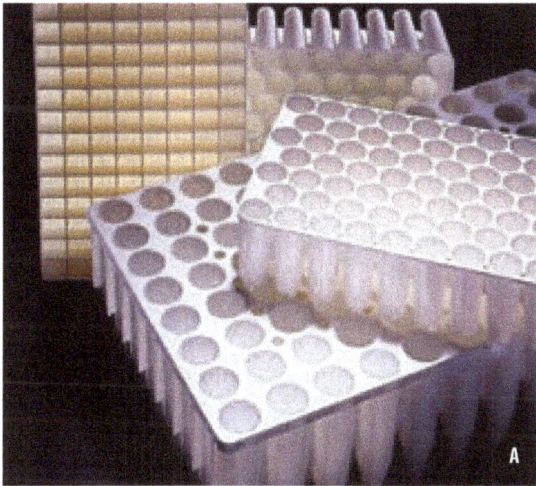

Figure 6.21—Hardwall plastic blocks are extremely durable containers made by several manufacturers. (A) Ropak® Multi-Pots, (B) IPL® Rigi-Pots, and (C) the Hiko™ Tray System. Photos A and C by Stuewe & Sons, Inc., B by Thomas D. Landis.

Miniplug Trays

Miniplug containers are used to start young seedlings that are transplanted into larger containers after establishment (figure 6.22). They are particularly useful for species with very small seeds that make precise seeding difficult. Multiple germinants can be thinned and plugs transplanted to larger containers. In these situations, the use of miniplug trays is much more space and labor efficient than direct seeding into larger cells. They require, however, constant attention because they dry out quickly. If you use miniplug trays, you may need to irrigate them several times a day, construct an automatic mist system, or use subirrigation.

CLEANING REUSABLE CONTAINERS

Reusable containers usually have some residual growing medium or pieces of roots that could contain pathogenic fungi. Seedling roots sometimes grow into the pores of containers with rough-textured walls, such as Styroblock™ containers, and remain after the seedling plug has been extracted (figure 6.23A). Liverworts, moss, and algae also grow on containers and are very difficult to remove from reusable containers. Used containers should be first washed to remove old growing media and other debris; a pressure washer is excellent for this purpose. The containers should then be sterilized with hot water, bleach, or other chemicals. Because many tribal nurseries choose not to use pesticides, how-

woody species. Cavity walls have vertical ribs to prevent root spiraling. IPL® Rigi-Pots™ are usually black but other colors can be obtained in large orders. They are available in a variety of block dimensions and cavity sizes and shapes including sideslit models to encourage air pruning of roots (figure 6.21B). The Hiko™ Tray System features a variety of block and cavity sizes and shapes (table 6.2). All cavities have vertical root training ribs and/or sideslits (figure 6.21C). The "Groove Tube" Growing System™ features grooves in the side walls and large drainage holes to promote root development.

Figure 6.22—
Miniplug containers are used to grow small seedlings that are transplanted into larger containers.
Photo by Tara Luna.

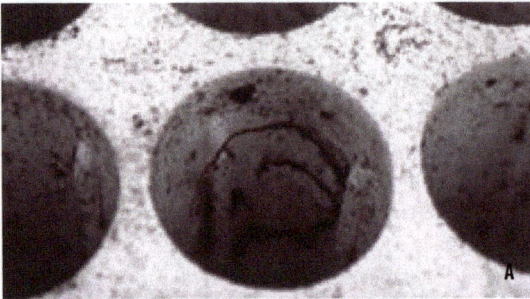

A

B

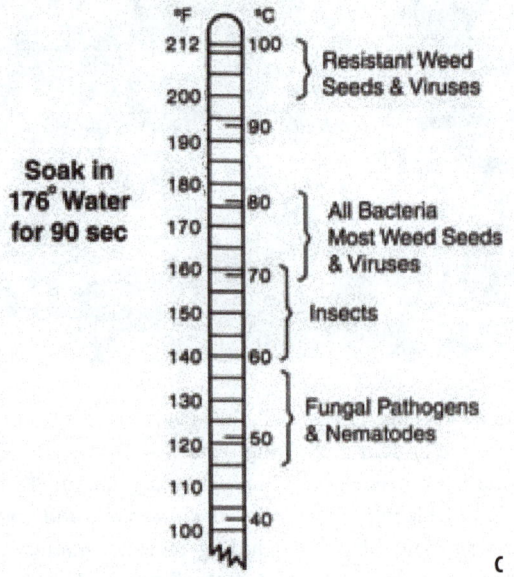

Soak in
176° Water
for 90 sec

°F   °C

212 — 100 } Resistant Weed Seeds & Viruses

200

190 — 90

180 — 80 } All Bacteria Most Weed Seeds & Viruses

170

160 — 70 }

150 —   Insects

140 — 60 }

130 —   Fungal Pathogens & Nematodes

120 — 50 }

110

100 — 40

C

Figure 6.23—(A) Used containers should be washed and sterilized before resowing, because residual growing media and seedling roots can contain disease organisms. (B) Hot-water dips are an effective treatment; (C) submersion in water of 165 to 185 °F (75 to 85 °C) for 30 to 90 seconds has been shown to be adequate for all types of containers.
Photos by Thomas D. Landis.

ever, hot-water dips are the most effective way to kill fungi and other pests in used containers (figure 6.23B). Most pathogens and weed seeds are killed when containers are held at 158 to 176 °F (40 to 60 °C) for more than three minutes (figure 6.23C). For containers, a good rule of thumb is to use a soaking temperature of 165 to 185 °F (75 to 85 °C) for 30 to 90 seconds for Styrofoam™ containers; 15 to 30 seconds is probably sufficient for hard plastic containers (Dumroese and others 2002). Soaking Styrofoam™ for longer durations at 185 °F (85 °C) or at hotter temperatures can cause the material to distort. Older Styrofoam™ containers often benefit from a longer soaking duration (up to three minutes). Commercial units are available, but many nurseries have built homemade container dipping systems that hold the containers under hot water in a dip tank.

SUMMARY

The types of containers selected for your nursery will have a great effect on seedling quality and will influence the horticultural practices used in the facility. Different species will require different types of containers based on the types of leaves and root systems they possess; small experiments may be needed to determine which containers are best. In some cases, custom containers may need to be designed to suit the species, outplanting objectives, and site. Container types currently on the market have advantages and disadvantages; however, the newest designs are focused on increasing seedling root quality and outplanting performance. Cost, available space, desired stock size, and the properties and dimensions of the container are among the most important factors to consider. Container selection is a key part of nursery planning.

LITERATURE CITED

Dumroese, R.K.; James, R.L.; Wenny, D.L. 2002. Hot water and copper coatings in reused containers decrease inoculum of *Fusarium* and *Cylindrocarpon* and increase Douglas-fir seedling growth. HortScience 37(6): 943-947.

Landis, T.D.; Tinus, R.W.; MacDonald, S.E.; Barnett, J.P. 1989. The container tree nursery manual: volume 2, seedling nutrition and irrigation. Agriculture Handbook 674. Washington (DC): U.S. Department of Agriculture, Forest Service. 119 p.

ADDITIONAL READINGS

Dumroese, R.K.; Wenny, D.L. 1997. An assessment of ponderosa pine seedlings grown in copper-coated polybags. Tree Planters' Notes 48(3): 60-64.

Landis, T.D.; Tinus, R.W.; MacDonald, S.E.; Barnett, J.P. 1990. The container tree nursery manual: volume 2, containers and growing media. Agriculture Handbook 674. Washington, DC: U.S. Department of Agriculture, Forest Service. 87 p.

Stuewe & Sons, Inc. 2008. Tree seedling nursery containers. http://www.stuewe.com/ (20 Oct 2008).

APPENDIX 6.A. PLANTS MENTIONED IN THIS CHAPTER

arrowleaf balsamroot, *Balsamorhiza sagittata*
lodgepole pine, *Pinus contorta*
oaks, *Quercus* species
pines, *Pinus* species
prairie turnip, *Pediomelum esculentum*

Collecting, Processing, and Storing Seeds

Tara Luna and Kim M. Wilkinson

7

Nurseries that work to strengthen and expand the presence of native species are concerned about fostering diverse populations that are strong and well adapted. For many native plants, however, the natural diversity of wild populations has been depleted. Habitat loss has reduced the range and sheer number of plants. For plants with commercial value, unsustainable harvesting practices may have reduced the number of plants with desirable characteristics while leaving behind inferior plants. This process of depleting a population of the best genetic properties so that future populations are weaker than the original is called "genetic degradation." Seed collection for plant propagation is an opportunity to reverse the trends of genetic degradation and species loss. Nurseries have a key role in conserving the gene pool of native plants.

GENETIC DIVERSITY AND SEED COLLECTION ETHICS

Today, seed collection ethics is an important consideration in native plant nurseries, and traditional indigenous practices serve as a valuable model. Collecting from a wide genetic base fosters a more diverse gene pool at the outplanting site. This practice can protect a planting against unforeseen biological and environmental stresses and also protect against inbreeding in future generations. For restoration and conservation projects, maintaining genetic diversity is a key part of project objectives. This important topic will be discussed in greater detail later in this chapter, but the first step is learning more about flowers and seeds.

A variety of dry native fruits and seeds by Tara Luna.

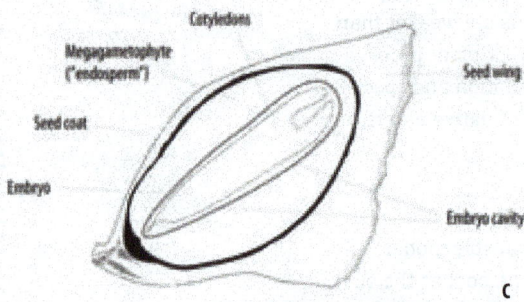

Figure 7.1—(A) Cones of gymnosperms can be dehiscent at maturity, as seen in Douglas-fir which (B) release winged seeds. (C) Gymnosperm seeds include an embryo, nutritive tissue, and seedcoat. Photos by Thomas D. Landis, illustration from Dumroese and others (1998).

UNDERSTANDING FLOWERS AND SEEDS

Seed collectors and growers need to be able to identify fruits and seeds to ensure collection of the right structure at the right stage of development. It is not uncommon for novices to collect open fruits after the seeds have dispersed or immature fruits or cones. In some species, what appears to be a seed is actually a thin-walled fruit containing the seed. Furthermore, a collector must be able to recognize mature seeds from immature or nonfilled seeds. Protocols describing flowering, fruits, seeds, and methods of seed collection, processing, and storage can be found in volume 2 of this handbook.

Plants are classified according to whether they produce spores or seeds. Spore-bearing plants such as ferns produce clusters of spores on the undersides of leaves that may or may not be covered with a papery covering. Spores can be collected like seeds just before they disperse, but they require special growing conditions to develop into plants. For further discussion on fern spore collection, see volume 2 of this handbook.

Seed-bearing plants are classified into two groups, gymnosperms and angiosperms, based on their flower types. Gymnosperms do not bear true flowers and are considered more primitive than angiosperms. Instead, gymnosperms bear male and female cones on the same tree. Male cones typically develop on the tips of branches and fall off after pollen is shed. Female cones enlarge and become more visible following pollination and fertilization, and seeds are borne naked on the mature scales. Gymnosperm cones can be dehiscent, indehiscent, or fleshy. Dehiscent cones have scales that open at maturity to release the seeds (figure 7.1A), whereas indehiscent cones rely on animals to pry them open and disperse the seeds. In dehiscent and indehiscent cones, the seeds are usually winged (figure 7.1B). Fleshy cones resemble berries and their seeds lack wings. Gymnosperm seeds include an embryo, nutritive tissue, and seedcoat (figure 7.1C).

Angiosperms bear true flowers, and seeds are enclosed in an ovary that develops and surrounds the seeds after fertilization. Pollen is transferred from the anthers (male reproductive structures) to the stigma surmounting the pistil (female reproductive structure). Following pollination and fertilization, the ovary enlarges into a fruit that contains one to many seeds. Thus, a fruit is a ripened ovary that develops and surrounds the seed after fertilization. It protects the seed and provides it with nutrition during development and helps with the dispersal of mature seeds. The seed is a ripened ovule consisting of a seedcoat, the nutritive tissue (endosperm), and the embryo, although embryo size varies widely among species (figures 7.2A–C).

Most species of angiosperms have perfect (bisexual) flowers, meaning they contain both the male and female reproductive structures in the same flower (figure 7.3A). Perfect flowers can be showy or very small and inconspicuous. Some species have imperfect flowers, meaning that separate male and female flowers are borne in single-sex flower clusters on the same plant (figure 7.3B). Some species are dioecious, meaning that individual plants are either male or female (figures 7.3C and D). Thus, only the female plants will bear fruits and seeds (figure 7.3E).

Because of the wide variety of flower types, resulting fruits also vary tremendously, and this variety greatly influences how seeds are collected, processed, cleaned, and stored. Dry, dehiscent fruits are those that are woody or papery and split open at maturity. Some examples include capsules (figure 7.4A), legumes or pods (figure 7.4B), and follicles (figure 7.4C). Dry, indehiscent fruits are those in which both the fruit and seed form an integrated part of the dispersal unit and do not split open at maturity. Seeds of these species are surrounded by thin shells that are fused with the outer layer of the fruit and are dispersed as single units that resemble a seed and often have winged appendages. Examples of dry, indehiscent fruits include achenes (figure 7.5A), schizocarps (figure 7.5B), samaras, and nuts (figure 7.5C). Fleshy fruits are those in which the tissue of the ovary is strongly differentiated. The pericarp is the part of a fruit formed by the ripening of the ovary wall. It is organized into three layers: the skin (exocarp), the often fleshy middle (mesocarp), and the membranous or stony inner layer (endocarp). These layers may become skin-like and leathery, fleshy, or stringy during development. Fleshy fruits such as berries, drupes, and pomes are indehiscent (figure 7.6). Berries contain a fleshy pericarp with many seeds, while drupes have a tough stony endocarp (known as the stone or pit) that encloses only a single seed. Furthermore, some fruits are known as aggregate fruits, which are actually clusters of many fruits that develop from single flowers, each bearing one seed.

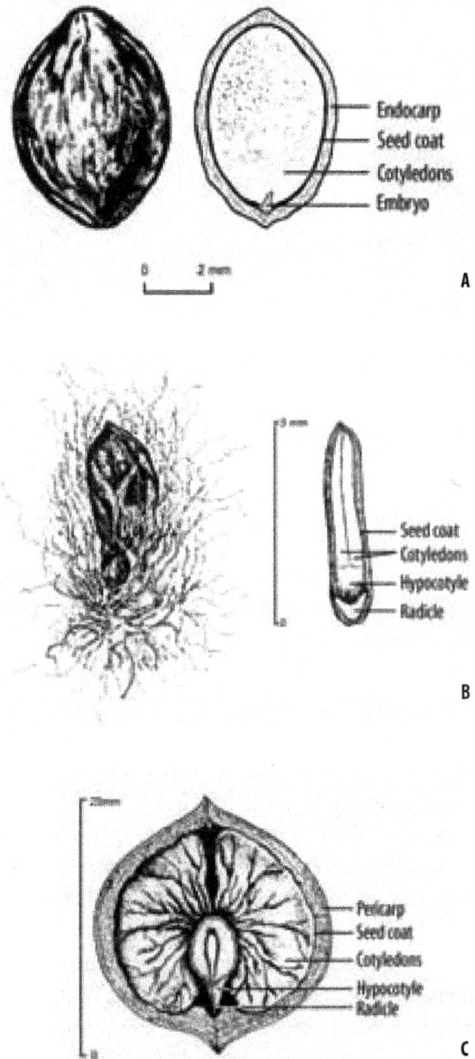

Figure 7.2—*Angiosperm fruits and seeds vary greatly among species: (A) American plum, (B) quaking aspen, and (C) shagbark hickory. In some species, the embryo is not visible to the naked eye.* Illustrations from Schopmeyer (1974).

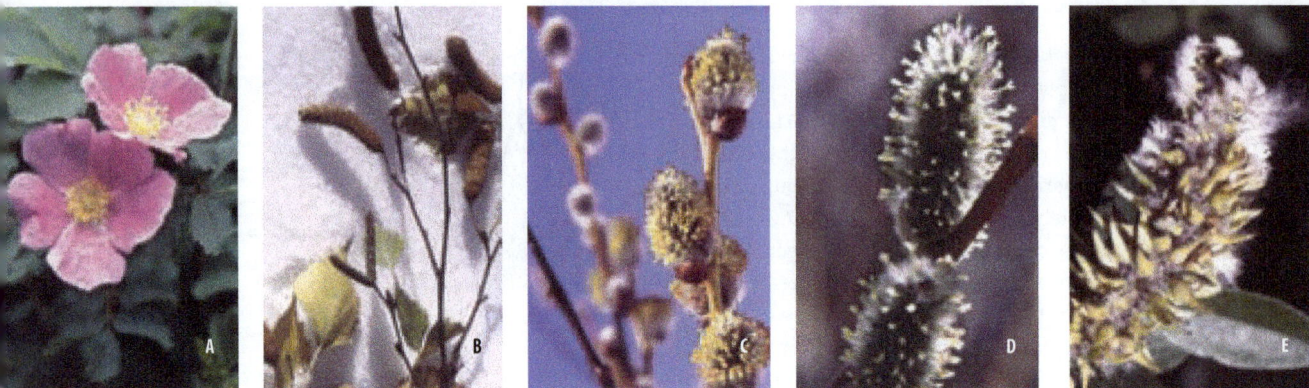

Figure 7.3—*Examples of flowers: (A) perfect bisexual flower of Woods' rose, (B) imperfect flowers borne in male and female clusters (catkins) on the same plant of paper birch, imperfect dioecious flowers borne on separate plants of willow (C) male flowers, (D) female flowers; (E) mature seed capsules on female plant.* Photos by Tara Luna

Concerning longevity, seeds can generally be classified into two groups: recalcitrant and orthodox. Recalcitrant seeds retain viability for only a few days, weeks, or months and, generally, for not more than 1 year. Examples include willow, cottonwood, and quaking aspen; nut-bearing trees such as white oaks, black walnut, and pecan; and many aquatic species such as wild rice. In general, large-seeded species that drop moist from perennial plants in habitats that dry out during the season are most likely recalcitrant (Hong and Ellis 1996). Recalcitrant species are usually sown in the nursery immediately after collection.

Orthodox seeds store easily for long periods of time because they tolerate desiccation. Most temperate species fall into this category. Orthodox seeds retain viability for periods longer than a year and they can be dried to low moisture levels and stored under lower temperatures. Species that grow in areas that are seasonally unsuited for germination usually produce orthodox seeds. Dry, hard seedcoats and small seeds from dry, dehiscent fruits are most likely to be orthodox. Medium-lived orthodox seeds remain viable from 2 to 15 years. This category includes many species of trees, shrubs, grasses, and forbs in temperate climates in a wide range of habitats. Long-lived orthodox seeds remain viable for 15 to 25 years or more. This category includes most conifers, shrubs, and legumes and other species that are adapted to fire ecology or desert environments.

COLLECTING SEEDS

Selecting a Proper Seed Source

Using the proper seed source is a key aspect of growing seedlings targeted to the needs of the outplanting site. Adaptation to site conditions, resistance to pests and diseases, desired characteristics of the plant, and growth rate of the plant are additional factors to consider. Ideally, seeds are collected from individuals currently thriving on the outplanting site. If the outplanting site has been heavily degraded, however, this may not be possible. In that case, seeds should be collected from populations similar to those of the outplanting site. Seed collection zones have been developed for some species, primarily commercial forest tree species, in some areas of the country, as described in Chapter 2, *The Target Plant Concept,* but are lacking for most other native plants. When collecting to ensure adaptation to the site, try to find seed sources having similar climate, soil type, elevation, precipitation, and environmental stresses such as wind and drought.

The presence of pests and diseases varies from site to site. Local plants that are thriving are excellent candidates. Plants showing obvious pest or disease problems may pass on those susceptibilities to their offspring. In addition, plants within a population can vary dramatically in their characteristics, such as the qualities and productivity of their fruit, wood, or medicinal products. Some characteristics may be difficult for seed collectors to discern. Collectors should ask for help from clients to choose parent plants with preferred properties.

Productivity is an important factor for some outplanting projects. For plants with food, wood, fiber, or other uses, high productivity in terms of abundant fruit, nuts, foliage, or fast growth rate for high-quality wood is often desired. Collectors should choose plants with high vigor and health. Many native species produce large seed crops only periodically. In good seed years, seed quality and quantity tends to be high. Heavy seed crops generally occur several years apart, and are often followed by light seed crops the following year. The interval between heavy seed-bearing years is referred to as periodicity.

For woody plants, growth form is also a key characteristic. For example, trees may range in form from small, multistemmed, shrubby individuals to large, straight-stemmed individuals. Depending on the preferred characteristics for the project needs, seed collectors can gather seeds from parents with the desired form.

Finally, the genetic quality of seeds used in the nursery can be a major factor in the success of an outplanting project. Seed selection is definitely an area in which "quality over quantity" should be the standard. Care in selecting seed sources can foster plants that are more productive, better adapted to local site conditions, and better suited to achieve the results planned for the project. The long-term ecological viability and sustainability of a planting is also at stake, because projects should contain enough diversity to reproduce healthy and productive offspring for future generations while remaining resilient to environmental stresses.

Ecology of the Species

To become an efficient seed collector, you need to learn as much as possible about each species. During the first few years, novice seed collectors need to spend as much time in the field as possible and keep good notes (figure 7.7A). Monitoring the development of the seed crop throughout the season is an important part of seed collection. Any observations may also provide clues on how to germinate the seeds. Good field experience has no substitute.

Be sure to record the time of flowering for each species. Flowering is easily observed in species with showy flowers but requires more attention for wind-pollinated species such as conifers, grasses, and willows. Most developing fruits become visible only a few weeks after flowering and pollination. Over time, recognizing the flowering sequence of the local flora allows the seed collection schedule to be simplified by keying it to the flowering period of a few index species. Familiarity with the species and local site conditions allows the development of a locale-specific seed collection schedule.

Figure 7.4—*Dry, dehiscent fruits such as (A) capsules (penstemon), (B) legumes or pods (lupine), and (C) follicles (milkweed) should be collected as soon as the sutures along the fruit wall begin to spilt open.* Photos by Tara Luna.

Figure 7.5—*Dry, indehiscent fruits such as (A) achenes (arrowleaf balsamroot), (B) schizocarps (biscuitroot), and (C) nuts (oak) are actually single units in which the fruit walls are fused to the seeds.* Photos by Tara Luna.

Figure 7.6—*Fleshy fruits of native plants, from top left: drupes of chokecherry, berries of snowberry, pomes of Cascade mountain-ash, berries of kinnikinnick and tall huckleberry, aggregate fruits of thimbleberry, berries of serviceberry, and hips of Woods' rose. Center: drupes of redosier dogwood.* Photo by Tara Luna.

Another important factor to consider is that each species has its own flower and fruit arrangement, pollination strategy, and mode of seed dispersal. Some species will flower and fruit over an extended period of time, while others will flower and fruit only once during the growing season. Other species, such as redosier dogwood, may produce several distinct fruit crops in a season on individual plants. If broad genetic representation is the collection goal, you will need to collect fruits at several times during the season.

Each species has different types of flower arrangements, which will have different blooming sequences. Within a single plant, there is often a range of fruit maturity stages. For example, species such as lupine have a flower stalk with a prolonged period of flowering and many different stages of fruit development. The seed collector will need to selectively harvest only the fully mature fruits and make repeated visits to the collection site (figure 7.7B).

Be aware of the dispersal strategy of the species before attempting to collect seeds. Fruits have developed many highly specialized devices that aid in the protection of seeds and the dispersal away from parent plants. Wind dispersal is very common; dispersal units are very small and light or equipped with specially formed devices for flying. Those devices include air-filled cavities, seeds or indehiscent fruits that are completely covered with hair like cotton, and seeds that are equipped with various kinds of wings or parachutes (figure 7.8). Collectors need to time collection before seed dispersal and windy days.

Other species, such as geranium, ceanothus, and lupine, disperse their seeds by force upon maturation. In these cases, you may need to bag the developing fruits with cloth to capture the seeds (figure 7.9). Use a fine-mesh cloth with a weave that allows light transmission but is small enough to prevent the seeds from falling through the cloth. Tie the bags over developing seed stalks so that seeds will be captured when they are dispersed.

Fleshy fruits are sources of food for animals, and collectors need to time their collection before the fruit is consumed. In some cases, it may be necessary to bag or cage fruits or indehiscent cones to obtain seeds. For example, the indehiscent cones of whitebark pine cones depend on birds and squirrels for seed dispersal, so they are caged with mesh wire to prevent animal predation.

Factors Affecting Seed Formation and Collection Timing

Environmental conditions during the growing season can be either beneficial or detrimental to flowering and seed development. Sudden frosts during flowering can eliminate fruit production for that year. Prolonged periods of cool weather will limit pollinator activity that can result in reduced fruit production. Periods of cold rains will also slow the rate of fruit maturation. Drought and high temperatures may promote flowering, but prolonged moisture stress may cause plants to abort developing fruits and seeds. In cases in which fruits do develop, overall seed viability may be low or seeds may be unfilled. For these reasons, perform a cut test on seeds just prior to collection, as discussed in detail in the following section.

Seed maturation is temperature dependent and is therefore affected by elevation, latitude, and aspect. Populations found on open, south-facing slopes will mature sooner than those on protected, north-facing slopes. Low-elevation populations usually mature first, and seed collectors can follow seed maturity upslope with increasing elevation. Collectors should use favorable microenvironments to their advantage. For example, populations growing in full sunlight tend to produce more seeds than those that are heavily shaded. Good soil nutrient status at the collection site also promotes good seed production. Locations that recently have been burned are good sources for herbaceous species because the overstory has been removed and a flush of nutrients has been made available to the plants during flowering, fruit, and seed formation. Older, burned areas are good collection sites for shrubs, which will flower and fruit for several years.

Sites intensely browsed by wildlife and livestock are poor choices for seed collection because animals often consume the current season's growth, which limits flowering and seed production. Certain insects and fungi also consume seeds. Fruits or seeds that have small exit holes, are discolored, or are misshaped should be avoided.

Correct timing is the most critical element in seed collection. Effective native seed collection involves a number of steps to ensure that seeds are collected at the right stage. If seeds are collected before they are fully mature, the effort often results in poor seed viability. Because most seeds are disseminated away from the parent plant upon maturation, arriving at the site too late may result

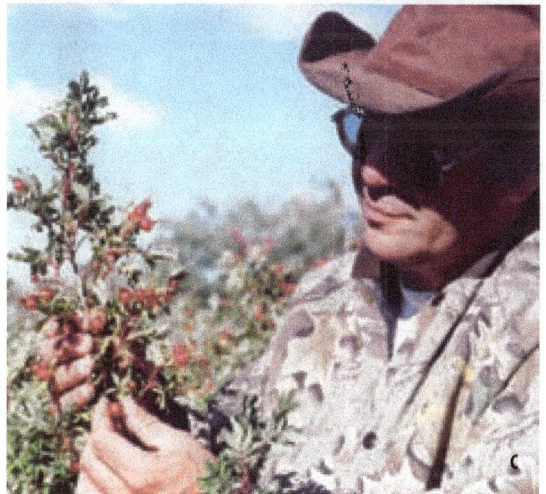

Figure 7.7—*Monitoring the development of fruits and seeds, such as (A) willow, is a necessary part of collecting native plant seeds. Collection should be timed just as the fruits fully mature, (B) as Linda Markins is doing with these panicled bulrush fruits and (C) Alex Gladstone with these rose hips, but before the fruits disseminate away from the plant.* Photo A by Dawn Thomas, B by Terrence Ashley, C by Tara Luna.

Figure 7.8—*Wind-dispersed seeds, like those of arctic dryad, can be challenging to collect because they must be gathered when they are fully mature but before the wind blows them away.* Photo by Tara Luna.

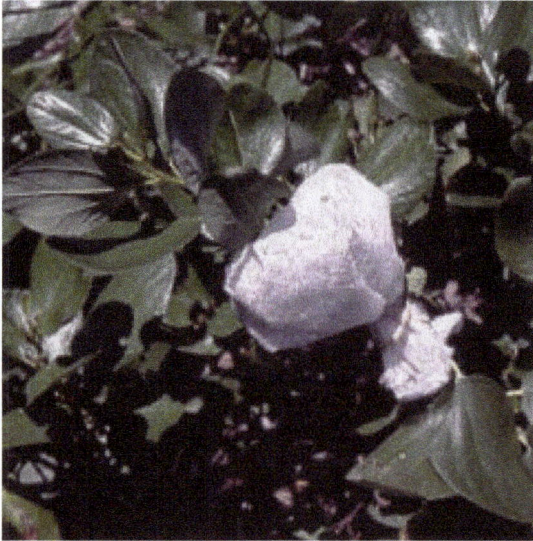

Figure 7.9—*Seeds that disseminate by force with explosive capsules, like snowbrush ceanothus, are best collected by tying cheesecloth bags over the developing fruits.* Photo by Tara Luna.

in the loss of seeds for the season. Proper seed collection timing requires the following practices:

— Locate populations of desired species early in the season.
— Monitor potential sites just after flowering when fruits are becoming visible.
— Record the dates of flowering, fruit, and cone formation. Cones are often a 2-year crop, so you can assess cone crop the year prior to collection.
— Observe carefully the weather patterns during pollination, fruit formation, and maturation.
— Visit the site frequently to monitor the development and quality of the seed crop.
— Use collection dates from previous years to predict target collection dates and other information.
— Use a cutting test of a few sample seeds to determine maturity prior to collection.
— Collect seeds during dry weather.

Ensuring That Seeds Are Healthy

The easiest way to ensure that seeds are healthy and ready for harvest is to use a cut test. A cut test allows inspection for mature, abnormal, infested, or empty seeds. Several seeds from several individuals within the population should be examined. The two essential tools are a hand lens and a safety razor, knife, or scalpel for cutting. With care, cut the fruit or seed along its longest axis. Inspect the seeds for their internal coloring, how completely the internal tissue fills the seedcoat cavity, and for the presence of an embryo. Depending on the species, the embryo may completely fill the cavity or be tiny and embedded in the endosperm. A microscope may be needed for examining very-small–seeded species. If the seedcoat is soft and the contents are watery and soft, the seed is immature. If the seedcoat is hard and the contents are firm and light tan to white in color, the seed is approaching maturity or is fully mature. Some species can be collected just prior to maturity if the entire inflorescence is cut and the seeds are allowed to cure properly before cleaning. Generally, the optimum time for seed collection is when fruits are splitting open at the top.

Seed Collection Tools

The choice of tools depends on the species to be collected. Select tools that will not damage the plant or the seeds. Some general collection methods include hand picking or hand stripping, cutting fruit clusters, raking or shaking branches over a canvas tarp, bagging or caging developing fruits or cones, and tying canvas tarps between large woody plants. The following tools and supplies are useful when collecting seeds from natural stands:

— Labels, permanent markers, pencils, and seed collection forms to attach to bags.
— Scissors, pruning shears, hand scythes, and extendible pruning poles for taller trees.
— A hand lens to examine seeds to ensure they are full.
— Safety razor blades or a sharp pocketknife for the cutting test of seeds and to examine fruits.
— Large paper bags for dry fruits.
— White plastic bags for fleshy fruits.
— Canvas tarps.
— Hand gloves.
— Wooden trays for collecting the seeds of low-growing plants.
— A storage box or cooler to keep collections from being overheated during transport.
— Binoculars for spotting fruits in taller trees.
— Fine-mesh bags, cages, fine-mesh cloth, and rubber bands for species with rapid dispersal.

Maintaining Genetic Diversity through the Ethical Harvest of Seeds

To maintain genetic diversity, follow these guidelines:

— Collect seeds from a minimum of 30 individual plants—50 to 100 individuals is better.
— Avoid collecting from single individuals or small populations with fewer than 30 individuals, unless you are propagating a rare species.
— Collect from distant individuals to reduce the chance of collecting only close relatives; however, this may not be possible with species that occur in small patches.
— Collect a few seeds from as many individuals as possible to ensure a good genetic representation of the population.
— As much as possible, collect seeds from habitat similar in elevation, aspect, and soils to that of the outplanting site to ensure genetic adaptation.
— For trees, gather seeds from throughout the canopy.

Some other ethical guidelines and common sense approaches to collecting include these practices:

— Do not collect all the seeds from a site. Suggestions vary, but most state that only 10 to 50 percent of the total seed crop should be collected from each site.
— Leave enough seeds as a food source for animals and to ensure the reproduction of the population.
— Avoid soil disturbance and plant damage while collecting seeds, especially in fragile habitats.
— If possible, allow an area to rest for at least two growing seasons between collections. Keep in mind that longer periods may be needed for some species and locations.
— Be absolutely certain of the identity of species being collected. If in doubt, collect a plant specimen for later identification.
— If avoidable, do not collect from weed-infested areas.
— Collect relatively even amounts from each plant; no single plant should be overrepresented in the collection.

PROCESSING SEEDS

Proper processing of fruits and seeds begins the moment the fruit or seed is removed from the parent plant. Proper processing includes temporary handling from the field to the nursery; temporary storage at the nursery; and prompt and proper seed extraction.

Temporary Handling from Field to Nursery

In general, it is best to transport material from the field to the nursery as quickly as possible, avoiding exposure to direct sun, high temperatures, and physical abuse. Dry fruits, seeds, and cones can be left inside their paper collection bags for short durations. Placing plastic bags filled with fleshy fruits inside coolers will help prevent them from fermenting and being damaged by high temperatures.

Temporary Storage at the Nursery
Recalcitrant Seeds

Recalcitrant seeds cannot withstand drying below a critical moisture level, so they are usually sown immediately after processing. During temporary storage before sowing, seeds must be kept fully hydrated by placing them in trays under moist burlap or in plastic bags filled with moistened sand or peat moss at cool temperatures. Relative humidity should be maintained at 80 to 90 percent.

Dry Fruits and Cones

At the nursery, small quantities of dry fruits and cones can be dried in paper bags or envelopes as long as the contents are loose. Large quantities must be dried immediately by spreading the material on a tarp or mesh screen, but the material will need to be turned several times a day. Turning the material prevents it from becoming too hot, drying in an uneven manner, or becoming moldy.

The best way to dry material is to spread the contents evenly on a drying rack. A drying rack consisting of a simple wooden frame with multiple screens can be constructed at low cost and will make efficient use of space in a seed drying room or greenhouse (figure 7.10). Drying racks should be made with mesh screens that allow air movement but prevent seed loss. Different mesh screens will be necessary for different seed sizes. Dry, dehiscent fruits should also be covered with a fine-mesh cloth to prevent the loss of seeds after fruits open. Good air movement, low relative humidity, and temperatures between 65 and 80 °F (18 and 27 °C) promote even drying and eliminate moisture buildup that can cause mold and damaging temperatures. A ventilated greenhouse or storage shed works well for this purpose. Temperature control is very important; you should use a shadecloth to keep temperatures from rising too high. Avoid rewetting dry fruits after collection. Also, make sure to exclude animals from the seed drying area.

Fleshy Fruits and Cones

Fleshy fruits and cones are very susceptible to fermentation and subsequent damaging overheating. On the other hand, it is important not to let the fruits dry out because this can make cleaning much more difficult. The best procedure is to temporarily store fleshy fruits in white plastic bags in a cool place or refrigerator until the seeds can be processed.

Processing Fruits and Cleaning Seeds

Seed cleaning is necessary before sowing or long-term storage. In some cases, seeds will fail to germinate if they are not removed from the fruits. Seeds of some species germinate more quickly and evenly if they are removed completely from their coverings. For example, Nebraska sedge has shown improved germination by removing the papery sac (perigynia) surrounding the seed (Hoag and others 2001). Germination is often improved if the seeds are extracted from stony endocarps (pits) prior to treatment and sowing.

Most native plant nurseries regularly deal with small seedlots. Seed cleaning and processing can be laborious and time consuming, and specialized cleaning equipment can be expensive. A variety of "low tech" methods and devices are easy to use, cheap, readily available, and work very well with a variety of fruit types. Some are described below. Whichever method of cleaning is chosen, the seed cleaning area of the nursery should be well ventilated. Some fruits can cause allergic reactions and fine dust can irritate skin, eyes, and lungs. It is important to wear gloves and dust masks during cleaning and to wash your hands afterward.

Cleaning Recalcitrant Seeds

Many species that have recalcitrant seeds, such as maple, often can be collected quite cleanly and are sown immediately. Others, such as members of the white oaks, need additional cleaning that is typically accomplished by flotation in water. Immediately after collection, acorns are placed in a bucket of water. Generally, the viable acorns sink whereas the nonviable acorns, trash, and debris float. As a side benefit, the soaking helps keep the acorns hydrated until they are sown. If acorns are collected in very dry conditions, viable acorns may also temporarily float. In this situation, a longer soak duration, perhaps even overnight, may be necessary to allow enough time for good acorns to hydrate and sink. Do a cut test to fine-tune the procedure.

Cleaning Dry Fruits and Cones

Separating seeds from dry, dehiscent fruits (figure 7.4) is usually the easiest cleaning procedure because the fruits split open at maturity. Small lots can be readily cleaned by shaking the fruits inside paper bags so that the seeds fall out; the woody capsules can then be removed from the bag. Modified kitchen blenders with rubber-coated blades are very useful for cleaning small lots of dry fruits (Thomas 2003). The ideal amount of dry fruit material to place in a blender varies with the blender's size, but one-quarter to one-third of the storage capacity of the blender works well (Scianna 2004).

Extracted seeds will be mixed with pitch globules, dry leaves, wings, and small pieces of cones or dry fruits. Screening is the easiest way to separate dry, indehiscent fruits and seeds. Screens can be constructed of hardware cloth and wooden frames. Commercial screens are also available in a range of sizes (figure 7.11). At least two screen sizes are needed. The top screen has openings large enough to allow the fruits and seeds to pass through and the bottom screen has smaller openings that allow fine chaff, but not the seeds, to pass through. By placing the collected material on the top screen and shaking, most of the trash can be removed. When separating other small-seeded plants, such as sedges, rushes, and some wildflowers, you will need screens with very fine mesh or kitchen sieves to properly separate seeds from other debris. Finally, finer chaff and empty seeds can be further removed by placing them in front of a fan on a low setting or running them through a series of fine kitchen sieves.

Larger quantities of dry fruits can be cleaned with a variety of commercial equipment, such as hammermills, clippers, dewingers, specific gravity separators, and air separators. The equipment of choice depends on the seeds being cleaned, the amount of debris in the collection, and the desired purity (cleanliness) requirements of the seedlot. A hammermill uses rotating hammers and stationary anvils to smash, crush, and tear dry fruits into smaller fragments to extract the seeds (figure 7.12A). Clippers are used to remove appendages and hulls from seeds and to separate out larger chaff material (figure 7.12B). Dewingers are used to remove winged appendages from seeds and fruits and are most commonly used with conifer seeds.

Conifer cones, after the scales open, can be resacked and shaken by hand or tumbled in a wire cage to dislodge additional seeds from the cone scales. Serotinous cones, such as jack pine or lodgepole pine, require exposure to heat before the scales open. Cones need to be exposed to 170 °F (77 °C) temperatures either by placing the cones in ovens for a period of a few minutes to a few hours or dipping them in hot water for a few minutes. If an oven is used, cones will need to be checked frequently during drying and removed when most have opened enough to allow the extraction of seeds. If the cones are

Figure 7.10—*Dry seeds require good ventilation to prevent mold development during post harvest handling and drying.* Photo by Dawn Thomas.

Figure 7.11—*Screens are used to separate large debris, seeds, and fine chaff from the seedlot and are available commercially in a wide range of screen holes sizes to complement any species.* Photo by Tara Luna.

Figure 7.12—Commercial seed cleaners such as (A) a hammermill or (B) clippers are useful for cleaning large lots of native seeds. Photos by J. Chris Hoag.

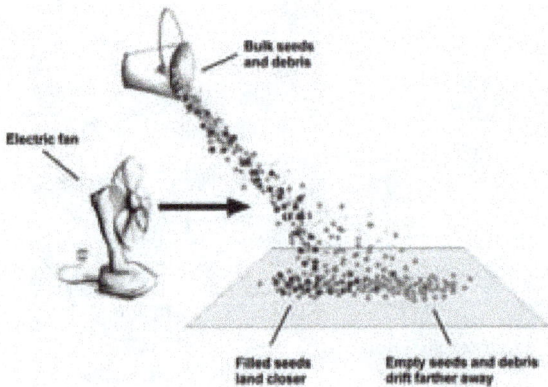

Figure 7.13—Use a small fan to winnow empty seeds and wings from filled seeds. The heavier filled seeds will land closer to the fan while lighter empty seeds and wings will land farther way. Illustration from Dumroese and others (1998).

dipped in hot water, the combination of heat and drying after the soak should be sufficient to open them. Seeds can then be removed from cones as described previously. Most conifer seeds are dewinged before sowing and this can be done by filling a burlap or cloth sack one-fourth full, tying or folding it shut, and gently kneading the seeds by squeezing and rubbing from outside the sack. Friction between the seeds and between the seeds and burlap will detach the wings. Remember to knead slowly and gently because too much friction might damage the seeds. This process requires only a few minutes. A few species, such as western redcedar, longleaf pine, and firs, have very tight wings that should be left on the seeds. Repeat the screening process again with a mesh size that retains seeds but allows the smallest debris to pass through (Dumroese and others 1998). Large seedlots are probably best processed by seed companies.

The last step is fanning or winnowing, which separates detached wings, hollow seeds, and seed-sized impurities from good seeds. The most efficient, high-tech, method is using an agricultural seed cleaning or fanning mill, but these machines require careful adjustment for each species to prevent retaining too many impurities or blowing away too many sound (full) seeds. Another method that works well is winnowing in front of a fan (figure 7.13). When seeds are poured slowly in front of a small electric fan, they separate according to weight from the base of the fan. Most heavy, sound seeds will come to rest near the base of the fan, and hollow seeds, wings, and lighter impurities will tend to blow farther away. Moving from the fan outward, periodically collect a small sample of seeds and cut them in half to check for soundness, determining where the hollow seeds are and discarding them. All species will probably require several successive separations to obtain the desired degree of seed purity. A good target for most species is 90 percent or more sound seeds (Dumroese and others 1998).

Cleaning Fleshy Fruits and Cones

Seeds in fleshy fruits should be processed soon after collection to avoid fermentation, mummification, excessive heating, or microbial infestation; all of which can damage seeds. The first step in cleaning is to soak fleshy fruits in water to soften the pulp. The soak may need to last a few hours to a few days, depending on the species, and the water should be changed every few hours. After the pulp is soft, flesh can be removed by hand squeezing or mashing using a wooden block, rolling pin, or other device. The flesh can also be removed by wet screening, which involves hand rubbing the fruits against screens using a steady stream of water to eliminate the pulp. Kitchen food processors and blenders with modified rubber-coated blades can be used for small lots of fleshy fruits. With both devices, run them for a minute or so to produce a puree of fruit and seeds. The puree should be placed in a bucket and water added slowly; viable seeds will sink and debris, including hollow seeds, will float to the surface. By slowly and continually adding water, most of the debris will float off, leaving clean seeds at the bottom of the bucket (Truscott 2004). If fleshy fruits of species with dormant seeds are being cleaned, they need to be washed with water to remove any remaining pulp and dried for several days before storage. Remember that the way in which seeds and fruits are handled during collection, temporary storage, postharvest handling, and cleaning can directly affect seed quality and viability, as well as seed storage life.

Cleaning Tools for Fleshy and Dry Fruits

Small, hobby-size rock tumblers are useful for dry, indehiscent fruits; rehydrating and cleaning fleshy fruits; or removing barbs or other appendages from seeds and fruits. Wet tumbling uses pea gravel or crushed stones and water in a rubber-lined tumbler vessel. Add just enough water so that the gravel and fruit makes a slurry. The tumbler can be run overnight and checked the following day. After a course of tumbling, the contents are dumped into a sieve and the pulp or debris is washed off, leaving clean seeds (Dreesen 2004) (figure 7.14). Another useful tool is the common kitchen blender with modified blades. Kitchen blenders can be used for small lots of fleshy and dry fruits after the impeller blades are coated with rubberized plastic coating (the material used to coat handtool handles) to prevent damage to the seeds (figure 7.15) (Thomas 2003).

Testing Cleaned Seeds

After the seeds are cleaned, it is a good idea to determine the quality of the seeds by testing seed viability, seed germination, or both. Seed viability and seed germination do not mean the same thing and growers need to know the difference between them. Seed via-

Figure 7.14—*A small hobby-size rock tumbler can be used to clean dry fruits with hard-to-remove appendages or fleshy fruits.* Photo by Tara Luna.

Figure 7.15—*Modifying kitchen blender blades by using rubber coating allows the grower to clean small lots of dry and fleshy fruits without damaging the seeds.* Photo by R. Kasten Dumroese.

bility tests estimate the potential for seeds to germinate and grow, whereas seed germination tests measure the actual germination percentage and rate. A seed germination test allows you to know the rate and percent germination to expect and thereby determine sowing rates so that seeds are used efficiently.

Seed Viability Tests

Cutting tests, described previously, are the simplest seed viability tests and are usually performed during seed collection and often just before treating seeds for sowing. Cutting tests should also be done on seedlots that have been stored for a long period of time to visu-

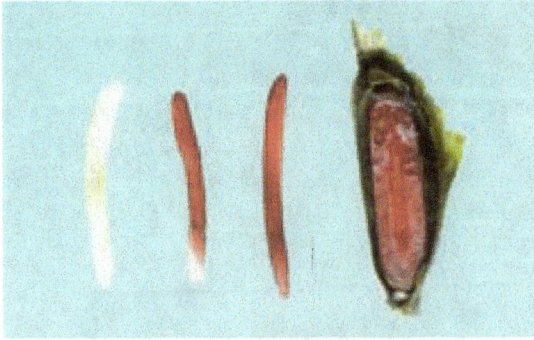

Figure 7.16—*Tetrazolium (TZ) tests stain living tissue red and can be used to estimate seed viability of a seedlot. Shown here is noble fir (left to right): dead embryo, damaged embryo, healthy seed.* From Stein and others (1986).

ally assess their condition. Cutting tests can reveal whether or not the seed is healthy, but really cannot tell anything about the potential for germination. A better test is the tetrazolium (TZ) test, a biochemical method in which seed viability is determined by a color change that occurs when particular seed enzymes react with a solution of triphenyltetrazolium chloride. Living tissue changes to red, while nonliving tissue remains uncolored (figure 7.16). The reaction takes place with dormant and nondormant seeds and results can be obtained within a couple of hours. Although the TZ test is easy to do, the interpretation of results requires experience. For this reason, most nurseries send their seed samples to seed analysts that have the necessary laboratory equipment and experience for testing. A third viability test is an excised embryo test. Embryos are carefully removed from seeds and allowed to grow independently of the seed tissue. Seeds often must be soaked for several days to remove hard seedcoats, and excision of the embryo is an exacting procedure that normally requires the aid of a microscope. As when doing TZ testing, most nurseries send their seed samples to seed analysts for excised embryo testing.

Germination Tests

Seed germination tests are regularly done by nurseries to determine seed sowing rates. If the species being tested has some type of seed dormancy, an appropriate treatment to remove dormancy will be needed before the germination test. Many nurseries will test dormant seedlots before and after the dormancy treatment to check its effectiveness. Testing measures the germination rate and percentage. The germination rate indicates how promptly seeds germinate, whereas the germination percentage indicates how many seeds germinate. Knowing the germination percentage is important for determining how many seeds to sow per container, and knowing the germination rate (speed) provides information on how long seeds will continue to germinate after sowing. See Chapter 8, *Seed Germination and Sowing Options*, for details about sowing rates and methods. Germination tests reflect the potential of a seed to germinate and actual germination in the nursery may vary greatly. This difference occurs because of the inherent variability of germination in most native species and also because of differences in the environmental conditions during testing and growing at the nursery.

Use the following steps to conduct a germination test:

— Select an area in the greenhouse or office that can be kept clean.
— Line the bottom of plastic trays, Petri dishes, or similar containers with paper towels. For large-seeded species, line the bottom with sterile sand (bake sand in an oven at 212 °F (100 °C) for at least 1 hour to sterilize it) or unmilled *Sphagnum* peat moss. Moisten the paper towels or other substrate with distilled water.
— Remove equally sized samples from each bag of the same seedlot, or, if there is only one container of the seedlot, from different portions of the container. Mix these samples together to form a representative sample (figure 7.17).
— Make four replicates of 100 seeds each and place them on the moist substrate in the container. The containers may be covered to reduce evaporation from the substrate.
— Use distilled water to remoisten the substrate as necessary, but never allow standing water to remain in the container.
— Place the containers under optimum germination conditions—ideally those in which light, temperature, and humidity can be controlled. Conditions similar to the nursery will yield more meaningful results.
— Count the number of germinants on a weekly basis for up to 4 weeks on herbaceous species and up to 3 months on woody species. Seed treatments that yield a high percentage of germination promptly are the best.

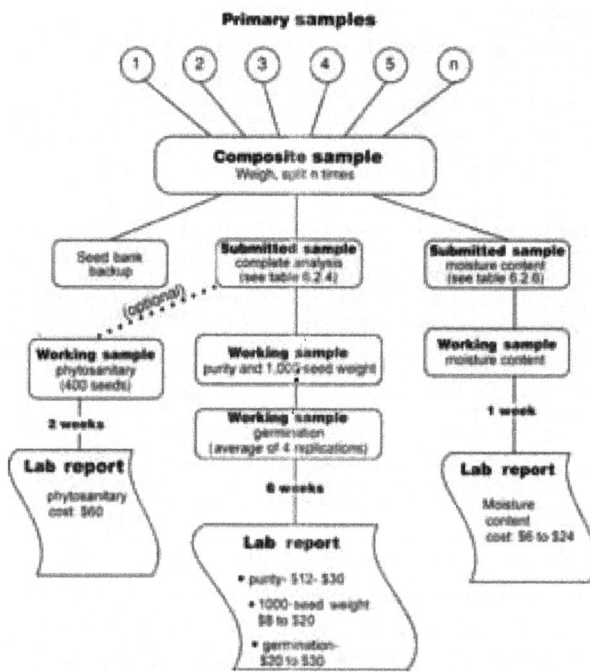

Figure 7.17—*Seeds cans be tested by collecting primary samples from an entire seedlot to make up a composite sample. The composite sample is further divided into samples tested at the nursery or submitted to a seed laboratory for testing.* Illustration by Jim Marin.

Purchasing Seeds from a Reputable Source

Knowing the original collection source of seeds is crucial to ensure that locally adapted, genetically appropriate materials are used. For example, some species, such as yarrow and prairie Junegrass, are native throughout the northern hemisphere so seeds from across the continent, or from Europe, are commonly sold. However, within each species are genetic variations; local populations have adapted to local climate, soils, and other site conditions. Local plant materials, collected from the same or similar area as the outplanting site, have been shown to perform better than non-local sources. Using locally-adapted seed sources is a key factor in ensuring the survival of the plants as well as the protection of local genetic resources. For some species, transfer guidelines may exist. For other species, people must decide on a case-by-case basis what makes sense, considering climate, soil type, elevation, and other site conditions (Withrow-Robinson and Johnson 2006).

Seed companies may collect seeds to order based on customer needs. If not, it is best to ask what sources the company has, rather than asking for a particular source—unscrupulous dealers may claim to have exactly what you want. Seeds must be high quality and free of weeds. When purchasing seeds, obtain and keep a certificate of the seed analysis for each seedlot. The seed analysis must have the scientific name of the species and cultivated variety (if applicable). It is also important to obtain information on the origin of the seeds; an estimate of viability; the percentage of pure live seeds (PLS); and the percentage of other crop seeds, weed seeds, and inert material.

Native cultivated varieties are available for some native grasses and forbs and are shown on the tag following the scientific name as a "named" variety. Cultivars are developed by selecting individual plants that undergo further selection for traits that allow for economical seed production. Cultivated varieties are typically used when wild sources are unavailable or when large quantities of seeds are needed for a restoration project.

The germination percentage reflects the germination potential of the seeds at a particular point in time. This potential is usually described as the percentage of 100 seeds that germinate between 0 and 28 days. Many native species, however, do not germinate within that timeframe. Instead, they are tested with dye (TZ) to determine the amount of living seeds. Often, seed distributors will provide a TZ estimate of viability instead of a germination percentage.

The percentage of PLS is a seed quality index that is calculated during seed testing (figure 7.18). PLS is a basis

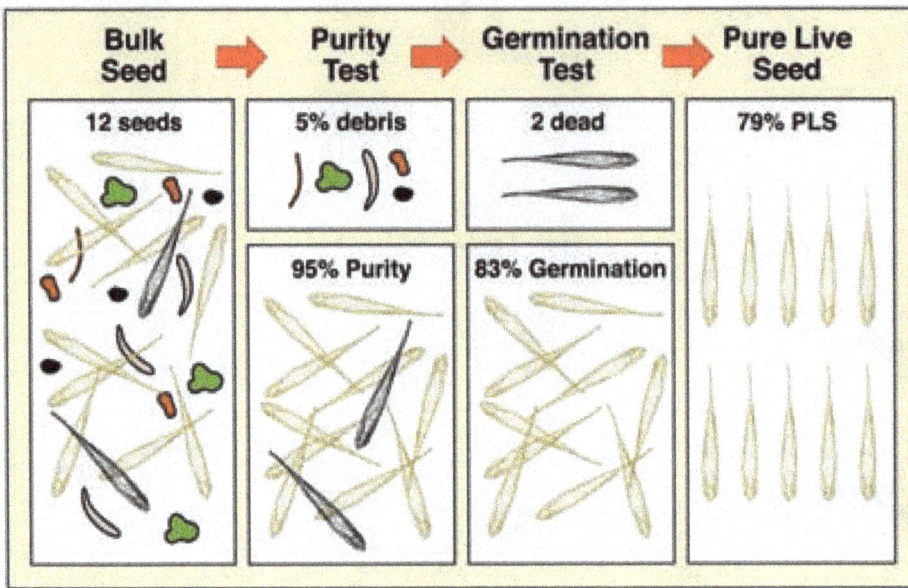

Figure 7.18—*Pure live seeds (PLS) is the percentage of the bulk seed weight that is composed of viable seeds. In this example, results of a purity test show 95 percent of the bulk weight is composed of seeds. The subsequent germination test indicates that 83 percent of the seeds germinated. Multiplying percentage purity by percentage germination yields 79 percent PLS.* Illustration from Steinfeld and others (2007).

for comparing seedlots that differ in purity (the percentage of clean seeds) and viability. Purchase only seeds with high PLS values and with very low percentages of weed seeds and other inert materials. It is often a good idea to ask about where the species was grown or collected to determine what weeds may be present in the seedlot.

Avoid purchasing generic wildflower seed mixtures. Often, these mixes include species that were originally native to another portion of the country or are foreign, nonnative species. Some species are aggressive and can displace native plant populations, and some mixes may contain noxious weeds. Purchase wildflower seed mixes only from reputable seed dealers that can provide the exact species composition with locally adapted seed sources.

STORING SEEDS

It can be quite beneficial to store seeds, especially for those species that yield seeds irregularly or to take advantage of a bumper crop of seeds. In addition, long-term seed storage is an important conservation method for threatened and endangered species. For proper seed storage, seeds must be mature and free of mechanical injury. The viability of seeds after storage depends on their viability at harvest, how they were handled during processing, and storage conditions.

Even under the best conditions, seeds degrade—the degree of longevity varies by species.

As previously discussed, recalcitrant seeds retain viability for only a few days or, at most, a year; they are usually stored only temporarily before sowing. Some species, however, can be stored for a few months as long as seeds retain high moisture content (35 to 50 percent) under high relative humidity conditions and are exposed to good air movement (stored in unsealed containers) and cool temperatures.

In nature, orthodox seeds of most conifers and hardwood trees usually survive less than 3 years. Under proper storage conditions, however, they may retain high viability after 25 years in storage (Landis and others 1999). Many hard-seeded species, such as lupine and American lotus, can remain viable under artificial storage conditions for even longer periods of time.

Storing orthodox seeds requires three basic principles: low relative humidity, low seed moisture content, and cool temperatures. These principles have been used for thousands of years. Indeed, the domestication of New World crops, such as corn, beans, and squash, by indigenous farmers was in part due to the storability of seeds (figures 7.19A and B).

The two most important factors affecting orthodox seed longevity under storage conditions are seed moisture content and temperature. A small change in seed moisture content has a large effect on the storage life of seeds. Therefore, it is important to know the moisture content to predict the possible storage life of a seedlot. With most orthodox species, the proper seed moisture content for storage is generally between 6 and 10 percent. An electronic moisture meter can be used to measure seed moisture content and is available from several suppliers.

After the seeds are clean, air-dry them in shallow trays for 2 to 4 weeks before storage to reduce the moisture content. Stir them once a week or often enough to prevent uneven drying. Put the seeds in an airtight container and label it well. Most species can be stored at temperatures slightly below to above freezing with good, long-term storage results.

Use the following guidelines to properly store orthodox seeds:

— Maintain low relative humidity and low temperatures in the storage environment to maximize the storage life of the seeds (figure 7.20A).

— Because relative humidity increases with a decrease in temperature, reduce relative humidity in the storage environment so dried seeds do not uptake additional moisture. For a small nursery, self-defrosting refrigerator will provide good results as long as the refrigerator is not used for other purposes.

— Use moisture-proof containers to maintain the proper seed moisture level. Small seedlots can be stored using sealed jars with rubber gaskets on the lids or envelopes kept in a sealed, thick-walled plastic tub with an airtight lid (figure 7.20B). Heat-sealed foil-lined plastic pouches used for food are effective and can be sealed and resealed with an ordinary clothes iron.

Storage Methods for Orthodox Seeds

Three methods of storage are used by small nurseries: freezer, cooler, and room temperature-low humidity storage. If freezer or cooler storage is being used and long power outages could occur, consider using a backup power supply; short-term fluctuations are generally not a problem. Storing seeds in a frozen condition is usually best for long-term storage. Most seeds of temperate species can be stored at temperatures at or slightly below freezing, although many species can be stored at 0 to -5 °F (-18 to -21 °C) in a small household freezer. Seeds are prepared by drying to low levels of seed moisture content. Seeds can be damaged by freezing if the seed moisture content is very high. Be sure to store seeds in airtight containers. When removing frozen seeds from the freezer, allow the container to reach room temperature before opening it. This practice prevents water condensation from forming on the seeds.

Some species will not tolerate freezing and can be cold-stored in a refrigerator. Seeds should be placed in an airtight container and kept at 38 to 41 °F (3 to 5 °C); use a self-defrosting refrigerator that maintains relative humidity between 10 and 40 percent. If the door is rarely opened, the humidity in a self-defrosting unit will maintain low relative humidity levels.

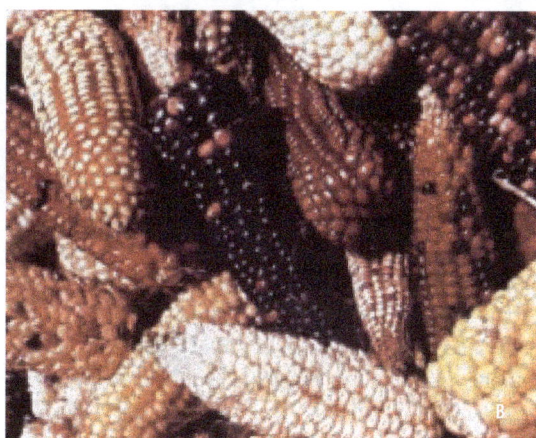

Figure 7.19— *The basic principles of seed storage, keeping seeds with low moisture contents stored in airtight containers at low relative humidity, has remained unchanged for thousands of years. (A) The indigenous farmers of the Southwestern United States practiced and continue to practice these techniques successfully to perpetuate many indigenous varieties of crops and native plants. (B) The domestication of New World crops such as corn was in part due to the storability of seeds.* Photos by Richard Hannan.

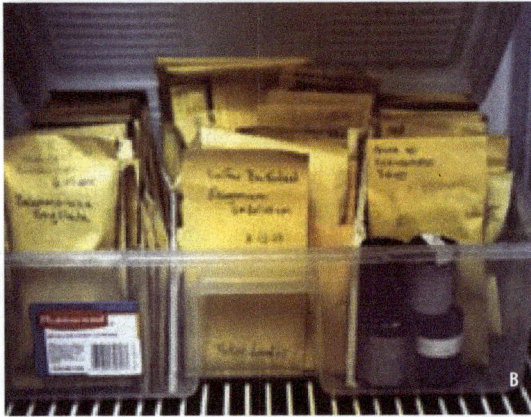

Figure 7.20—(A) Orthodox seeds should be properly dried before storage and kept in moisture-proof containers under cool conditions with low humidity. (B) Each seedlot should be labeled noting origin, date, and the viability percentage. Small lots can be stored in envelopes as long as they are kept in a moisture proof container (B). Photo A by R. Kasten Dumroese, B by Tara Luna.

it inside a tightly sealed jar for every 2 ounces (57 g) of seeds that need to be stored. The silica gel will remove water vapor and ensure that seeds remain at the proper storage moisture. To recharge them, the gels can be baked in an oven (150 °F [66 °C]) for an hour or so.

Storage Methods for Recalcitrant Seeds

Some nut- and acorn-bearing species can be stored for several months as long as the seeds have high seed moisture content (35 to 50 percent) and are stored under cool and moist conditions. Nondormant seeds need to have constant gas exchange, so they are usually stored in unsealed containers in plastic bags filled with moist peat moss in the refrigerator.

Sowing Seeds after Long-Term Storage

In some cases, seeds of large-seeded species that have been dried to low moisture levels may be damaged by absorbing water too quickly. Therefore, when rehydrating these seeds, remove them from storage and spread them evenly in a sealed plastic tub. Place moistened paper towels in the tub so that the towels do not touch the seeds directly. Water vapor released from the towels will be slowly absorbed by the seeds; after a couple of days, the seeds will be able to handle water uptake without injury.

SUMMARY

Most native plant species are propagated by seeds to preserve wide genetic variability that is needed for successful seedling establishment and survival in the natural environment. Growers must become familiar with the type of fruits and seeds they plan to collect and propagate: they should know the species. Seed source is critical because it affects seedling growth in the nursery and also is important to the adaptability of the seedling to the outplanting site. Maintaining genetic diversity using proper, ethical collection techniques is very important. The manner in which seeds are handled during collection and postharvest handling and cleaning can greatly affect their viability. Seeds of some species are inherently short lived and cannot be stored for periods longer than a few months. Most temperate species, however, have orthodox seeds and can be successfully stored if optimum conditions are provided. Seed quality and testing are necessary to plan and produce a high-quality seedling crop in a timely manner.

Although orthodox seeds can be stored at room temperature as well, they will deteriorate faster than those stored at lower temperatures. Ideally, room temperature storage should be used only on seedlots that are held for a short time. The seed moisture content at the time of storage should be at the low end of the range—6 to 8 percent. Seeds must be placed in airtight containers and stored in a room or area with low relative humidity. This storage method works best in the more arid portions of the country.

Silica gels, available from hobby shops and florists, can be used to maintain low seed moisture content. They have been used on short-lived native grass seeds placed into long-term storage to enhance longevity and should be tried with other short-lived native species (Dremann 2003). A good rule of thumb is to pour about a teaspoon (about 5 ml) of silica gel into a paper envelope and place

LITERATURE CITED

Dreesen, D. 2004. Tumbling for seed cleaning and conditioning. Native Plants Journal 5: 52-54.

Dremann, C. 2003. Observations on *Bromus carinatus* and *Elymus glaucus* seed storage and longevity. Native Plants Journal 4: 61-64.

Dumroese, R.K.; Landis, T.D.; Wenny, D.L. 1998. Raising forest tree seedlings at home: simple methods for growing conifers of the Pacific Northwest from seeds. Contribution No. 860. Moscow, ID: University of Idaho, Idaho Forest, Wildlife, and Range Experiment Station. 56 p.

Hoag, J.C.; Dumroese, R.K.; Sellers, M.E. 2001. Perigynium removal and cold moist stratification improve germination of *Carex nebrascensis* (Nebraska sedge). Native Plants Journal 2: 63-66.

Hong, T.D.; Ellis, R.H. 1996. A protocol to determine seed storage behavior. IPGRI Technical Bulletin No. 1. Rome, Italy: International Plant Genetic Resources Institute.

Landis, T.; Tinus, R.W.; Barnett, J.P. 1999. The container tree nursery manual: volume 6, seedling propagation. Agriculture Handbook 674. Washington, DC: U.S. Department of Agriculture, Forest Service. 167 p.

Schopmeyer, C.S., technical coordinator. 1974. Seeds of Woody Plants in the United States. Agriculture Handbook 450. Washington, D.C.: U.S. Department of Agriculture, Forest Service. 883 p.

Scianna, J.D. 2004. Blending dry seeds clean. Native Plants Journal 5: 47-48.

Stein, W.I., Danielson, R., Shaw, N., Wolff, S., Gerdes, D. 1986. Users guide for seeds of western trees and shrubs. General Technical Report PNW-193. Corvallis, OR: U.S. Department of Agriculture, Pacific Northwest Station. 45 p.

Steinfeld, D.E.; Riley, S.A.; Wilkinson, K.M.; Landis, T.D.; Riley, L.E. 2007. Roadside revegetation: an integrated approach to establishing native plants. Vancouver, WA: Federal Highway Administration, U.S. Department of Transportation, Technology Development Report. FHWA-WFL/TD-07-005. 413 p.

Thomas, D. 2003. Modifying blender blades for seed cleaning. Native Plants Journal 4: 72-73.

Truscott, M. 2004. Cuisinart for cleaning elderberry (*Sambucus* spp. L. [Caprifoliaceae]). Native Plants Journal 5: 46.

Withrow-Robinson, B.; Johnson, R. 2006. Selecting native plant materials for restoration projects: insuring local adaptation and maintaining genetic diversity. Corvallis, OR: Oregon State University Extension Service. EM-8885-E. 10 p.

APPENDIX 7.A. PLANTS MENTIONED IN THIS CHAPTER

American lotus, *Nelumbo lutea*

American plum, *Prunus americana*

arctic dryad, *Dryas octopetala*

arrowleaf balsamroot, *Balsamorhiza sagittata*

biscuitroot, *Lomatium* species

black walnut, *Juglans nigra*

Cascade mountain-ash, *Sorbus scopulina*

ceanothus, *Ceanothus* species

chokecherry, *Prunus virginiana*

cottonwood, *Populus* species

Douglas-fir, *Pseudotsuga menziesii*

fir, *Abies* species

geranium, *Geranium* species

jack pine, *Pinus banksiana*

kinnikinnick, *Arctostaphylos uva-ursi*

lodgepole pine, *Pinus contorta*

longleaf pine, *Pinus palustris*

lupine, *Lupinus* species

maple, *Acer* species

milkweed, *Asclepias* species

Nebraska sedge, *Carex nebrascensis*

noble fir, *Abies procera*

oaks, *Quercus* species

panicled bulrush, *Scirpus microcarpus*

paper birch, *Betula papyrifera*

pecan, *Carya illinoinensis*

penstemon, *Penstemon* species

prairie Junegrass, *Koeleria macrantha*

quaking aspen, *Populus tremuloides*

redosier dogwood, *Cornus sericea*

rose, *Rosa* species

serviceberry, *Amelanchier alnifolia*

shagbark hickory, *Carya ovata*

snowberry, *Symphoricarpos albus*

snowbrush ceanothus, *Ceanothus velutinus*

tall huckleberry, *Vaccinium membranaceum*

thimbleberry, *Rubus parviflorus*

western redcedar, *Thuja plicata*

white oaks, *Quercus* species

whitebark pine, *Pinus albicaulis*

wild rice, *Zizania palustris*

willow, *Salix* species

Woods' rose, *Rosa woodsii*

yarrow, *Achillea millefolium*

Seed Germination and Sowing Options

Tara Luna, Kim M. Wilkinson, and R. Kasten Dumroese

8

Seeds of many native species are challenging to germinate. One important thing a grower can do is learn as much as possible about the life history, ecology, and habitat of the species they wish to grow. What processes do seeds of this species go through in nature? Any observations will be valuable when trying to germinate and grow species that have little or no published information available. How seeds are handled, treated, and sown will affect the quality of the seedling produced. Several sowing options are best suited to seeds with certain characteristics. In this chapter, we discuss seed dormancy treatments that can be used to stimulate germination, and different types of sowing options.

SEED DORMANCY

Dormancy is an adaptation that ensures seeds will germinate only when environmental conditions are favorable for survival. The conditions necessary to allow seeds to "break" dormancy and germinate can be highly variable among species, within a species, or among seed sources of the same species. This degree of variability is advantageous because seeds will germinate at different times over a period of days, weeks, months, or even years, ensuring that some offspring will be exposed to favorable environmental conditions for survival. Horticultural practices may tend to discourage dormancy either intentionally through breeding programs or unintentionally by favoring seedlings that germinate more quickly under nursery conditions. There are several types of seed dormancy. Before attempting to grow a plant, it is important to know the seed

Hand-sowing by Tara Luna.

dormancy type. A simple key to determine the type of seed dormancy is provided in the following sections. Knowing the ecology and life history of the species will help you develop treatments and provide conditions to dissipate, or "break," seed dormancy and achieve good rates of germination.

Types of Seed Dormancy

Nondormant Seeds

Nondormant seeds can germinate immediately after maturation and dispersal from the mother plant. The length of time, however, required for the initiation of germination is variable. Some species may germinate immediately (most willows, quaking aspen, and cottonwoods), whereas others may take up to a month to germinate after sowing (some species of white oaks).

Dormant Seeds

Dormant seeds will not germinate immediately even when ideal environmental conditions exist. Dormant seeds may take several months or even years before they germinate. Dormancy may be caused by factors inside (internal) or outside (external) the seeds. Some species have a combination of internal and external dormancy, a condition known as double dormancy.

Internal dormancy may be physiological, morphological, or both (Baskin and Baskin 1998). Physiological dormancy is the most common type seen in temperate and arctic native plants. Seeds are permeable to water but certain environmental conditions are necessary to modify the internal chemistry of the seed and thus allow germination. Usually a period of cold, moist conditions or holding seeds in dry storage overcomes physiological dormancy. Seeds with morphological dormancy have an underdeveloped embryo when dispersed from the mother plant. A period of after-ripening (usually warm and moist conditions) is needed for the embryo to fully mature before the seed is capable of germination. Seeds with morphological-physiological dormancy usually require a combination of warm and cold conditions, often over an extended period of time, before they are capable of germination.

External seed dormancy may be physical or physical-physiological (Baskin and Baskin 1998). Seeds that have hard, thick seedcoats that physically prevent water or oxygen movement into seeds have physical dormancy. These seeds normally germinate over a period of sev-

eral years. Depending on species and habitat, various environmental factors cause these seeds to become permeable over time. Seeds that require additional exposure to particular temperatures after they become permeable have physical-physiological dormancy.

Determining Seed Dormancy Type

Knowing the type of seed dormancy is essential for successful propagation. The following key to dormancy types is based on the permeability of seeds to water, the size and characteristics of the embryo (which can often be obtained from other literature sources), and whether seeds germinate in 30 to 45 days at temperatures similar to those found in the natural habitat at the time of seed maturation (Baskin and Baskin 2004).

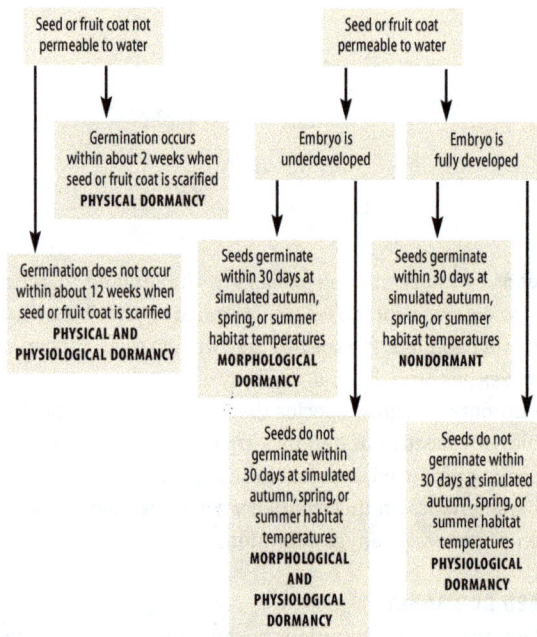

Seed or fruit coat not permeable to water

→ Germination occurs within about 2 weeks when seed or fruit coat is scarified **PHYSICAL DORMANCY**

→ Germination does not occur within about 12 weeks when seed or fruit coat is scarified **PHYSICAL AND PHYSIOLOGICAL DORMANCY**

Seed or fruit coat permeable to water

Embryo is underdeveloped

→ Seeds germinate within 30 days at simulated autumn, spring, or summer habitat temperatures **MORPHOLOGICAL DORMANCY**

→ Seeds do not germinate within 30 days at simulated autumn, spring, or summer habitat temperatures **MORPHOLOGICAL AND PHYSIOLOGICAL DORMANCY**

Embryo is fully developed

→ Seeds germinate within 30 days at simulated autumn, spring, or summer habitat temperatures **NONDORMANT**

→ Seeds do not germinate within 30 days at simulated autumn, spring, or summer habitat temperatures **PHYSIOLOGICAL DORMANCY**

Move-Along Experiment

Information collected from a "move-along" experiment can be very useful for learning about the dormancy breaking and germination requirements of a species. This technique allows a grower to determine if summer only, winter only, or a summer-to-winter sequence of temperatures is required for breaking dormancy in seeds with permeable seedcoats (or fruitcoats) (Baskin and Baskin 2003, 2004). Two temperature profiles simulating 1-year cycles from winter to winter or summer to summer and control treatments are run concurrently. Under this experiment, seeds are exposed to 14 hours of light per day and are subjected to alternating temperatures. The following information provides general guidelines for simulating temperatures, moving from summer through autumn, winter, and spring:

— Summer: 86 °F daytime and 59 °F nighttime (30 and 15 °C).
— Early autumn: 68 °F daytime and 50 °F nighttime (20 and 10 °C).
— Late autumn: 59 °F daytime and 43 °F nighttime (15 and 6 °C).
— Winter: 34 to 41 °F daytime and nighttime (1 and 5 °C).
— Early spring: 59 °F daytime and 43 °F nighttime (15 and 6 °C).
— Late spring: 68 °F daytime and 50 °F nighttime (20 and 10 °C).

Temperature regimes can be modified to reflect more specific conditions. Controls for the experiment are incubated continuously at each temperature regime.

Seeds can be placed in a refrigerator with a thermometer to simulate the winter treatment. Seeds can be placed in trays under a grow light in a heated nursery office to simulate the summer, spring, and autumn temperatures. Ideally, you will need to adjust the temperature of the refrigerator and office in the evening to reflect cooler nighttime temperatures. By moving seeds through this experiment a grower can determine whether only a warm, moist treatment is needed or if only a cold, moist treatment (stratification; see the following paragraphs) is needed or if both are required to break dormancy (figure 8.1).

TREATMENTS TO OVERCOME DORMANCY AND ENHANCE GERMINATION

A variety of treatments are used by native plant nurseries to prevent seed diseases and to break dormancy. It is important to remember that a degree of variability in dormancy can occur within a species, among seedlots of that species, or even between the same seed sources from year to year. Thus, seed treatments may need to be adjusted to compensate for these differ-

Figure 8.1—*The move-along experiment to determine seed dormancy type can be employed by using simple equipment such as placing seed flats under grow lights for the warm period and a refrigerator for the cold period.* Photo by Tara Luna.

ences. For example, seeds from high elevations or far northern latitudes may require a much longer cold treatment to break seed dormancy than do seeds of the same species collected at lower elevations or from more southerly latitudes.

A variety of seed treatments has been developed in response to the diversity of plants grown in nurseries. Before treating seeds, be sure to consult available references to see what treatments have been used on that species; see the literature cited at the end of this chapter and the Native Plant Network (http://www.native plantnetwork.org). If no information is available, check references for closely related species. Any personal observations made on the species in the habitat may also provide some clues on how to germinate the seeds. In general, however, the process of treating seeds follows the fairly standard progression outlined below. Some steps are optional or mandatory depending on whether the seeds are nondormant or have internal or external dormancy.

1. Seed Cleansing

Any seeds can be cleansed of bacterial and fungal infestation; this treatment is necessary for species that easily mold (figure 8.2). Often, molding can be related to the most common disease seen in nurseries, damping-off. Seed cleansing is especially important in humid climates and for species that take a long time to germinate. Often, without treatment, seeds can be lost to pathogens before they are planted in the nursery.

One of the best cleansing treatments is to simply soak seeds in a stream of running water for 24 to 48

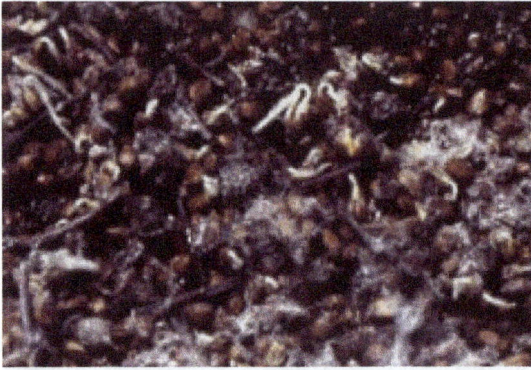

Figure 8.2—*Seeds that are not cleansed before treatment or sowing can easily mold or be susceptible to serious pathogens such as damping-off disease.* Photo by Thomas D. Landis.

hours. The running water flushes bacterial and fungal spores from the seeds (James and Genz 1981), and this treatment can be used to satisfy the soaking requirement described in the next section. Seeds can also be cleansed with several chemicals, and some of these also act to stimulate germination. Bleach is the most common chemical used to cleanse seeds; depending on the species, the solutions range between one part bleach (5.25 percent sodium hypochlorite) in eight parts water to two parts bleach in three parts water. With most species, treatment duration is 10 minutes or less. Species with very thin seedcoats should not be cleansed with bleach. Hydrogen peroxide can be an effective cleanser and can sometimes enhance germination (Narimanov 2000). The usual treatment is one part peroxide (3 percent solution) in three parts water solution. Many native species of the rose family, such as serviceberry and Woods' rose, benefit from this treatment.

2. Scarification

Scarification is any method of disrupting an impermeable seedcoat so that water and oxygen can enter seeds with external dormancy. In nature, hard seedcoats are cracked or softened by fire, extreme temperatures, digestive acids in the stomachs of animals, or by the abrasion of blowing sand or ice. After the seedcoat has been disrupted, oxygen and water pass into the seeds and germination can proceed. Species with external dormancy include many of the legumes, mallows, and other species that are adapted to fire or inhabit dry or desert environments.

Seeds can be scarified many ways. How well the method works depends on the species and the thickness of the seedcoats. Whichever method is chosen, it is very important not to damage the endosperm, cotyledons, or embryo during the treatment. Taking time to learn seed anatomy of the species is helpful. Trying several methods and recording the results will help determine the best method for that species and seed source.

Mechanical scarification includes filing or nicking seeds by hand and is most often used on large-seeded species such as locust, acacia, mesquite, and whitebark pine (figure 8.3). Be sure to scarify on the side of the seed opposite the embryo. This method is time consuming and requires precision to adequately modify the seedcoat without damaging the internal portions of the seed. Sandpaper can be used on smaller-seeded species such as sedges; placing seeds into a shallow wooden box and then rubbing them under a block of wood covered in sandpaper is the simplest technique. Often, however, the degree of scarification achieved with sandpaper can be variable.

Many species, especially those from fire-adapted ecosystems, respond to germination cues from heat. In nurseries, this response can be simulated by using either wet or dry heat to scarify the seeds. Using wet heat is an effective method for many small-seeded species because it provides a rapid, uniform treatment that can be assessed within a few hours. Native legumes (lupine, milkvetch, Indian breadroot, and wild licorice), ceanothus, buckthorn, and wild hollyhock can be scarified by wet heat (figure 8.4). Because the thickness of the seedcoat may vary among sources, it is wise to dissect a few seeds and examine the thickness of their seedcoats to help determine treatment duration . Seeds are added to boiling water for just 5 to 10 seconds and then immediately transferred to a vat of cold water so that they cool quickly to prevent embryo damage. Seeds imbibe the cool water for 1 day and are ready for sowing or for stratification. Some species cannot tolerate excessively high temperatures, so you may want to heat the water to only 158 °F (70 °C) and monitor your results.

Dry heat is most commonly used on fire-adapted species such as laurel sumac and ceanothus. Seeds are placed in an oven at temperatures ranging from 175 to 250 °F (80 to 120 °C) from a few minutes to 1 hour, depending on the species. The seedcoat cracks open in response to the heat. To avoid damaging seeds, this treatment should be monitored closely.

Figure 8.3—*Hand-scarified seeds of American lotus.* Photo by JF New Nursery.

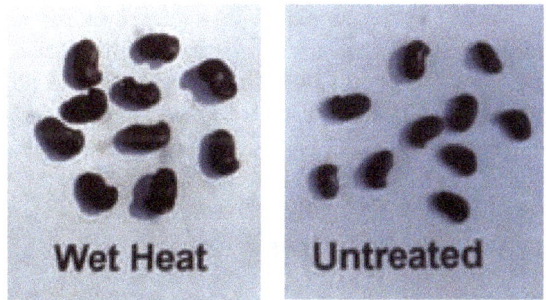

Figure 8.4—*Seeds, such as these of New Mexico locust, that have been scarified by hot water are visibly larger than untreated seeds since the seed coat has been breached and seeds can then absorb water, increasing their size.* Photo by Tara Luna.

Figure 8.5—*Smooth sumac seeds that have been treated with sulfuric acid.* Photo by Nancy Shaw.

Sulfuric acid is most commonly used on species with very thick seedcoats and with stony endocarps that surround the embryo. Blackberry, kinnikinnick, manzanita, and skunkbush sumac are species with very thick seedcoats and have been scarified with sulfuric acid. Treatment length varies with the species and often among seed sources, and it must be carefully monitored because seeds can be destroyed if the treatment is too long. A simple way to monitor the process is by removing seeds at regular intervals and cutting them with a sharp knife. When the seeds are still firm but can be cut fairly easily, the treatment is probably sufficient. Another way is to run a pilot test on a subsample of seeds. Again, remove some seeds periodically and evaluate how well they germinate. Once the best duration is known, the entire seedlot can be treated. Sulfuric acid is very dangerous to handle and requires special equipment, personal protective gear, and proper disposal after use (figure 8.5). Some species, such as wild raspberry and salmonberry, have thick seedcoats but can easily be damaged by sulfuric acid. Instead, citric acid or sodium or calcium hypochlorite baths with longer treatment durations may be used, especially for species with thinner seedcoats.

The safe use of sulfuric acid requires the following criteria and procedures:

— Treat seeds that are dry and at room temperature.
— Require workers to wear safety equipment, including face shield, goggles, thick rubber gloves, and full protective clothing.
— Add acid to water, never water to acid.

— Immerse seeds in an acid-resistant container, such as a glass, for the duration required.
— Stir seeds carefully in the acid bath; a glass rod works well.
— Immerse the container with seeds and acid in an ice bath to keep temperatures at a safe level for the embryos (this temperature depends on the species; many do not need this step).
— Remove seeds from the acid by slowly pouring the seed-acid solution into a larger volume of cool water, ideally one in which new, fresh water is continually being added.
— Stir seeds during water rinsing to make sure all surfaces are thoroughly rinsed clean.

Hobby-size rock tumblers can be used to scarify seeds and to avoid seed destruction that can occur with sulfuric acid or heat scarification (figure 8.6). Dry tumbling involves placing seeds, a coarse carborundum grit (sold by rock tumbler dealers), and pea gravel in the tumbler. The duration is usually for several days, but this method is an effective and safe way of scarifying many species. Wet tumbling, a method in which water is added to the grit and gravel, has been an effective treatment for redosier dogwood, golden currant, and wolfberry (Dreesen 2004). A benefit of wet tumbling is that seeds are soaked in well-aerated water and chemical inhibitors may be leached.

3. Soaking

After cleansing and/or scarification, seeds must have exposure to water and oxygen before germination can occur. The standard procedure is to soak seeds in running water for 1 to several days until they are fully hydrated. This condition can be checked by weighing a subsample: pull a sample; allow it to dry until the seedcoat is still wet but dull, not glossy; and weigh it. When the weight no longer increases substantially with additional soaking time, the seeds have absorbed sufficient water. Scarified seeds will be more obvious; the seeds will enlarge drastically during the soak. Seeds that only had physical dormancy can be immediately planted. As mentioned previously, running water rinses are good seed cleaning treatments and are effective in reducing the need for fungicides in nurseries (Dumroese and others 1990). Running water soaks also help to remove any chemical inhibitors present on or within the seeds, and an aquarium pump can be used to agitate the seeds to improve the cleaning effect and keep the water well aerated. If seeds are not soaked with running water, change the water often (at least a couple of times each day).

4. Germination Stimulators

Several chemicals are known to increase germination of many native plants. These chemicals are usually applied after seeds are fully hydrated. As with the other treatments already discussed, species, seed source, and other factors will affect how well the treatment works. In general, only seeds with internal dormancy receive this treatment. Germination stimulators include gibberellic acid, ethylene, smoke, and potassium hydroxide.

Gibberellic Acid. Gibberellic acid is the most important plant hormone for the regulation of internal seed dormancy and is often used on seeds with complex

Figure 8.6—*Hobby-size rock tumblers can be used to scarify seeds. This method is an effective alternative to acid or heat scarification.* Photo by Tara Luna.

internal dormancy and with those species having underdeveloped embryos. In some cases, it has been used to substitute for a warm, moist treatment and to hasten embryo after-ripening. Gibberellic acid can be purchased from horticultural suppliers. A stock solution of 1,000 parts per million (ppm) is prepared by dissolving gibberellic acid in distilled water at the rate of 1 mg in 1 ml. A 100-mg packet is dissolved into 100 ml (about one-half cup) of water. Preferred concentrations vary, but most nurseries use 500 to 1,000 ppm. High concentrations can cause seeds to germinate, but the resulting seedlings may be of poor quality. Therefore, it is best to experiment with low concentrations first.

Ethylene. This gas occurs naturally in plants and is known to stimulate the germination of some species. Ethylene gas is released from ethephon, a commercially available product. Ethephon, used either alone or in combination with gibbrellic acid, has enhanced the germination of species such as purple coneflower and arrowleaf balsamroot (Chambers and others 2006; Feghahati and Reese 1994; Sari and others 2001). It may inhibit germination in other species, so consult the literature and/or experiment.

Smoke Treatments. Smoke stimulates germination in many fire-adapted native species from the California chaparral, especially those that have thin, permeable seedcoats that allow entry of smoke into the seeds (Keeley and Fotheringham 1998). Seeds can be treated with smoke fumigation, a method in which smoke is piped into a specially constructed smoke tent containing seeds sown in trays (figure 8.7A), or with smoke

TIPS FOR USING GIBBERELLIC ACID

—Because gibberellic acid takes a long time to dissolve, stir it constantly or prepare it the day before use.

—Store unused solution away from direct sunlight.

—Using unbleached coffee filters, cut the filters into squares and fold them diagonally to form a container.

—Pour gibberellic acid solution evenly into an ice cube tray to a depth sufficient to cover the seeds.

—Place each folded coffee filter containing the seeds into the wells of the tray.

—After 24 hours, remove the seeds, rinse them with water and either sow them or prepare them for stratification.

Potassium Hydroxide Rinses. Potassium hydroxide has been used to stimulate germination in several native plant species. Optimum concentration varies from 5.3 to 7.6 Molar for 1 to 10 minutes depending on the species, but longer soaks at higher concentrations were found to be detrimental (Gao and others 1998).

5. Stratification

Fully soaked seeds with internal dormancy are treated with stratification. Historically, stratification was the process of alternating layers of moist soil and seeds in barrels and allowing these "strata" to be exposed to winter temperatures. For centuries, people have known that this treatment causes seeds to germinate because it mimics what occurs naturally during winter. Nowadays, stratification is often used more generically to describe the combined use of moisture

water. Smoke water is an aqueous solution of smoke extract made by burning vegetation and piping the smoke through distilled water or allowing the smoke to infuse into a container of water. Seeds are then soaked in the treated water (figure 8.7B). Conversely, growers can experiment with commercially available smoke products such as liquid smoke, smoke-infused paper discs, and ash that is added to growing media. Many variables, such as the material used for combustion, the combustion temperature, and the duration of exposure, will need to be determined on a species-by-species basis. Experiments performed by Keeley and Fotheringham (1998) found that the length of exposure to smoke was very important in some species; a 3-minute difference in exposure resulted in seed mortality. Some fire species did not germinate under heat or smoke treatments alone. With some species, such as beargrass, seed burial for 1 year or stratification is required in addition to smoke exposure.

All these factors can have an effect on germination and should be considered when using smoke or chemicals in smoke to induce germination. Some native species that responded favorably to smoke treatments include antelope bitterbrush, big sagebrush, Great Plains tobacco, Indian ricegrass, white sage, beargrass, scarlet bugler, and big sagebrush (Blank and Young 1998; Landis 2000). Success with this novel treatment will require trials; keep good records.

Figure 8.7—Smoke treatments have been used to overcome seed dormancy and enhance germination rates for many native species inhabiting fire-dependent ecosystems. (A) A smoke tent for treating seeds. (B) Smoke water-treated seeds of angelica.
Photo A by Kingsley Dixon, B by Tara Luna.

and any temperature to overcome seed dormancy. We use the term "stratification" to refer to only cold, moist treatments and the term "warm, moist treatment" instead of warm, moist stratification.

Many native species with double internal seed dormancy require a combination of a warm, moist treatment for a period of time followed by stratification. Remember that some species or seedlots may require only a few days or weeks of stratification, while other species or seedlots may require several months. Therefore, as a general rule, it is best to use the maximum recommended treatment. Also keep in mind that what works well at one nursery may not necessarily work well at another nursery because of differences in seed source, handling, processing, cleaning, and storage.

Another valuable advantage to stratifying seeds is that it speeds up germination and makes it more uniform, which is desirable in a container nursery. Therefore, instead of having germination occur sporadically over several months, it all occurs within a few days or weeks, making it much easier to care for the crop.

Stratification Techniques

Seeds sown in flats or containers in late summer or autumn and left outdoors during winter undergo "natural" stratification. This technique may be preferred if the species has double dormancy (requires both a warm, moist treatment and stratification), requires a very long stratification, and/or requires low temperatures or fluctuating temperatures for a long period of time (figure 8.8A). Conversely, "artificial" stratification involves placing seeds, sometimes within a moist medium such as peat moss, inside permeable bags or containers under refrigeration for a period of time. Artificial stratification has several advantages: (1) it allows for a routine check of seeds to ensure they are moist and not moldy, (2) a large number of seeds can be stratified in a small space, and (3) seeds or seedlots that begin to germinate can be removed from the treatment and planted in the nursery as they become available. Artificial stratification is preferred over natural stratification unless the natural treatment provides higher rates of germination (figure 8.8B).

Artificial stratification can be accomplished a couple of ways. For small seedlots and/or small seeds, seeds can be placed between sheets of moistened paper towels and inserted in an opened plastic bag or sown on a

SEED TREATMENT TEMPERATURES

Stratification—34 to 38 °F (1 to 3 °C)
Warm, moist treatment—72 to 86 °F (22 to 30 °C)

medium in flats with drainage holes (figures 8.9A and B). For paper towels, first moisten them with clean water and let excess water drain away from the towel by holding it up by one corner. The paper towel should remain moist but not waterlogged. Second, be sure to distribute the seeds evenly across the moist paper towel or the flat to help prevent the spread of mold to other seeds. If necessary, seeds can then be first exposed to warm temperatures before exposure to cold temperatures. The paper towel method also works well for those species that require only a few weeks of stratification.

Another technique is "naked" stratification. Most conifer seeds, for example, are stratified this way (figures 8.10A and B). Seeds are placed in mesh bags and then soaked in running water as described previously. After the seeds are hydrated, the bag is pulled from the soak, allowed to drip dry for 30 to 90 seconds, and then suspended in a plastic bag. Make sure the seeds are not in contact with standing water in the bag and hang the bags in the refrigerator. If naked seeds need a warm, moist treatment before stratification, it is easiest to first spread the seeds onto moistened paper towels enclosed in large plastic bags. After the warm treatment, the seeds can be returned to the mesh bags for stratification. One other hint: if a particular species or seedlot has a tendency to begin germinating during stratification, surface-dry the seedcoats—seeds should be moist and dull, not shiny—and then put the seeds into the bag for refrigeration. The seeds should still have enough moisture for chemical processes that dissipate dormancy to occur but not enough moisture to allow germination.

Many wetland and aquatic species can be treated with naked stratification in water. In general, these species can be easily stratified in Ziploc®-type bags filled with water. Insert a soda straw into the bag, ensuring that the end is sticking out of the bag, and seal the rest of the bag securely. Place under refrigeration for the stratification period.

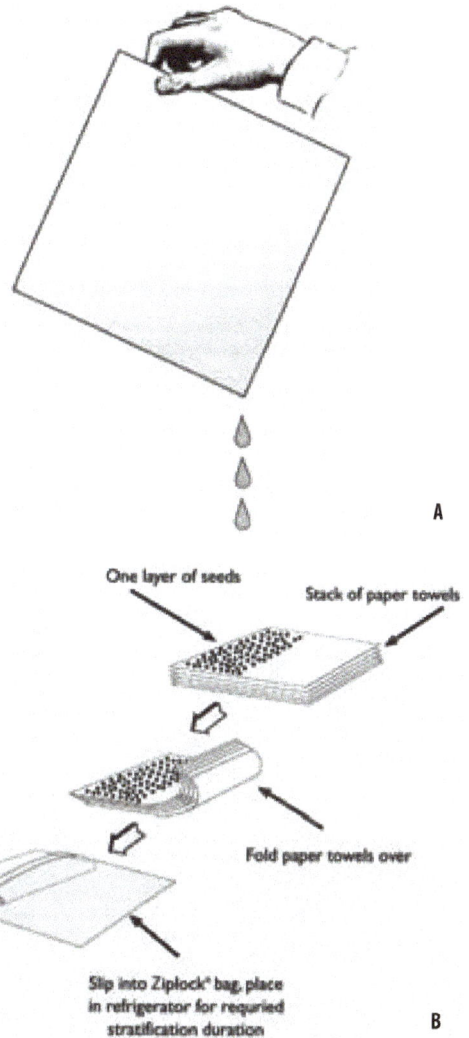

Figure 8.9—*(A) Small seeds requiring only a few weeks of stratification can be stratified on moistened paper towels within a plastic bag inside a refrigerator. The paper towels should be moist; hold them by an end to allow excess water to drain away. (B) Place the seeds evenly on the towels and insert them into a Ziploc®-type bag.* Illustration from Dumroese and others (1998).

Figure 8.8—*(A) Natural stratification works well for seeds with a long stratification requirement or those requiring cool or fluctuating temperatures for germination. (B) Artificial stratification enables the inspection of seeds during treatment. Although some seeds will not show any signs that the stratification duration was sufficient others will begin to crack open or sprout when sufficient chilling has occurred.* Photos by Tara Luna.

ENVIRONMENTAL FACTORS INFLUENCING GERMINATION

Four environmental factors, light, water, oxygen, and temperature, are normally required for germination. Native plants often have specific germination requirements, however, so enviornmental conditions will vary from species to species. These result from their specific ecological adaptations and the environmental cues that trigger germination. For example, light quality and duration can influence germination. Some species, such as sedges, bulrushes, and rushes, have very small dust-like seeds that require light for germination, so they should be sown on the surface of the

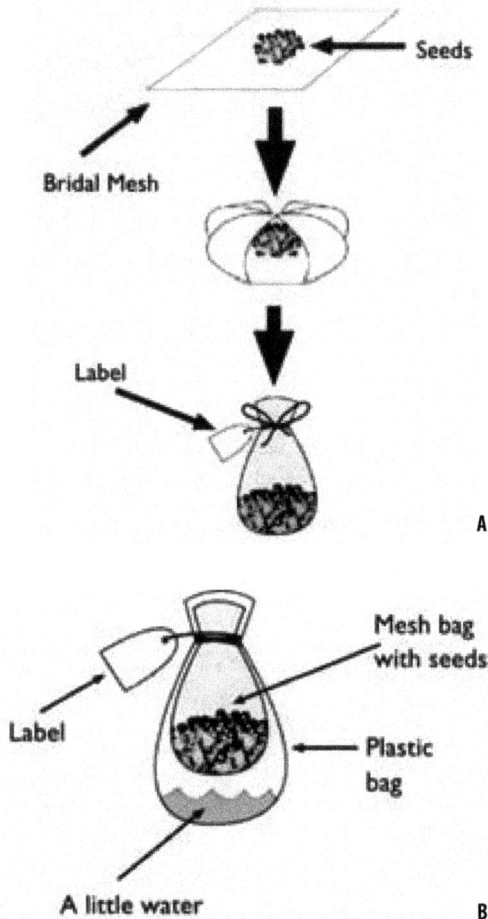

Figure 8.10—*(A) This is an easy way to handle seeds under naked stratification: Cut a square piece of mesh. Place the seeds in the center and fold up the corners of the mesh to make a bag. Make sure there is plenty of room in the bag for the seeds to expand when they absorb water. Place the bag full of seeds into a running water soak for 24 to 48 hours. Remove the bag and allow the seeds to drip dry for a minute or two before suspending the mesh bag inside a plastic bag. (B) Make sure the seeds hang above any water that collects in the bottom of the plastic bag and hang the bags in a refrigerator.* Illustrations from Dumroese and others (1998).

medium and left uncovered. Some species are light sensitive and will fail to germinate if they are buried too deeply. Many native species fall into this category. Species with larger-sized seeds are conditioned to germinate only if they are buried in the soil. Wetness is also important. Overwatering seeds during germination results in reduced levels of oxygen in the medium and promotes tissue breakdown and disease whereas underwatering delays or prevents germination. Therefore, seeds should be kept evenly moist during germination. Although oxygen is needed for respiratory processes in germinating seeds, some aquatic species may require low oxygen levels for germination.

Temperature plays an important role in germination. Species that require "cool" temperatures for germination will germinate only when temperatures are below 77 °F (25 °C). In this case, flats or seeded containers should be left outside the greenhouse to germinate under the natural fluctuating temperatures during spring.

Species that tolerate cool temperatures will germinate over a wide range of temperatures from 41 to 86 °F (5 to 30 °C). These species can probably be germinated outside or inside a warm greenhouse. Species requiring warm temperatures will germinate only if temperatures are above 50 °F (10 °C) and should be kept in a warm greenhouse. Some species germinate better when exposed to alternating temperatures. The fluctuation of day and night temperatures often yields better germination than do constant temperatures for seeds such as antelope bitterbrush, mountain mahogany, and cliffrose. The most effective alternating temperatures have a difference of at least 18 °F (10 °C) between the daytime and nighttime temperature. Native species that require alternating temperatures are probably best germinated directly outdoors (autumn seeding), where they are exposed to the naturally fluctuating temperatures of the seasons.

SEED SOWING OPTIONS

Native plant growers often work with seeds of species that have not yet been propagated in nurseries. Usually, little literature or experience is available to answer questions about seed dormancy-breaking requirements, germination percentages, and other factors. "Standard" sowing options, such as direct seeding, are often not ideal when working with unfamiliar species. Even when key questions about seed performance have been answered, the actual process of sowing seeds into containers will vary with the species, type of seed, and seed quality. Nurseries have several options for improving the efficiency and effectiveness of seed sowing. Several sowing techniques have been used for native plants (Table 8.1) and are described below.

Direct Sowing

Direct sowing is fast, easy, and economical because it minimizes seed handling and labor. It can be mechanized when done on a large scale. For direct sowing to be efficient, the seeds must be uniform in size and shape, easy to handle, abundant in supply; have simple dormancy treatments; and have a known, high germination rate (figures 8.11A–C). The success of direct seeding depends on the accuracy of seed germination information. Growers must realize that actual seedling emergence may be different from the results of laboratory germination tests that are conducted under ideal environmental conditions. Nursery managers must adjust for this discrepancy based on their own operational experience. Growers should conduct a small germination test of each seedlot to determine the percentage of germination that they will obtain in "real life" and use those percentages when deciding the number of seeds to direct sow (see table 8.2). Follow these steps for successful direct sowing:

— Determine how many seeds must be sown to obtain the production target.
— Treat seeds as necessary to break dormancy.
— Sow seeds, ideally centering the seeds in each container. Some seeds require a specific orientation for optimal growth and development; if so, make sure seeds are sown in the correct orientation.

Table 8.1—Options for sowing seeds (modified from Landis and others 1999)

Propagation Method	Good Method for Seeds with the Following Characteristics	Advantages	Disadvantages
Direct sowing: Seeds are sown into containers	—Have a known, high-percentage germination —Are inexpensive —Are in abundant supply —Have uniform, smooth shapes	—Fast and easy —Economical —Minimizes seed handling —Seeds are all sown at once	—Less efficient use of space, seeds, and/or growing medium —Causes of poor germination are difficult to track —May require thinning and/or consolidation and associated labor costs —Not good for large or irregularly shaped seeds
Planting germinants: Seeds sprouting or germinating in trays or bags are sown into containers while roots are just beginning to emerge	—Are of unknown viability —Are valuable or rare —Have unknown germination requirements —Germinate over a period of time during stratification	—Efficient use of seeds —Efficient use of nursery space —Can adjust for unknown seed quality or performance	—Labor intensive —May result in nonuniform crop development —Root deformation possible —Requires frequent, skilled monitoring
Transplanting emergents: Seeds are sown into flats or seedbeds for germination; once germinated and leaves appear, seedlings are transplanted to containers	—Are being tested but will not be transplanted to produce a crop —Do not respond well to other sowing methods —Have long or unknown dormancy —Good for trials to observe seed performance	—Useful with fibrous rooted species —Efficient use of seeds —Efficient use of nursery space —Can adjust for unknown seed quality or performance	—Not recommended for woody and/or taprooted species because of problems with transplant shock and/or root deformation —Requires skilled labor
Miniplug transplants: Seeds are sown directly into small containers. After germination, they are transplanted into larger containers	—Are of unknown quality —Are valuable or rare —Have unknown germination requirements —Have very tiny seeds —Will be transplanted into large containers	—Efficient use of space —Uniform crop development —Low risk of transplant injury	—Requires two sets of containers —Timing is critical —Transplanting by hand is labor intensive

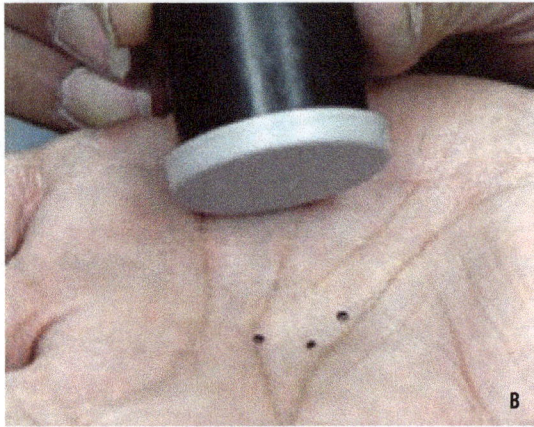

Figure 8.11—(A) Direct sowing works well for seeds that have a known dormancy, are easy to handle, and are in abundant supply. (B) Simple tools such as a film canister or (C) a folded envelope can be used to accurately sow small seeds of native plants.
Photo A by Tara Luna, B and C by Dawn Thomas.

— Depending on the light requirements of the species, cover seeds with the correct amount of mulch.
— Gently water the seeds with a fine watering head to press them into the growing media.

Seeds can be direct sown in one of two ways in containers: multiple seeds or single seeds.

Multiple-Seed Sowing and Thinning

"Multiple sowing," the most common direct sowing practice, is a method in which several seeds are sown into each container with the expectation that at least one will germinate. The number of seeds to sow can be calculated based on the percentage of germination of the seeds. Commonly, two to five seeds are sown per container. As a general rule, seeds with less than 50 percent germination are not recommended for direct sowing because the high density of nonviable seeds in the container may cause disease problems, more containers will need to be thinned, and many plants will be wasted (figure 8.12). Table 8.2 provides general recommendations for the number of seeds to sow per container based on the germination percentage. At some point, adding more seeds per container does not really increase the number of containers with plants but does drastically increase the number of containers with too many plants (table 8.3). Sometimes it may be better to single-sow a few containers than thin extra seedlings from many containers. For example, sowing a single seed per container of a seedlot with 85 percent germination yields 15 percent empty containers whereas sowing two seeds per container yields only 2 percent empty containers, but sowing the extra seed requires thinning 72 percent of the containers. The nursery manager may have been better off, in terms of seed use efficiency and labor, to have just oversown 10 percent more containers rather than pay for the labor to thin. Therefore, the amount of seeds to sow per container is a function of germination, seed availability, nursery space, thinning costs, and so on.

If more than one plant emerges per container, the extra(s) must be removed or clipped ("thinned"). When several seedlings emerge in the same container, they compete for light, water, and nutrients. The result is lower initial growth rates of seedlings until they are thinned. For this reason, thinning should be done as soon as possible after seedlings emerge. Thinning is a labor-intensive practice that can damage remaining seedlings if done improperly. Train workers to thin plants carefully and to follow these guidelines:

— Thin germinants as soon as possible; the more developed the root system becomes, the more difficult it is to thin.
— Retain the strongest seedling closest to the center of the container; thinning is an opportunity for selecting the healthiest seedling while removing inferior plants.
— Pull or cut extra plants. For species with a long, slender taproot at germination (for example, pines), extra seedlings can be easily pulled until they develop secondary roots. For species with vigorous, fibrous root systems, cutting extra plants with scissors or nipping them with fingernails may be better.
— Discard removed plants into compost or waste.
— Check the remaining seedling. If thinning disrupted the mulch, adjust it so the seedling has the best environment possible.

Single-Seed Sowing

Sometimes, particularly when seeds are scarce or costly, single seeds can be directly sown into containers. This practice ensures that every seed has the potential to become a plant. If a particular number of

Figure 8.12— *Calculating and testing germination rates help reduce costs and problems associated with thinning.* Photo by Thomas D. Landis.

Table 8.2—For a given seed germination percentage, increasing the number of seeds sown per container increases the number of filled containers. Generally, a target of 90 to 95 percent filled containers is reasonable (from Dumroese and others 1998)

Seed Germination Percentage	Seeds To Sow per Container	Percentage of Containers with at Least One Seedling
90 +	1 to 2	90–100
80–89	2	96–99
70–79	2	91–96
60–69	3	94–97
50–59	4	94–97
40–49	5	92–97

Table 8.3—A sowing example for a seedlot of chokecherry having 65 percent germination. Assuming 1,000 seedlings are desired and that many containers are sown, notice that adding more than three seeds per container really does not improve the number of containers with seedlings, but does use (waste) many seeds and increases the number of containers that will require thinning (modified from Dumroese and others 1998)

Seeds Sown per Container	Total Seeds Sown	Containers with at Least One Seedling	Empty Containers	Additional Seedlings Produced per Additional 1,000 Seeds Sown	Containers Requiring Thinning To Remove Extra Seedlings
1	1,000	650	350		
2	2,000	880	120	230	420
3	3,000	960	40	80	720
4	4,000	990	10	30	870
5	5,000	1,000	0	0	945

plants are required, then extra containers are planted, often referred to as "oversowing," to make up for any empty cells. The number of extra containers to sow can be calculated based on the percentage of germination. If a seedlot has only 78 percent germination, for 100 plants you must sow at least an extra 28 containers (100 desired seedlings/0.78 success rate = 128 containers required). The number of oversown containers may need to be increased to account for seedling losses during the growing cycle.

Oversowing works best if the nursery has extra space and is using containers with individual, exchangeable cells because, containers with live plants can be consolidated and the extra containers can be removed (see Chapter 6, *Containers*, for example). Single sowing is efficient because no seeds are wasted and plants that do emerge are not subjected to competition or the stresses of thinning as they are with the multiple-sowing technique. Oversowing does, however, waste potting materials and bench space. Consolidating the empty containers, however, can be a labor-intensive and expensive process.

Planting Germinants

Germinant sowing ("sowing sprouts") is the practice of sowing seeds that are germinating (or "sprouting") into the container just as their young root emerges (figure 8.13). When done properly, germinant sowing ensures that one viable seed is placed in each container. The resulting seedlings are often larger because they can begin to grow immediately without competition. This technique results in minimal waste of materials, labor, and space, and works best for seeds that:

— Are from a rare or valuable seedlot.
— Have a low or unknown germination percentage.
— Are large or irregularly shaped.
— Germinate in stratification
— Have deep dormancy and germinate over a long period of time.
— Rapidly produce a long root after germination (such as many desert and semidesert species).

Germinant sowing is a relatively simple process. Seeds are treated as necessary, but, rather than being direct sown into containers, they are germinated in trays or placed in bags. Trays can be very simple; Styro-foam™ meat trays, cake pans with clear plastic lids, or layers of paper towels or fabric on a sheet of plastic or cardboard all work well. Ideally, seeds are dispersed enough on the trays to prevent mold. Larger seeds are sometimes placed in plastic bags filled with a moist medium such as *Sphagnum* peat moss. Seeds are routinely checked every few days or weeks. After seeds begin to germinate, they should be checked daily. Germinated seeds are removed daily and planted directly into their containers. Larger seeds can be planted by hand; smaller seeds must be sown with tweezers.

Two factors are critical when sowing germinants: timing and root orientation. Seeds should be sown into containers as soon as the root emerges. The embryonic root, often called a "radicle," should be short, ideally no longer than 0.4 in (1 cm). If the radicle becomes too long, it may be difficult to plant without causing root deformation (figures 8.14A and B). Some growers like to prune the radicle of taprooted species, such as oaks, prior to planting to ensure a more fibrous root system. No more than the very tip (up to 0.1 in [3 mm]) is trimmed with clean scissors or clipped with the thumbnail. The germinating seed is carefully placed in the container, either sown on its side or with the radicle extending downward. After the seeds are properly planted, the medium should be firmed around the root and the seed covered with mulch.

Planting germinants is highly effective because it makes efficient use of space and seeds. All containers are filled with one growing plant and subsequently losses are minimal. Another advantage is that the germination process is more visible to the growers than when seeds are direct sown. This means that germination timing can be better monitored and the causes of germination problems are easier to track. On the other hand, because seeds in trays or bags are very close together, a mold or pathogen can contaminate all the seeds. Labor is required to routinely check for germination, skill is required to achieve proper planting orientation of the seeds, and planting must be done in a timely fashion. Because germinants may emerge over several weeks or longer for some species, crop development wil be more variable and require special cultural treatments.

Transplanting Emergents

Transplanting emergents ("pricking out") is a common practice, but is not recommended for taprooted

Figure 8.13—*Planting germinants works well for species that require cool temperatures for germination and break dormancy over a prolonged period during stratification.* Photo by Tara Luna.

Figure 8.14—*When planting germinants, seeds must be sown as soon as the radicle is visible and must be oriented correctly when planted. Incorrect orientation leads to severe root deformation in woody species such as (A) bitterbrush and herbaceous species such as (B) arrowleaf balsamroot.* Photo A by Thomas D. Landis, B by R. Kasten Dumroese.

woody plants because root problems often result (table 8.1). Seeds are hand-sown in shallow trays that are usually filled with about 2 in (5 cm) of peat moss-vermiculite growing medium (figure 8.15). After the seeds germinate, they are "pricked out" of the tray and transplanted into a container.

Transplanting emergents works best when:

— Species have a fibrous root system that recovers well from transplanting (herbaceous forbs without a taproot, grasses, sedges, and rushes).
— Tests or trials are being used to observe germination timing, seed treatments, observations of root development, or early growth.
— Seeds are too small or fragile to be sown by any other method.
— Seeds have very complex dormancy and/or germinate over an extended period of time.
— Limited nursery growing space makes direct seeding uneconomical.

Some key disadvantages include the following:

— Disease potential is very high in densely planted trays.
— Root orientation and timing is critical; poor root form and other problems can result.
— Transplanting is very skill and labor intensive.

Great care and proper technique must be used during sowing and transplanting. Larger seeds are scattered by hand over the surface of the moistened medium. Smaller seeds can be sown with a salt shaker with enlarged holes. Sown seeds are covered with a light application of fine-textured mulch, irrigated, and placed in the greenhouse. Although the exact size or age to transplant the germinating seedlings varies by species, it is usually done at the primary leaf stage. Emergents are carefully removed from the tray, usually by gently loosening the medium around them (figure 8.16A). A small hole is made in the medium of the container and the germinant is carefully transplanted, ensuring proper root orientation (figure 8.13). Some species benefit from root pruning prior to transplanting. The potting medium is then firmed around the root and stem (figure 8.16B). When done improperly on tap-rooted woody species, this practice can produce a "J-root" or kink in the seedling stem that can reduce growth in the nursery and cause mechanical weakness or mortality after outplanting. Therefore, unless no other sowing method works, transplanting emergents of woody plants is discouraged.

Transplanting Miniplugs

A miniplug is a small-volume container or expanded peat pellet in which seeds are direct sown (see Chapter 6, *Containers*). After the seedlings are well established, they are transplanted into a larger container (figure 8.17). Transplanting miniplugs has a number of bene-fits. The miniplug container preserves healthy root form because damage to roots during transplanting is eliminated. Planting miniplugs also makes efficient use of growing space. Large numbers of miniplugs can be started in a very small area and managed intensively during germination and early growth, avoiding the expense associated with operating the entire nursery. For nurseries that produce large container stock, plant-ing miniplugs can result in a more efficient use of pot-ting materials and space than other sowing methods.

Seeds are direct seeded into the miniplug containers or peat pellets. Timing this practice is very important because plants in miniplug containers must have a firm enough root plug to withstand the transplanting process, but they must not have so many roots that they are rootbound or that the roots may become deformed after transplanting. If peat pellets are used, too few roots are not a problem because the entire pellet can be transplanted. Seedlings with enough roots to hold the mini-plug together are carefully extracted. A hole large enough to accept the plug is made in the medium of the larger container, and the miniplug-grown seedling is carefully inserted. Planters should ensure that the roots go straight down and are not deformed during trans-planting. The medium is gently firmed around the root system, mulch is applied, and the plant is watered.

Transplanting miniplugs is labor intensive and requires skill. Another drawback with miniplugs is the need for two sets of containers; one for miniplugs and one for the product. The savings of space and climate control in the greenhouse, however, may compensate for the extra expense. Before investing in miniplugs on a large scale, a small trial is advised.

Figure 8.15—*Pricking out trays for transplanting emergents.* Photo by Tara Luna.

SEED COVERINGS (MULCH)

Regardless of the seed sowing method, a seed cover or "mulch" is necessary to create an optimal environment for germinating seeds. The only exception is for species that require light to germinate. Mulch is usually a light-colored, nonorganic material spread thinly over the seeds. Examples of mulches include granite grit (such as poultry grit) (figure 8.18A), pumice, perlite (figure 8.18B), coarse sand, or vermiculite (figure 8.18C). When properly applied, mulches:

— Create an ideal "moist but not saturated" environ-ment around germinating seeds by making a break in the texture of the potting medium (water will not move from the medium into the mulch).

Figure 8.17—*Miniplugs are a viable option for growing seedlings that will later be transplanted to a larger container. Miniplugs work very well with species with very tiny seeds.* Photo by Tara Luna.

— Keep seeds in place. This practice improves contact with the medium and minimizes the number of seeds washed out of the containers by irrigation or rainfall.

— Reflect heat when mulches are light colored, so seeds do not get too hot on bright, sunny days.

— Reduce the development of moss, algae, and liverworts (figure 8.18D).

The recommended depth of the seed covering varies by species; a general rule is to cover the seed twice as deep as the seed is wide. If mulch is too shallow, seeds may float away in the irrigation water. If the mulch is too deep, small plants may not be able to emerge above it (figure 8.19).

Seeds requiring light should be left uncovered. Very small seeds should be left uncovered or barely covered with a fine-textured material such as fine-grade perlite or milled *Sphagnum* peat moss. Uncovered and barely covered seeds must be misted frequently to prevent them from drying out. After light-requiring and light-sensitive species have emerged and are well established, mulch can be applied to prevent moss and liverwort growth and to help keep the medium moist.

TRY DIFFERENT SOWING TECHNIQUES AND KEEP DETAILED RECORDS

Germinating and growing native plants is directly tied to the natural processes they go through in nature. Understanding the biology and ecology of native plants

Figure 8.16—*Transplanting emergents works well for fibrous rooted shrubs, forbs, and grasses. Great care must be taken to lift the emergent from the pricking out tray without damaging the roots and to carefully and properly transplant it into a the new container filled with moistened growing media.* Photos by Tara Luna.

Figure 8.18—*Seed mulches are important to hold the seeds in place and to moderate the surface temperature of the medium during germination. Common mulches include (A) grit, (B) perlite, and (C) vermiculite. (D) Mulches help to prevent the development of mosses and liverworts, which can compete with the seedling.* Photos by Thomas D. Landis.

will provide important clues on how to overcome seed dormancy (if any) and provide the correct environmental conditions needed for germination.

It is important to develop a good recordkeeping system to refine and improve results over time and prevent the loss of valuable information. Keep details on the general information of the species and seedlot and the seed treatments and resulting germination. Because growers have a number of options for sowing seeds, it may be a good idea to do small trials of several of the methods described; see Chapter 17, *Discovering Ways to Improve Crop Production and Plant Quality*, for proper ways of conducting trials. Although several methods may "work"—that is, result in a viable plant produced—the question during the trials should be: Which method is optimal?

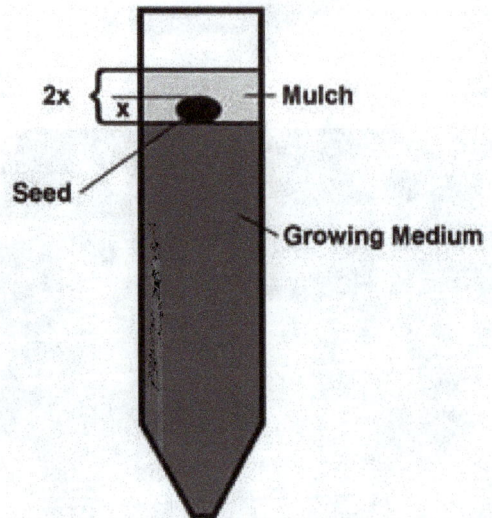

Figure 8.19—*A general rule of thumb for covering seeds with mulch is to cover the seed twice as deep as the seed is wide. Species requiring light for germination should never be covered with mulch, although mulch can be added after germination to reduce the growth of moss and liverworts.* Illustration by Jim Marin.

LITERATURE CITED

Baskin, C.C.; Baskin, J.M. 1998. Seeds: ecology, biogeography and evolution in dormancy and germination. San Diego, CA: Academic Press. 666 p.

Baskin, C.C.; Baskin, J.M. 2003. When breaking seed dormancy is a problem try a move along experiment. Native Plants Journal 4: 17-21.

Baskin, C.C.; Baskin, J.M. 2004. Determining dormancy-breaking and germination requirements from the fewest numbers of seeds. In: Guerrant, E.O., Jr.; Havens, K.; Maunder, M., editors. *Ex situ* plant conservation: supporting species survival in the wild. Washington, DC: Island Press: 162-179.

Blank, R.R.; Young, J.A. 1998. Heated substrate and smoke: influence on seed emergence and plant growth. Journal of Range Management 51: 577-583.

Chambers, K.J.; Bowen, P.; Turner, N.J.; Keller, P.C. 2006. Ethylene improves germination of arrow-leaved balsamroot seeds. Native Plants Journal 7: 108-113.

Dreesen, D. 2004. Tumbling for seed cleaning and conditioning. Native Plants Journal. 5: 52-54.

Dumroese, R.K.; Wenny, D.L.; Quick, K.E. 1990. Reducing pesticide use without reducing yield. Tree Planters' Notes 41(4): 28-32.

Dumroese, R.K.; Landis, T.D.; Wenny, D.L. 1998. Raising forest tree seedlings at home: simple methods for growing conifers of the Pacific Northwest from seeds. Moscow, ID: Idaho Forest, Wildlife and Range Experiment Station. Contribution 860. 56 p.

Feghahati, S.M.J.; Reese, R.N. 1994. Ethylene-, light-, and prechill-enhanced germination of *Echinacea angustifolia* seeds. Journal of American Society of Horticultural Science 119(4): 853-858.

Gao, Y.P.; Zheng, G.H.; Gusta, L.V. 1998. Potassium hydroxide improves seed germination and emergence in five native plant species. HortScience 33: 274-276.

Hartman, H.T.; Kester, D.E.; Davies, F.T.; Geneve, R.L. 1997. Plant propagation: principles and practices. 5th ed. Upper Saddle River, NJ: Prentice Hall Press. 770 p.

James, R.L.; Genz, D. 1981. Evaluation of ponderosa pine seed treatments: effects on seed germination and disease incidence. Forest Pest Management Report 81-16. Missoula, MT: U.S. Department Agriculture, Forest Service, Northern Region. 13 p.

Keeley, J.E.; Fotheringham, C.J. 1998. Smoke induced seed germination in California chaparral. Ecology 79: 2320-2336.

Landis, T. 2000. Where's there's smoke there's germination? Native Plants Journal 1: 25-29.

Narimanov, A.A. 2000. Presowing treatment of seeds with hydrogen peroxide promotes germination and development in plants. Biologia 55: 425-428.

Sari, A.O.; Morales, M.R.; Simon, J.E. 2001. Ethephon can overcome seed dormancy and improve seed germination in purple coneflower species *Echinacea angustifolia* and *E. pallida*. HortTechnology 11(2): 202-205.

ADDITIONAL READING

Landis, T.D.; Tinus, R.W.; McDonald, S.E.; Barnett, J.P. 1999. The container tree nursery manual: volume 6, seedling propagation. Agriculture Handbook 674. Washington, DC: U.S. Department of Agriculture, Forest Service. 166 p.

APPENDIX 8.A. PLANTS MENTIONED IN THIS CHAPTER

acacia, *Acacia* species

American licorice, *Glycyrrhiza lepidota*

American lotus, *Nelumbo lutea*

angelica, *Angelica* species

antelope bitterbrush, *Purshia tridentata*

arrowleaf balsamroot, *Balsamorhiza sagittata*

beargrass, *Xerophyllum tenax*

big sagebrush, *Artemisia tridentata*

blackberry, *Rubus* species

buckthorn, *Rhamnus* species

bulrush, *Schoenoplectus* species

ceanothus, *Ceanothus* species

cliffrose, *Purshia stansburiana*

cottonwood, *Populus* species

golden currant, *Ribes aureum*

Great Plains tobacco, *Nicotiana attenuata*

Indian breadroot, *Pediomelum esculentum*

Indian ricegrass, *Achnatherum hymenoides*

kinnikinnick, *Arctostaphylos uva-ursi*

laurel sumac, *Malosma laurina*

locust, *Robinia* species

lupine, *Lupinus* species

mallow, *Sphaeralcea* species

manzanita, *Arctostaphylos* species

mesquite, *Prosopis* species

milkvetch, *Astragalus* species

mountain mahogany, *Cercocarpus ledifolius*

New Mexico locust, *Robinia neomexicana*

purple coneflower, *Echinacea angustifolia*

quaking aspen, *Populus tremuloides*

redosier dogwood, *Cornus sericea*

rushes, *Juncus* species

salmonberry, *Rubus spectabilis*

scarlet bugler, *Penstemon centranthifolius*

sedges, *Carex* species

serviceberry, *Amelanchier alnifolia*

skunkbush sumac, *Rhus trilobata*

smooth sumac, *Rhus glabra*

white oaks, *Quercus* species

white sage, *Salvia apiana*

whitebark pine, *Pinus albicaulis*

wild hollyhock , *Iliamna* species

wild raspberry, *Rubus idaeus*

willow, *Salix* species

wolfberry, *Lycium* species

Woods' rose, *Rosa woodsii*

9

Vegetative Propagation
Tara Luna

For the past 30 years, interest in the propagation of native plants has been growing. Many desirable and ecologically important species, however, are difficult or very time consuming to propagate by seeds. Thus, nursery growers may want to investigate how to propagate a species of interest by vegetative propagation. This can be done by combining classic horticultural propagation techniques with an understanding of the ecological and reproductive characteristics of the species. By investigating how a species perpetuates under natural conditions, nursery growers may be able to vegetatively propagate the species and produce nursery stock in situations when there are constraints on using seed propagation.

Many native plants naturally propagate vegetatively (that is, without seeds or spores) as a method of ensuring reproduction. Vegetative propagation is commonly found with species that have short seed life, low seed viability, or complex or delayed seed dormancy strategies. Species that inhabit ecosystems with drastic weather patterns, short growing seasons, and endure fires and other disturbances often reproduce vegetatively. All new daughter plants that arise from vegetative propagation are genetically identical to the mother (donor) plant, and these resulting individuals are known as "clones" (figure 9.1).

Nursery managers can make use of a plant's ability to regenerate vegetatively. The following situations favor vegetative propagation over seed propagation:

Joanne Bigcrane of the Confederated Salish and Kootenai Tribes in Montana by Tara Luna.

Figure 9.1—*Pacific yew is a species that is propagated by stem cuttings: (1) because it has very complex seed dormancy; (2) to obtain a larger plant in a shorter period of time; and (3) to perpetuate certain genotypes or individuals exhibiting high levels of taxol, a medicinal product found within the bark.* Photo by Thomas D. Landis.

- Seed propagation is difficult, very time consuming, or few viable seeds are produced.
- Larger nursery stock is needed in a shorter period of time.
- An individual, unique plant needs to be propagated.
- There is a need to shorten time to flower for seed production.
- A uniform stock type is needed.
- Specific genotypes are desired.
- Disease-free nursery stock is required.

Some disadvantages of using vegetative propagation include:

- Greater production costs than seed propagation, usually because of increased labor.
- Reduced genetic diversity.
- Specialized propagation structures may be required, depending on the species or time of year.

In general, vegetative propagation can be done with pieces of stems, leaves, roots, bulbs, corms, tubers, and rhizomes. Many factors, however, contribute to successful vegetative propagation of native plants. The type of vegetative material used, the time of year that material is collected, how it is handled and manipulated to induce rooting, and proper application of the correct environmental conditions all affect vegetative propagation. In addition, how plants are handled after rooting also plays an important role.

Because vegetative propagation is more costly than growing seedlings, the production system must be efficient. A general rule of thumb is that at least 50 percent rooting must be obtained to produce cuttings economically. If rare species or individual plants are being propagated, however, costs may be less important. Consider these methods to reduce production costs:

- Develop a smooth production line, from the collection of material to the final product.
- Train nursery staff how to properly collect, pro-cess, plant, and grow material.
- Build a dibble for making holes in the rooting medium.
- Control waste caused by poor propagation or growing practices.
- Lift and harden cuttings properly to reduce mortality.
- Develop a good system for overwintering cuttings.
- Keep good records to improve your results and to document production costs.

The following discussion will provide a broad overview of vegetative propagation. More specific details on vegetative propagation of particular plants can be found in Landis and others (1999) and volume 2 of this handbook.

STRIKING CUTTINGS

A cutting is the portion of a plant that is collected, treated, and planted to develop into a new intact plant complete with stems, leaves, and roots. Cuttings can be collected from mother plants in the wild, or special donor plants can be cultured in the nursery. Selection of mother plants, whether in the nursery or the wild, must be done carefully; it is just as important as the origin of seeds to ensure that nursery stock is well adapted to the outplanting environment. Collection of cuttings should follow the same ethical guidelines as collection of seeds to establish proper genetic diversity and sustainability of wild populations. See Chapter 7, *Collecting, Processing, and Storing Seeds*, for guidelines. In addition, the ability of cuttings to root is often clone specific, so it is important to record the origin of cuttings and subsequent rooting success.

Striking is the process of placing the cutting into soil or a rooting substrate. Often, propagators will say that cuttings have been "struck" to indicate that the cuttings have been placed in the rooting substrate.

SHOOT OR STEM CUTTINGS

Shoot cuttings, also referred to as stem cuttings, are the most common type. These cuttings can be broadly placed into three categories depending on the time of year they are collected (figure 9.2). Hardwood cuttings are collected when plants are dormant, from late autumn through early spring. Softwood cuttings are collected in late spring and early summer when stems and leaves are actively growing. Semihardwood (greenwood) cuttings are collected in late summer and early autumn when stem tissues have hardened and terminal buds have formed. Within each category, several cutting types are possible. Both deciduous and coniferous (evergreen) species can be propagated with these types of cuttings.

Hardwood Cuttings

Deciduous Species

Deciduous hardwood stem cuttings are the easiest, least expensive type of cuttings because they are easy to prepare, are not as perishable as softwood or semihardwood cuttings, can be stored in coolers or shipped if necessary, and require little or no special equipment during rooting. They are sometimes struck directly on the outplanting site or brought back to the nursery to grow as bareroot or container stock.

If hardwood cuttings are struck directly on the outplanting site, they can be live stakes (12 to 16 in [30 to 40 cm] long), poles (12 to 16 ft [3.6 to 4.9 m] long), or branched cuttings (2 to 6 ft [0.6 to 1.8 m] long). These cuttings are collected and outplanted during late autumn to early spring when the cutting is dormant and the soil at the outplanting site is wet. Live stakes and branched cuttings need to be long enough to reach moisture in the soil profile and are usually driven into the ground with a mallet with only three to four nodes (buds) above ground. Poles are much longer and are driven deep enough so they can remain in contact with the water table during the driest part of the year. Hardwood cuttings of willows and cottonwoods are commonly used this way in restoring riparian areas.

If hardwood cuttings are struck in the nursery, they can be straight, heel, or mallet cuttings (figure 9.3). Straight cuttings are made from straight hardwood

Figure 9.2— *"Hardwood" cuttings are dormant wood collected in winter. In late spring and early summer, new growth that occurs from the hardwood and bends but does not snap when bent is considered "softwood." In late summer and early autumn, as softwood matures, subsequent cuttings from that wood are termed "semihardwood." In late autumn, semihardwood further hardens for winter, becoming dormant wood from which hardwood cuttings can be made.* Illustration by Steve Morrison.

Figure 9.3— *Straight, heel, and mallet cuttings. Straight cuttings are used on easy-to-root species, while heel and mallet cuttings are commonly used on more difficult-to-root species.* Photo by Tara Luna.

stems and are the most common type for easy-to-root species. Heel cuttings are made from side shoots on stems that are 2 years old. To make a heel cutting, pull the side shoot away from the tip so that there is a section of older wood at the base of the cutting. Mallet cuttings include a cross-section of older stem at the base of the side shoot (figure 9.3).

All hardwood stem cuttings have an inherent polarity and will produce shoots on the distal end (nearest the bud) and roots on the proximal end (nearest the main stem or root system). If planted upside down, the cutting will not root. When using straight or live stake deciduous cuttings, the tops and bottoms of the stems need to be distinguished. The absence of leaves can make it difficult for nursery workers to discern. The solution is to cut the bottoms straight across the base of a node and cut the tops at an angle (figure 9.4).

Coniferous Species

Hardwood cuttings of evergreen conifers are usually taken in late winter to early spring. Unlike hardwood cuttings, evergreen cuttings must be struck into a special rooting environment (see Chapter 4, *Propagation Environments*) as soon as possible because they cannot be stored for any length of time. Evergreens are best rooted in special rooting environments after being wounded or treated with rooting hormone (described in the following paragraphs). Usually, cuttings are 4 to 8 in (10 to 20 cm) long with all leaves removed from the lower half. Straight, mallet, and heel cuttings are also used with evergreen species (figure 9.3).

Softwood Cuttings

Prepared from the new growth of deciduous or evergreen species, softwood cuttings generally root easier than other types of cuttings but require more attention and a special rooting environment. The best cutting material has some degree of flexibility but is mature enough to break when bent sharply (figure 9.5). Extremely fast-growing tender shoots are not desirable.

Herbaceous stem cuttings are softwood cuttings made from nonwoody plants. They are handled in the same way as softwood cuttings (figure 9.6A). Many succulent desert plant cuttings, such as those from cacti, are easily propagated; cuttings should be allowed to develop callus for a week before inserting the cutting into rooting media. They root readily without misting or high humidity (figure 9.6B).

Semihardwood Cuttings

Semihardwood (greenwood) stem cuttings are those made from leafy broad-leaved evergreen plants and leafy summer and early autumn wood from deciduous plants. Cuttings are taken during the late summer and autumn just after a flush of growth has taken place and the wood is partially matured. In many cases, the terminal bud has formed for the next growing season (figure 9.7).

ROOT CUTTINGS

Although not used as much as other types of cuttings, root cuttings can be made by dividing roots into indi-

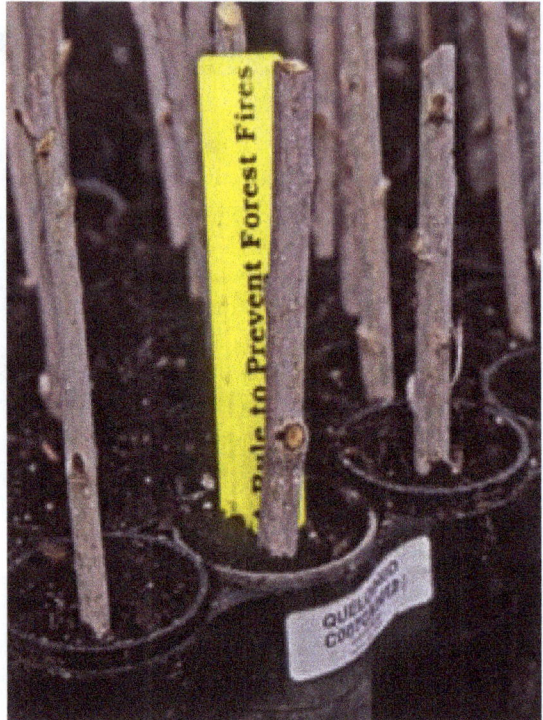

Figure 9.4—*Polarity means cuttings will produce shoots on the distal end (nearest the bud) and roots on the proximal end (nearest the main stem or root system). It is important that the cutting is oriented properly when planted; the absence of leaves can make this an easy but important point to miss. The solution is to cut the bottoms straight across the base of a node and cut the tops at an angle.* Photo by Tara Luna.

vidual segments containing dormant shoot buds capable of developing into new plants. Root sections are collected from late autumn to early spring before new tissue emerges from buds.

Root cuttings are planted horizontally in containers with the dormant leaf buds on the upper side. Some root cuttings are also planted in the containers vertically, but it is important to maintain the correct polarity (figure 9.8). To ensure that root cuttings are planted correctly, cut the upper end of the cutting horizontally and cut the basal end diagonally. Root cuttings generally do not require a special rooting environment unless shoots are cut from the root piece and treated as a stem cutting.

WHAT TO CONSIDER WHEN SELECTING CUTTINGS FROM MOTHER PLANTS

A variety of factors can greatly influence the rooting success of cuttings. Collectors need to be aware of these factors and, with experience, will be able to discern the right type of cutting material to collect. Important factors include seasonal timing, juvenility, plagiotropism, species, and cutting size and quality (figure 9.9).

Some species can be readily propagated from cuttings collected in any season of the year, while others have very specific seasonal trends when they will form roots. For any given species, small experiments are required to determine the optimum time to take cuttings, which is related to the physiological condition of the donor plant at collection time rather than any given calendar date. Recordkeeping is important to improve rooting results from year to year.

All plants progress from a juvenile phase (incapable of producing flowers) to a mature or adult flowering phase. Different parts of the plant, however, can be at different stages of maturity at the same time. Sometimes the juvenile phase can be distinguished from the adult phase by differences in leaf shape or color or by the overall habit of the plant. The juvenile phase is easily seen in junipers in which juvenile leaves are feathery and needle-like and often differ in color from mature leaves that are more rounded at the tips. In other conifers, juvenile wood is usually found on the lower portion of the tree crown and the adult, cone-bearing wood is located in the upper crown. In deciduous plants, juvenile wood is found near the stem base or root crown and can be discerned as the long, non-flowering shoots (sucker shoots). Cuttings collected from this region of the plant root more easily than those from older, mature wood. In some cases, many difficult-to-root species will root only from stems collected from young seedlings. Hedging or coppicing is the practice of regularly cutting back donor plants to maintain juvenile wood and is an efficient means of generating many long, straight cuttings from a limited number of plants. Donor plants in natural stands can be selected for hedging on an annual basis if cuttings will be collected from the area for several years (figure 9.10). Otherwise, mother plants can be held in the nursery and used as a source of cuttings.

Plagiotropism is the habit of a cutting to continue to grow in the direction it was growing on the donor plant. In some species plagiotropism is strong but in other species it is weak. Similarly, plagiotropism can be strong or weak depending on the original position of the cutting on the donor plant. Often, plants produced from cuttings from lateral shoots will maintain a later-

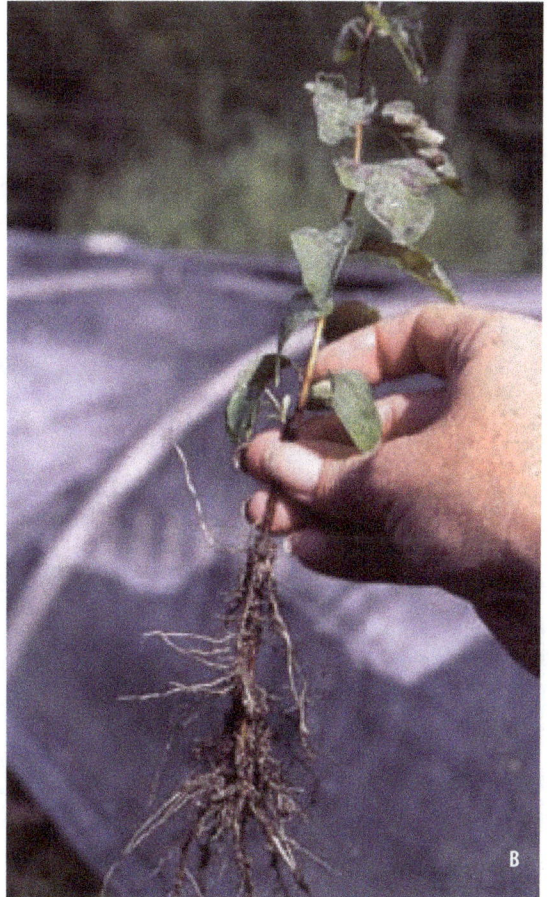

Figure 9.5—*(A) Softwood stem cutting material has some degree of flexibility but is mature enough to break when bent sharply. Tender softwood shoots that do not break should not be used. (B) Rooted softwood snowberry stem cutting.* Photos by Tara Luna.

Figure 9.7—*Semihardwood cuttings of Cascade mountain ash collected in late summer from a lateral branch with a maturing terminal bud.* Photo by Tara Luna.

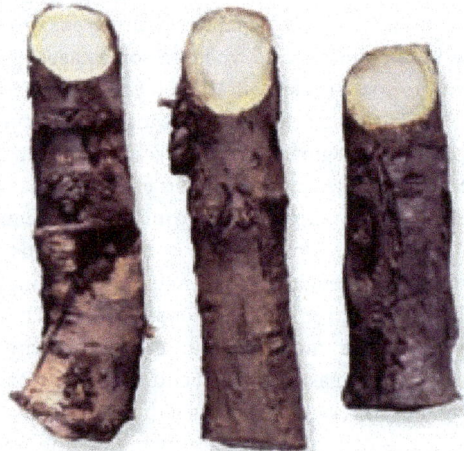

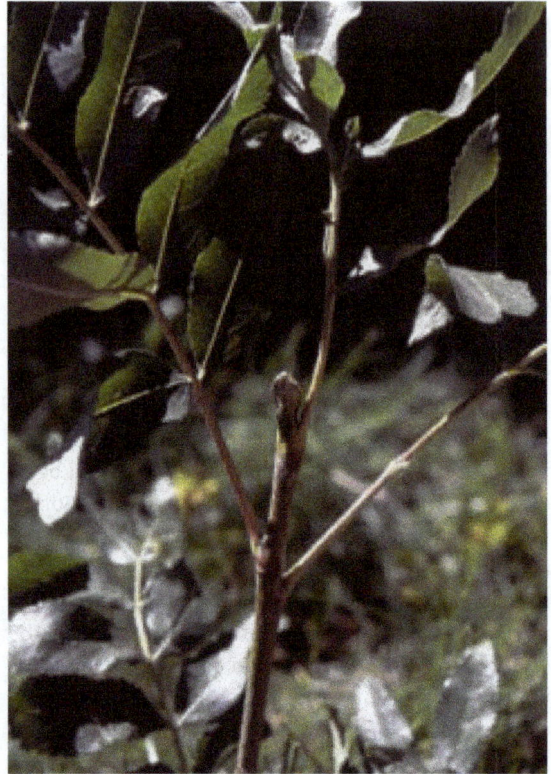

Figure 9.6—*Softwood cuttings of many herbaceous native perennials, such as (A) Alberta pentsemon, can be easily rooted using mist and application of a rooting hormone. (B) Cacti cuttings must air-dry for several days before sticking into containers; unlike other cuttings, they do not need mist or high humidity to root.* Photos by Tara Luna.

Figure 9.8—*Root cuttings, such as these from quaking aspen, are used when stem cuttings do not root well.* Photo by Tara Luna.

al habit, whereas plants produced from terminal shoots will grow vertically. This habit can create problems with the growth habit of the nursery stock (figure 9.11).

Some species of plants are dioecious, meaning that male and female flowers are borne on separate plants. In such cases, collectors may not realize they have collected cuttings of only one sex. Outplanting plants of only one sex onto the restoration site may compromise project objectives because seed production over the long term would be impossible (Landis and others 2003). Therefore, be sure to collect both male and female cutting material (see Landis and others 2003).

The size and quality of the cutting are important. Cutting size varies from species to species and by seasonal timing. Easily rooted species such as willow can be collected as long poles for rooting or made into small microcuttings. Microcuttings consist of one bud and a small section of internode stem and are typically less than 2 in (5 cm) long (figure 9.12). Hardwood cuttings vary in length from 4 to 30 in (10 to 76 cm). At least two nodes are included in the cutting. The basal cut is made just below a node and the top cut is made above a node. If more than one cutting is being made from a stem, be certain that nursery workers handling the cuttings maintain the correct polarity. The very tip portions of the shoot, which are usually low in carbohydrates, are usually discarded. Central and basal portions of the stem cutting usually make the best cuttings, but there are exceptions. Good cutting wood has some stored carbohydrates that will supply the cutting with food reserves until roots form. Very thin or elongated shoots are not desirable. If cuttings are collected from natural stands, harvest from individuals that are growing in full sun to partial shade and avoid those in deep shade. Often, the ability of a cutting to produce new roots changes from the base of the cutting to the tip. Softwood stem cuttings are usually straight, 3 to 6 in (7.5 to 15 cm) long with two or more nodes. Generally, softwood cuttings root better from terminal shoots. Semihardwood cuttings are usually 3 to 6 in (7.5 to 15 cm) long with leaves retained in the upper end. Semihardwood cuttings usually root best from lateral shoots.

COMMON DIOECIOUS PLANTS

ash
buffaloberry
cottonwood
fourwing saltbush
joint fir
maple
silverberry
willow

Figure 9.9—*Cuttings should be collected only from healthy donor plants, such as these kinnikinnick, and preferably from juvenile wood.* Photo by Tara Luna.

Figure 9.10—*Hedging plants such as redosier dogwood on an annual basis provides many straight, juvenile woody shoots that can be used for cuttings.* Photo by Tara Luna.

Figure 9.11—*Plagiotropism is the effect of the position of the branch utilized for a cutting on the growth habit of the progeny. The terminal shoot on the juniper cutting on the right, collected as a lateral shoot, still exhibits lateral growth tendencies.* Photo by John Edson.

Figure 9.12—*Microcuttings are small one- to two-node cuttings that can be made from a single stem of willow.* Photo by Tara Luna.

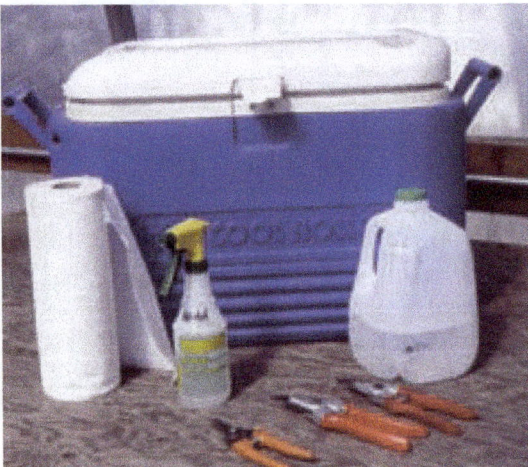

Figure 9.13—*Equipment used for the collection of cuttings should include sharp tools, a cleaning agent for tools, and a cooler to keep cuttings from drying out during transport.* Photo by Tara Luna.

COLLECTING, TRANSPORTING, AND TEMPORARY STORAGE OF CUTTINGS

Some basic equipment and supplies are necessary to efficiently collect cuttings and ensure their health until they are used in the nursery (figure 9.13). The following items are recommended:

— High-quality, sharp pruning shears and pruning poles for collecting from trees.
— Spray bottles filled with disinfectant (1 part bleach (5.25 percent sodium hypochlorite) in 10 parts water) for pruning shears.
— Permanent labels and marking pens for noting origin of collection.
— Large, white plastic bags with ties for bulk collections.
— Spray bottles filled with water to keep cuttings moist in the plastic bags after collection.
— Portable, insulated coolers for transport back to the nursery.

When collecting and handling cuttings, it is important to:

— Collect only from healthy donor plants.
— Keep cuttings cool to avoid wilting and desiccation.
— Handle cuttings carefully so that tissues are not bruised.
— Make sure that some buds or leaves are present on stem cuttings.
— Collect from nonflowering shoots. In general, cuttings root better before or after flowering.
— Place cuttings in the same direction when bundling to avoid mix-ups with polarity during preparation back at the nursery.

Although hardwood cuttings collected during the dormant season are quite tolerant of handling and can be stored in refrigeration for weeks or months before striking (figure 9.14A), softwood and semihardwood cuttings should be collected during the early part of the day and storage should be kept to a minimum. Ideally, softwood and semihardwood cuttings are made on cloudy, cool days or during the early morning. All cuttings, even hardwood cuttings, should be handled with care to avoid water loss and physical damage. Techniques used for collecting and handling cuttings can

Figure 9.14—*Hardwood and softwood cuttings are handled very differently during collection: (A) hardwood cuttings can be collected during the dormant season and can be stored for several weeks in a cooler, (B) while softwood cuttings require more attention so that the stems and leaves do not wilt and desiccate before they are taken back to the nursery, and they are usually placed in the rooting chamber the same day of collection.* Photo A by Joyce Lapp, B by Tara Luna.

greatly affect rooting results. Cuttings should be kept cool and shaded during collection and transport back to the nursery. Never lay cuttings on the ground in full sun. Place cuttings into white plastic bags, mist them, and label with origin information and the date (figure 9.14B). When collecting from mother plants, make a proper cut that facilitates healing of the mother plant. Take the cutting just above a node, ensuring that you do not leave a stub. Then trim the base of the cutting to just below the node where rooting is more likely to occur. Between collection sites, disinfect the pruning shears with a solution of 1 part bleach (5.25 percent sodium hypochlorite) to 10 parts water to avoid the spread of disease.

At the nursery, refrigerated storage should be available to hold cuttings if they are not struck immediately. Deciduous hardwood cuttings can be stored for several days or weeks but generally no longer than 4 to 8 weeks. Wrap deciduous hardwood cuttings in moist peat moss or burlap before placing them into storage. Inspect stored cuttings frequently to make certain that tissues are slightly moist and free from fungal diseases. Hardwood and softwood evergreen cuttings, deciduous softwood cuttings and semihardwood cuttings should not be stored for longer than 1 day and preferably should be struck in propagation beds the same day of collection.

TYPES OF ROOTING AND PROPAGATION ENVIRONMENTS

The development of new roots on a shoot is known as "adventitious root formation." Two types of roots occur depending on whether buds capable of producing new roots are present. If buds are present, the resulting roots are termed "preformed" or "latent." Native species such as willow and cottonwood have preformed or latent root initials. In the nursery, cuttings of these species are usually struck directly into containers because they do not require a special rooting environment. This method is the easiest and most economical way to produce these species because no additional transplanting is needed.

If no buds are present, then the roots are termed "wound-induced" and new roots form only in response to the wound caused by preparing the cutting (figure 9.15A). Species requiring wounds can vary considerably in their ability to form new roots. After a root is wounded, callus tissue forms at the base of a cutting, primarily from the vascular tissue (figure 9.15B), but callus formation is not always essential to rooting. In easy-to-root species, callus formation and root formation are independent processes that occur at the same time because of similar environmental triggers (figures 9.15B and C). In difficult-to-root species, adventitious roots arise from the callus mass. In some cases, excessive callus can hinder rooting and is a signal to use a lower concentration of rooting hormone. Often, excess callus should be scraped away and the cutting replaced in the rooting environment.

In general, all species with wound-induced roots must first be rooted in a special propagation environment in which the temperature of air and medium are tightly controlled. High relative humidity is encouraged, light levels are often reduced, and the medium is kept "moist but not wet." See Chapter 4, *Propagation*

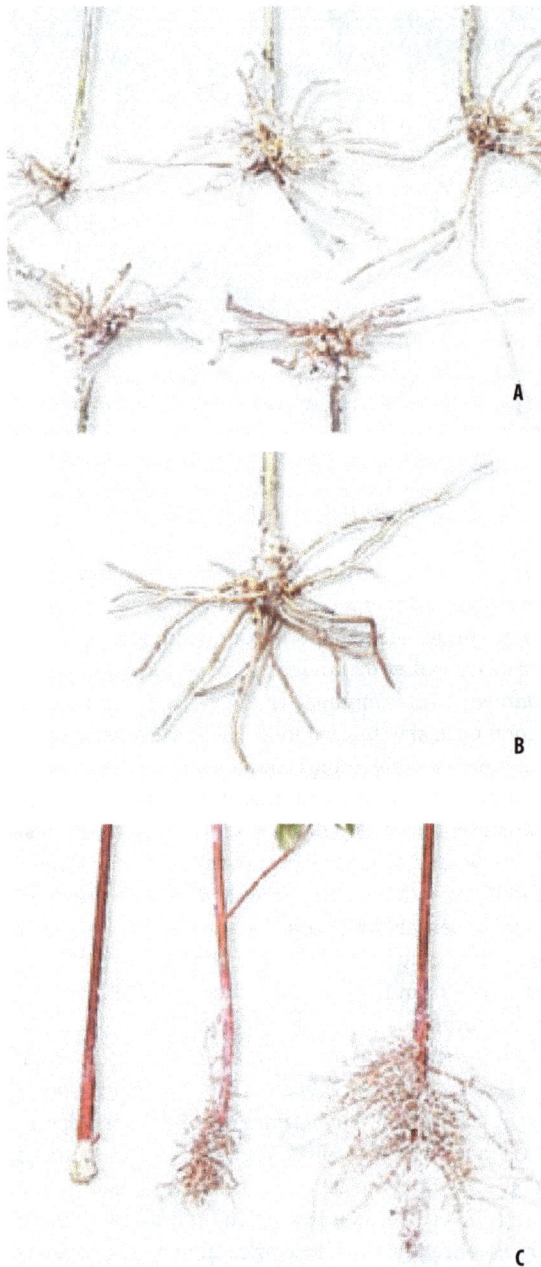

Figure 9.16—*Developing an efficient system for producing cuttings, employing experienced propagators, and keeping the work area clean are key aspects to reducing production costs.* Photo by William Pink.

Figure 9.15—*(A) Note adventitious roots of a cutting, (B) callus and roots forming at the base of a cutting, and (C) the development of adventitious roots over a 6-week period.* Photos by Tara Luna.

Environments, for more details on propagation environments. Easy-to-root species are often struck directly into containers filled with regular growing medium and, once rooted in the special propagation environment, are moved into the regular nursery. Hard-to-root species are often struck into a special rooting medium and, after roots form, are transplanted into containers to continue their growth.

Cutting Preparation

While preparing cuttings, it is important to keep the work area clean (figure 9.16). Use sharp, well-maintained shears and knives to make clean cuts and disinfect them often to reduce the possible spread of disease. Preparing cuttings standardizes their size and shape, promotes side shoots, and eliminates shoot tips that often die back. It is important to maintain polarity during this process, especially for deciduous hardwood cuttings. Cuttings that will require hormone treatment to encourage rooting, such as those of hardwood narrowleaf evergreens or any softwood or semihardwood cuttings, should have one-third to one-half of the leaves and buds removed to reduce the amount of water loss from the cutting. Any flower buds should also be removed. It is important, however, to retain some buds or leaves on the cutting so that the cutting can manufacture food during rooting.

Figure 9.17—*Wounding the lower end of the stem often increases rooting results and root mass, especially on cuttings that are difficult to root.* Photo by Tara Luna.

Figure 9.18—*Willow water solution is a homemade rooting hormone that can be used on easy-to-root species.* Photo by Tara Luna.

Wounding Cuttings

Wounding, used on species that are difficult to root, increases rooting percentages and improves the quantity and quality of roots produced. Wounding exposes more cells to rooting hormone, encourages callus formation, and, in some cases, removes thick woody tissue that can be a barrier to root formation (figure 9.17). Cuttings are commonly wounded by hand-stripping small lower stems and leaves to create wounded areas along the basal portion of the cutting, scraping the base of the stem with a small, sharp knife or potato peeler (figure 9.17), or slicing one or two long, shallow slivers (0.75 in to 1.25 in [2 to 3.2 cm] long) of tissue from the base of the stem, making sure to penetrate the cambium layer of the stem. Slicing requires precision and experience so that cuttings are not excessively damaged.

Rooting Hormones

Auxins are natural plant hormones that encourage root formation on cuttings and are available from natural and synthetic sources. In practice, auxins are commonly referred to as rooting hormones. Willows are a natural source of auxins (Leclerc and Chong 1984). "Willow water" is a rooting hormone solution that can be made by cutting green, actively growing willow stems into 1-in (2.5-cm) pieces, mashing them, placing them in water brought to a boil, and then removing them from the heat to cool and steep overnight. After removing the willow stems, cuttings can be soaked overnight in the willow water and then planted (figure 9.18).

Most cuttings, however, are treated with synthetic hormones that are available in powder and liquid form, and some preparations may contain chemical fungicides (figure 9.19). Synthetic hormones can be purchased ready to use or can be mixed by growers to specific concentrations. Indole-3-butyric acid (IBA) and naphthaleneacidic acid (NAA) are the most widely used synthetic auxins for rooting. Often, mixtures of IBA and NAA are more effective than either component alone. The effect of rooting hormones varies widely between species and, in some cases, between genotypes. The concentration of a rooting hormone is expressed in either parts per million (ppm) or as a percentage. In general, rooting hormone powders are expressed as a percentage, while liquid solutions are expressed as ppm.

Although nursery workers can either purchase synthetic rooting hormones in liquid or powder forms or prepare their own from ingredients purchased from horticultural suppliers, it is generally easiest to purchase ready-to-use formulations. It is important to remember that all rooting hormones have a limited shelf life of 18 to 24 months. Therefore, when purchasing or mixing hormones:

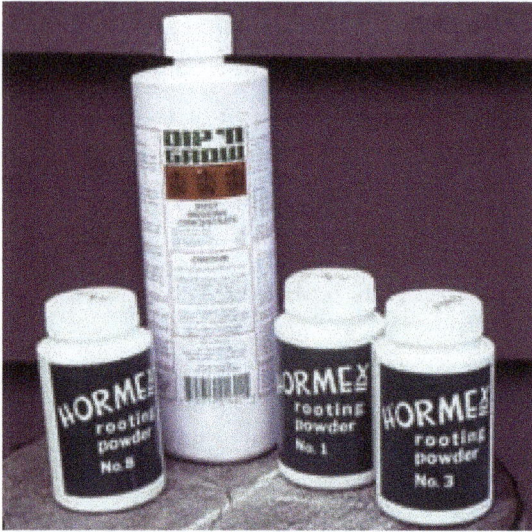

Figure 9.19—*Advantages of using rooting hormones on cuttings include (1) increased overall rooting percentages if applied correctly and at an effective concentration, (2) more rapid root initiation, (3) an increase in the total number and quality of roots, and (4) more uniform rooting.* Photo by Tara Luna.

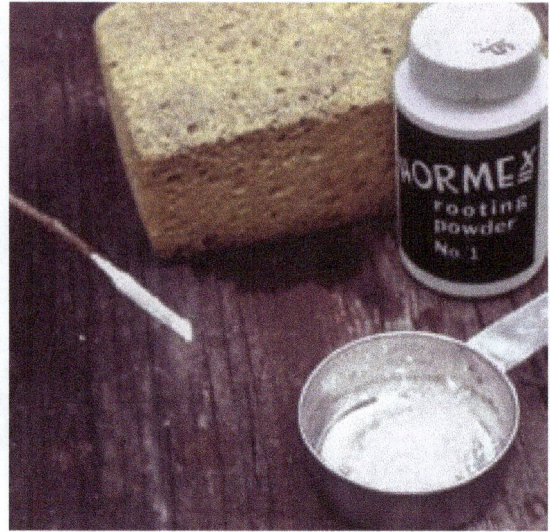

Figure 9.20—*Although powders are preferred to liquid hormones by many growers, care should be taken to apply powder hormones evenly and consistently.* Photo by Tara Luna.

- Record the date of purchase on the container.
- Order only what you plan to use within 18 to 24 months. Order smaller quantities more often to ensure that the rooting hormone remains effective.
- Keep containers sealed and refrigerated when not in use to preserve the activity of the root hormone.
- Always pour a little into a separate container when treating cuttings.

Many growers prefer powders because a number of prepared commercial products of varying strengths are available, they are easy to use, and large quantities of cuttings can be treated quickly.

However, powder must be applied uniformly to all cuttings; variable amounts of rooting powder adhere to the base of a cutting, which can affect rooting results (figure 9.20). The following precautions and special techniques are needed when using powders:

- Wear gloves during application.
- Transfer enough hormone to a smaller container from the main stock container for use. Never transfer unused hormone back to main stock container.
- Apply the hormone uniformly; make sure the base of the cutting is moist so that the powder adheres. Pressing cuttings lightly onto a moist sponge is a good idea.

- Ensure cuttings are dipped into the powder to a depth of 0.2 to 0.4 in (5 to 10 mm). Make certain that cut surfaces and other wounds are also covered with rooting hormone.
- Remove excess powder by lightly tapping the cuttings on the side of the dish.

Liquid products are formulated with alcohols and often must be diluted with great care to the desired strength. Some of the advantages of using solutions are a wide range of commercial preparations is available, specific concentrations can be formulated at the nursery, and they can be stored for longer periods under the right conditions. Some growers believe that using solutions is more accurate than powders are in regard to the amount of rooting hormone entering the stem tissue. The most common procedure for treating cuttings is using the concentrated-solution-dip method (quick-dip method) in which the base of the cutting is dipped into the solution for 3 to 10 seconds. Whole bundles of cuttings can be treated at once (figure 9.21). Alternately, cuttings can be soaked for a longer period of time in a more dilute hormone solution. When using liquid rooting hormones, it is important to:

Figure 9.21—*Using the liquid hormone "quick dip" method is preferred by many growers because bundles of cuttings can be treated at the same time with a consistent uniformity of application.* Photo by Tara Luna.

— Wear gloves during mixing, preparation, and application.
— Make certain that the solution was diluted to the right concentration correctly and precisely.
— Place the solution in a clean jar.
— Ensure that the treatment time is constant for a uniform application rate and to avoid damaging the plant tissue (phytotoxicity).
— Make certain that the basal ends are even to obtain uniform depth of dipping in the solution if bundles of cuttings are dipped.
— Allow the alcohol to evaporate from the stem of the cutting before striking cuttings into the propagation bed, a process that usually takes only a couple of minutes.
— Properly discard any remaining solution, because it is contaminated with plant material.

STRIKING, MONITORING, AND GROWING CUTTINGS

As mentioned previously, cuttings may be struck directly into containers or struck into special rooting environments. Direct striking into containers is more efficient and therefore more economical than striking into a special rooting environment because the cuttings are handled only once and expensive transplanting is avoided. Easy-to-root hardwood cuttings, such as those from redosier dogwood, willow, and cottonwood (figure 9.22), should always be direct struck. Often, a dibble of the same diameter and depth of the stem of the cutting is a useful tool for making openings in the medium into which the cutting can be struck. If using powdered rooting hormones, this practice will help keep the hormone from being brushed off. The following should be encouraged when striking cuttings:

— Wear gloves if the cuttings were treated with rooting hormones.
— Maintain polarity (keep the correct end of the cutting up).
— When using stem cuttings, make certain that at least two nodes are below the surface of the rooting medium.
— If cuttings were wounded, make certain that wounded tissue is adequately covered with rooting hormone and is below the surface of the rooting medium.
— Strike cuttings firmly in the rooting medium. Make certain to avoid air pockets around the base of the stem.
— Try to strike cuttings within 1 to 2 days so that all the plants will have the same level of root development and thus can be hardened off properly prior to lifting.

After cuttings are struck, maintain a clean rooting environment (figure 9.23); routinely inspect cuttings for proper temperature, humidity, and moisture levels and adjust as necessary to match daily weather conditions. Check to ensure that all equipment (including bottom heat) is working properly.

Environmental Conditions for Direct Struck Cuttings

In general, easy-to-root hardwood cuttings directly struck into containers can be treated similar to seedlings. Details on growing container seedlings are presented in chapters 10 through 13.

Figure 9.22—*Easy-to-root hardwood cuttings can be directly struck in containers; this is the most economical way of producing cuttings.* Photo by Tara Luna.

Figure 9.23—*The propagation environment must be carefully monitored to ensure that the mist system, the timers that control the frequency of mist, bottom heat, and other equipment are working properly. Equipment controls for outdoor mist systems need to be adjusted to accommodate daily changes in wind, temperature, and rain.* Photo by Tara Luna.

Environmental Conditions in Special Rooting Environments

Achieving successful rooting requires attention to sanitation, relative humidity, temperature, light, rooting medium, and sometimes mycorrhizae and mineral nutrition. See Chapter 4, *Propagation Environments*, for information about equipment necessary to regulate humidity, temperature, and light.

Sanitation

Always keep the propagation environment as clean as possible. Strike cuttings into a sterilized rooting medium. Routinely inspect for and remove dead leaves or cuttings that could be a source of disease infection.

Humidity

Until the root system forms, high levels of relative humidity must be provided to slow the rate of water loss from the cutting.

Temperature

The optimum air temperature for rooting cuttings is 65 to 75 °F (19 to 24 °C). The rooting medium temperature should be 10° F (5.5 °C) higher than the air temperature; this heat is generated by bottom heating cables beneath the rooting media or flats.

Light

Providing light for photosynthesis is necessary so that cuttings can continue to manufacture food during rooting, but too much sunlight can cause excessive air temperatures. Shadecloths of 30 to 50 percent shade cover are most effective to reduce air temperature while providing sufficient light.

Rooting Medium

A good rooting medium provides aeration and moisture and physically supports the cuttings. A pH of 5.5 to 6.5 is optimum for most plants, but acid-loving plants prefer 4.0 to 5.0. Some common components of rooting media generally include a combination of two or more of the following: large-grade perlite, pumice, *Sphagnum* peat moss, sawdust, sand, and fine bark chips.

Different combinations of the components are used depending on the species being propagated. Selection of the rooting medium components influences rooting percentages and the quality of roots produced. Using very fine- or very coarse-grade sands tends to discourage the development of secondary roots. Roots that do form tend to be brittle and break off during the process of transplanting the cuttings into containers for further plant development. A good rooting medium promotes the development of fibrous root systems that retain rooting medium during transplanting, which reduces "transplant shock."

Mycorrhizal Fungi

Some growers inoculate the rooting medium with mycorrhizal fungi or other symbiotic organisms, which has improved rooting results with some plants (Scagel and others 2003). This practice may be especially important for those species that take a long time to form roots, such as Pacific yew, blueberries, cranberries, and rhododendrons. See Chapter 14, *Beneficial Microorganisms*, for more information on mycorrhizae.

Nutrient Mist

Some difficult-to-root cuttings may remain in a special rooting environment for a long period of time. Over time, the cuttings can become weakened, resulting in yellowing of the leaves or leaf and needle drop. Nutrients can be leached from the leaves by the long exposure to overhead misting. In these cases, the application of a dilute, complete foliar fertilizer through the mist line can improve cutting vigor and may aid in rooting. Because some species respond favorably to nutrient mist while others are adversely affected, you will need to do some preliminary trials before treating all the cuttings.

Transplanting Cuttings from Special Rooting Environments

A few weeks after striking cuttings into the rooting environment, they should be inspected for root development. Using a trowel, carefully lift a few cuttings by digging well below the end of the cutting. After most cuttings have initiated roots, turn off the bottom heat to encourage the development of secondary roots. When cuttings have developed adequate root systems, they need to be hardened for life outside the rooting environment. See Chapter 12, *Hardening*, for more information. The goal is to condition stem and leaf tissues and promote secondary root development before transplanting. Cuttings can be hardened by following these guidelines:

— Gradually reduce the misting frequency over a period of 3 to 4 weeks.
— Increase the frequency and duration of ventilation in enclosed propagation systems.
— Do not let the rooting medium dry out completely.

After cuttings have hardened, transplant them into containers and transfer them to the nursery for additional growth. Because cuttings are more expensive to produce than seedlings, it is important to handle them carefully at this stage (figure 9.24). It is essential to avoid root damage by following these guidelines:

— Examine each cutting to ensure it has a sufficient root system capable of sustaining the cutting after transplanting. Cuttings with only a few slender roots or very short roots should remain in the propagation bed for further root development (figure 9.24A.)

— Transplant only on cool, overcast days or during early morning hours to avoid transplant shock.
— Transplant cuttings in an area of the nursery protected from wind and sunlight.
— Prepare containers, medium, labels, and transplanting tools before removing cuttings from the rooting medium.
— Moisten the growing media prior to transplanting to prevent tender roots from drying out.
— Remove cuttings from the rooting medium carefully and remove only a few at a time so roots will not dry out. Loosely wrap a moist paper towel around the root systems until they are transplanted.
— Handle cuttings carefully by holding the cutting by the stem and by leaving any rooting medium still attached to the root mass. Do not shake medium off the root system.
— Partially fill the container with moistened medium before inserting the cutting. Then add additional moistened medium and gently firm the medium with fingers without breaking the roots (figure 9.24B).
— Do not transplant the cuttings too deep or too shallow.

After transplanting the cuttings, they should be placed in a shadehouse or protected from full sun and wind for at least 2 weeks. When the cuttings appear to be well established, gradually increase the level of sunlight by moving them to a different area of the nursery or by exchanging the shadecloth for one with a more open weave. After a couple of weeks, move the sun-requiring species into full sun. Cuttings should be closely monitored for any sign of stress. Adequate sunlight is needed for new shoot growth and adequate accumulation of carbohydrates prior to winter. Cuttings should put on as much growth as possible to minimize winter mortality.

OVERWINTERING CUTTINGS

Sometimes cuttings will not have a sufficient root mass for transplanting by the end of summer. In this case, they should be left undisturbed and transplanted the following spring. Be sure to begin hardening them at least 6 weeks before the first frost. See Chapter 12, *Hardening,* for procedures. Newly rooted cuttings need extra protection during overwinter storage. Ideally, the root temperature should be kept at 34 to 41 °F (1 to 5 °C).

Transplant

Retain in propagation environment

A

B

Figure 9.24—*(A) Cuttings should have enough developed roots to support the cutting once it is lifted and planted outside the mist system. Cuttings with underdeveloped roots should be left in the propagation environment longer to develop an adequate root mass. (B) Cuttings should be handled carefully during lifting and potting by keeping the tender roots undamaged and moist.* Photos by Tara Luna.

C

Figure 9.25— *Many native plants, such as antelope bitterbrush on the Hopi Reservation, will layer naturally when lower branches come in contact with the soil.* Photo by Tara Luna.

The medium surface can freeze but not to a depth that injures young roots. Loss of cuttings due to improper overwintering storage is a common occurrence and contributes significantly to costs. For a full discussion on overwintering plants, including cuttings, see Chapter 13, *Harvesting, Storing, and Shipping.*

OTHER METHODS OF VEGETATIVE PROPAGATION

Besides stems and roots, several other portions of mother plants can be used to vegetatively propagate new daughters, and stems can be used in ways other than the traditional cutting described previously.

Layering

Layering is a technique by which adventitious roots are formed on a stem while still attached to the plant. Layering often occurs naturally without the assistance of a propagator (figure 9.25). It is mostly used by nurseries with a long growing season and on those species that fail to root from stem or root cuttings. Layering is started when plants are dormant. Four types of layering can be used by propagators: simple, French, mound, and drop. **Simple layering** is used on species that produce many shoots annually. Long, low-growing flexible shoots are pegged down 6 to 9 in (15 to 23 cm) from the shoot tip, forming a "U" (figure 9.26A). The bottom of the U stem is girdled with a sharp knife and is covered with soil or sawdust, leaving the tip exposed. After a sufficient root system is formed, the new plant can be severed from the donor plant (figure 9.26B). **French layering** is similar to simple layering but uses a long, single branch that is pegged down to the soil surface. The following spring, pegs are removed and the branch is laid into a trench and buried up to the tips of the shoots with well-aerated soil (figure 9.27A) and sawdust or mulch (figure 9.27B). After burying repeatedly, each shoot along the stem will form roots by autumn of the second year. **Mound layering** or stooling involves selecting a young stock plant (figures 9.28A and B) and cutting back shoots to a couple of inches above ground level (figure 9.28C). Numerous shoots develop in consecutive growing seasons and are covered to half their height with sawdust (figure 9.28D). This procedure is repeated three times as the shoots grow so that, by the end of the second or third growing season, the well-rooted shoots are unburied and are ready to plant as individuals (figures 9.28E and F). **Drop**

layering is very similar to mounding. Drop layering involves planting well-branched container plants deeply in the ground with only the tips of the branches exposed. New growth forms from the exposed branch tips, but the buried portions of the stems form roots along the stems.

Stacked layering is a new vegetative propagation method for quaking aspen and other rhizomatous species (Landis and others 2006). This technique takes advantage of the rapid and extensive root growth of seedlings and the fact that severed roots will form new shoots. In the spring, a stack of Styrofoam™ containers is created with a 1-gallon pot containing a seedling inserted in the top block. The lower Styrofoam™ containers are filled with growing medium with a thin layer of medium sandwiched between the blocks (figure 9.29A). By next spring, the roots of the mother plant will have grown down through and colonized the cavities in the lower blocks. Running a sharp knife between the Styrofoam™ containers severs the roots, which then form new shoots (figure 9.29B). After a few months, the new plants can be transplanted into larger pots. Another set of filled Styrofoam™ containers can be situated below the block with the mother plant to start another propagation cycle. Stacked layering is ideal for propagating species where seeds are rare or other vegetative techniques will not work.

Bulbs, Corms, Tubers, Rhizomes, and Crown Division

Bulbs, tubers, corms, and rhizomes are specialized plant structures that function in the storage of food, nutrients, and water. Many culturally important native species not easily grown from seeds have these structures.

A *bulb* is an underground storage organ consisting of a short, fleshy stem surrounded by fleshy modified leaves (scales). Tunicate and scaly are two types of bulbs. Tunicate bulbs have outer scales that are dry and membranous (figure 9.30A). Tunicate bulbs can be propagated by using offsets, scoring, scooping, coring, sectioning, and cuttage. **Offsets** of tunicate bulbs are the main method used to increase bulb stock. Any species that readily produces offsets can be propagated this way, but a large amount of bulbs is required to produce many plants (figure 9.30B). Camas and mariposa lily may be propagated by offsets. **Basal scoring** is when three incisions are made at the base of the bulb,

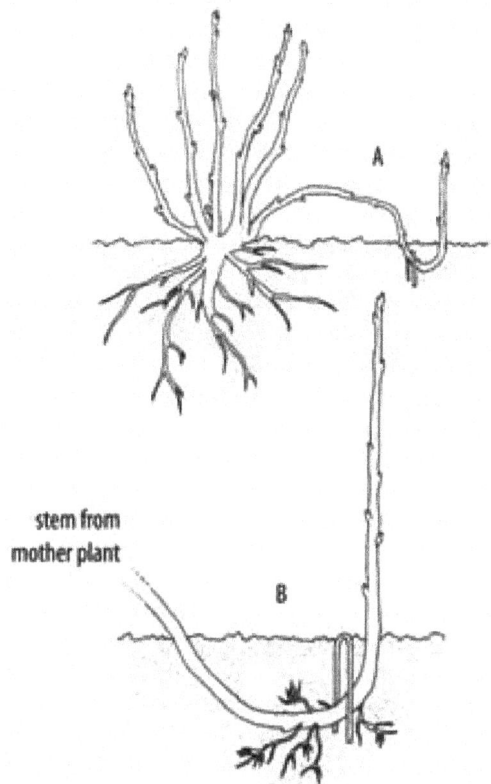

Figure 9.26—*Simple layering.* Illustration by Bruce McDonald and Timber Press Inc.

deep enough to go through the basal plate and the growing point. Growing points in the axils of the scales grow into bulblets. The bulbs can be placed in a warm, dark place at high humidity for a few months or planted upside-down (basal plate up) in clean dry sand, vermiculite, or perlite. Basal scooping is the removal of the entire basal plate to remove the shoot and flower bud at the center of the bulb. This exposes the fleshy leaf bases from which small bulblets will develop. **Coring** involves removal of the center portion of the basal plate and the main growing point of the bulb. Cored bulbs can be treated as described for scoring. **Sectioning** involves cutting a mature bulb into five to ten pie-shaped sections, each with a portion of the basal plate attached. These sections are treated and handled as described for scoring. **Cuttage** involves cutting a mature bulb vertically into six or eight sections. Next, the basal plate is cut so there are one to four scale segments attached on each piece of basal plate. These are treated with a fungicide and planted in perlite or vermiculite with the segment tips exposed above the rooting medium.

Scaly bulbs (also known as nontunicate bulbs) lack a tunic and are characterized by the scales being sepa-

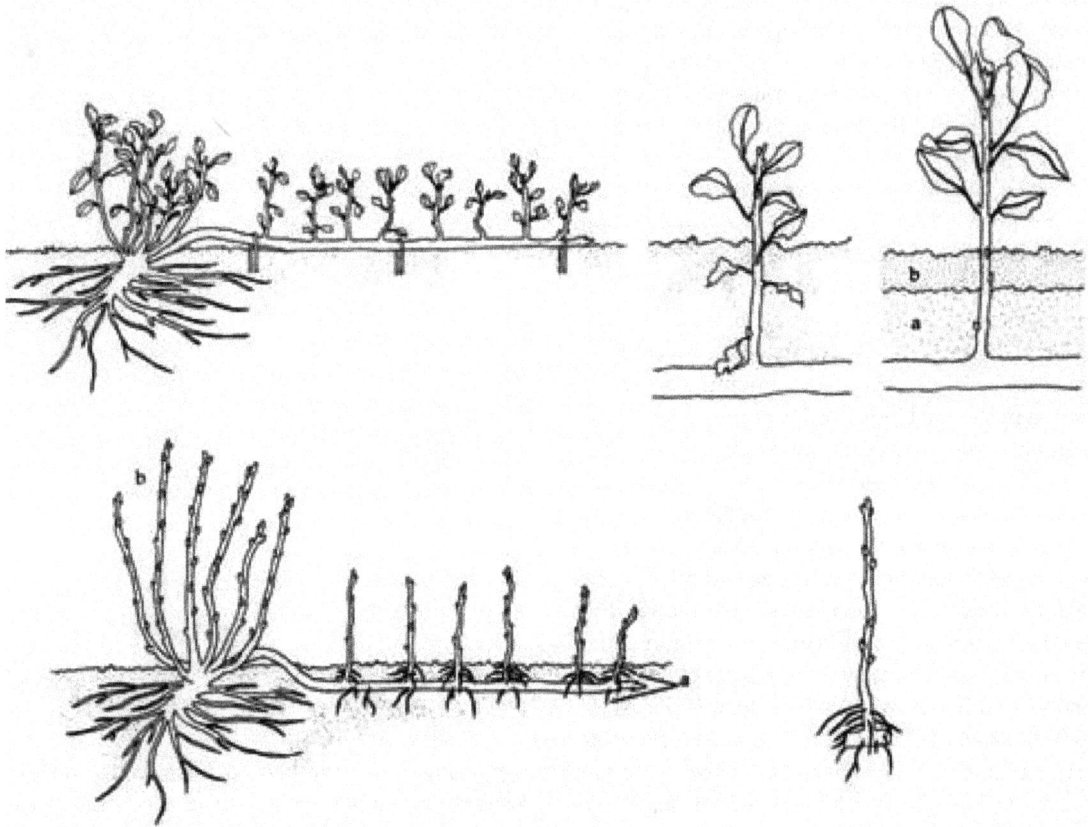

Figure 9.27— *French layering.* Illustration by Bruce McDonald and Timber Press, Inc.

rate and attached to the basal plate (figure 9.31). They are easily damaged and will dry out quickly without the protective tunic, so it is necessary to handle them with more care. These bulbs are propagated by scaling, which is done after flowering. The outer two layers of scales are removed from the mother bulb and treated with rooting hormone to induce bulblet formation. Scales are inserted vertically (about half their length) in propagation flats with moist sand and peat moss. Two or three growing seasons are needed to reach flowering size. Individual bulb scales are removed from the mother bulb and placed in growing conditions so that bulblets form at the base of each scale. Usually, three to five bulblets form per scale. Native lilies and fritillaries are propagated by scaling.

A *corm* is very similar to a tunicate bulb and consists of a swollen stem base enclosed by the dry, scale-like leaves (figure 9.32). It differs from a bulb in being a solid stem structure consisting of nodes and internodes that are compressed. Cormels are miniature corms that form between the old and new corms. One to 2 years of growth is usually required for them to reach flowering size. Trout lilies and some shooting stars produce corms.

Tubers are swollen modified stems that serve as underground storage organs. The most common tuber is the potato. "Eyes" are actually nodes containing buds. Propagation by tubers involves planting the entire tuber or dividing it into sections containing at least one eye or bud. Wapato is a native plant that produces tubers.

Rhizomes are specialized stems in which the main axis of the plant grows horizontally or vertically at or below the soil surface. Many native species, such as iris, repro-

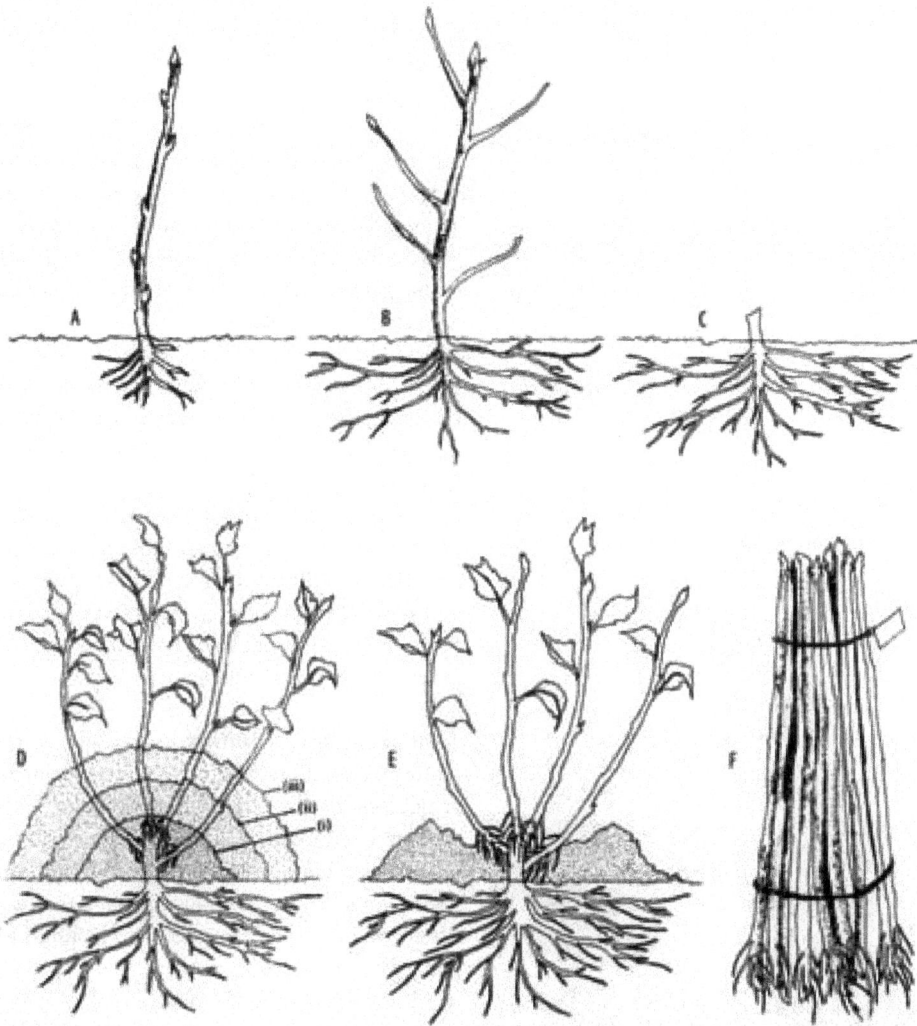

Figure 2.28—*Mound layering*. Illustration by Bruce McDonald and Timber Press, Inc.

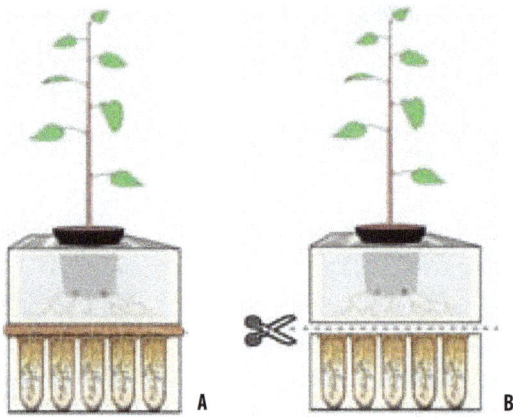

Figure 9.29—*(A) Roots from the mother plant grow downward through the cavities of stacked Styrofoam™ containers. (B) The roots are then severed, after which they develop new shoots.* Illustration by Jim Marin.

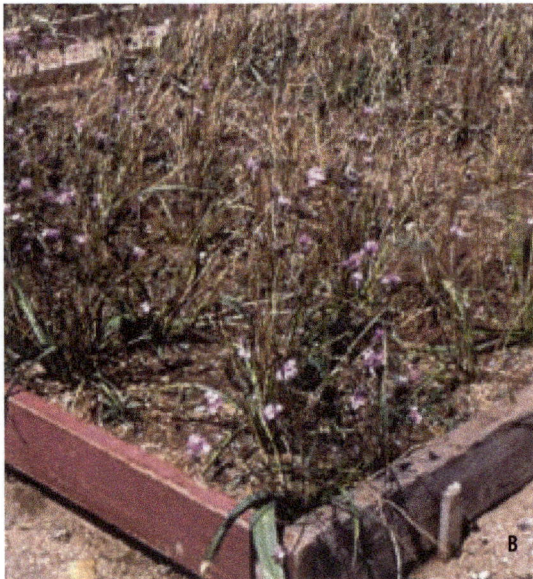

Figure 9.30—*(A) Camas is a native cultural plant with a tunicate bulb. (B) Native bulbs are often grown in raised beds; offsets from the mother bulbs are harvested and planted when they are dormant.* Photos by Tara Luna.

duce by rhizomes and they are easily propagated into larger numbers from a few nursery plants by divisions (figure 9.33).

Rhizomes vary in length and size according to species. Rhizomes are cut into sections with each containing at least one shoot bud or active shoot; some roots are attached to the bottoms of the rhizomes and are planted into containers individually. Rhizomes can also be planted in nursery beds and used as a source for bareroot stock for planting or for cultural uses such as basketry (figure 9.34).

Another method of propagation by division that differs slightly from dividing rhizomes is crown division. It is an important method for propagating many native herbaceous perennials that produce multiple offshoots from the crown. Crown divisions are usually done in early spring just before growth begins (species that flower and grow in spring and summer) or in late summer (species that bloom in late summer and early autumn). Plants are dug up and cut into sections with a sharp knife, each with a substantial portion of the root system, and transplanted individually.

Stolons, Runners, and Offsets

Stolons, runners, and offsets are specialized plant structures that facilitate propagation by layering. Stolons are modified stems that grow horizontally above the soil line and produce a mass of stems (figure 9.35A). Runners are specialized stems that arise from the crown of the plant and grow horizontally along the ground and produce a plantlet at one of the nodes (figure 9.35B). Raised beds planted with species with stolons or runners can be an endless source of material, and plants can be dug and potted individually or transplanted as bareroot stock.

Plants with rosettes often reproduce by forming new shoots, called offsets, at the base of the main stem or in the leaf axils. Offsets are cut close to the main stem of the plant with a sharp knife. If well rooted, an offset can be potted individually. Sever the new shoots from the mother plant after they have developed their own root systems. Nonrooted offsets of some species may be removed and placed in a rooting medium. Some of these offsets must be cut off, while others may simply be lifted from the parent stem.

Micropropagation

Micropropagation is a process used to propagate plants using very specialized tissue culture techniques. Tissue culture is the procedure for maintaining and growing plant tissues and organs in an aseptic culture in which the environment, nutrient, and hormone levels are tightly controlled. A small piece of vegetative material called the explant is used to create a new, entire plant. Rare or greatly endangered native species have been micropropagated to increase the number of individuals for restoration projects when other methods of propagation have been limiting or failed. Micropropagation has also been used as a method to offer plants in the nursery trade in order to preserve them from poaching and eventual extirpation from wild populations. Micropropagation works well for some species and poorly for others. For some native plants, such as orchids, it is one of the only options for successfully germinating seeds. Most native plant nurseries do not have an elaborate tissue culture facility because of the high cost, although small-scale micropropagation can be done with minimal equipment in a clean room. Micropropagation continues to play a role in the conservation of native plants and is included here as a viable propagation option when other methods fail. See volume 2 of this handbook for species that have been micropropagated.

ESTABLISHING MOTHER PLANTS AT THE NURSERY

Some nursery managers find it advantageous to maintain donor plants at the nursery as a continual source of cutting material. This practice can be more efficient than collecting from wild populations year after year, especially if the same ecotypes will be used for a long-term restoration project. Usually, mother plants are planted in field beds at the nursery or, in some cases, are kept in large containers. Regardless, mother plants must be clearly labeled as to species and origin. If mother plants are in field beds, an accurate map should be kept. Mother plants should be hedged on an annual basis to maintain wood juvenility and to produce numerous straight shoots to use as cutting material. One disadvantage to using mother plants grown at the nursery is that they require nursery space and must be intensively managed.

Figure 9.31—*Cormels form between the old and new corms and can be separated and planted individually.* Photo by Tara Luna.

Figure 9.32—*Scaly bulbs, such as those of yellowbells, can be propagated by (A) scaling and by (B) removing and planting the small rice-like bulblets individually.* Photos by Tara Luna.

Figure 9.33—*Many native plants have rhizomes. Rhizomes vary in thickness and length and can be used to propagate many more plants from existing nursery stock. (A) Missouri iris and (B) and (C) water sedge propagated from divisions.* Photo A by Tara Luna, B and C by Thomas D. Landis.

Figure 9.34—*(A) Dogbane rhizomes can be raised in field beds for cultural uses such as basketry. (B) White-rooted sedge being raised in field beds for basketry.* Photo A by William Pink, B by Chuck Williams.

SUMMARY

Vegetative propagation is the production of daughter plants from the stems, leaves, roots, or other portions of a single mother (donor) plant. Daughter plants contain the exact genetic characteristics of the donor. Vegetative propagation techniques can be used on many species if seeds are unavailable or difficult to germinate. Producing plants from cuttings is more labor intensive and expensive, and production may require special propagation structures. Experience is required to determine the best collection time, hormone treatment, and medium necessary to root cuttings. Some native species cannot be successfully rooted from stem cuttings, while others can be propagated from root cuttings, divisions, or layering. Micropropagation is an option for propagating rare and endangered species or others that are very difficult to propagate by other methods. Keeping good records will improve the success of rooting cuttings and reduce nursery costs.

LITERATURE CITED

Landis, T.D.; Tinus, R.W.; Barnett, J.P. 1999. The container tree nursery manual: volume 6, seedling propagation. Agriculture Handbook 674. Washington, DC: U.S. Department of Agriculture, Forest Service. 167 p.

Landis, T.D.; Dreesen, D.R.; Dumroese, R.K. 2003. Sex and the single *Salix*: considerations for riparian restoration. Native Plants Journal 4: 111-117.

Landis, T.D.; Dreesen, D.R.; Pinto, J.R.; Dumroese, R.K. 2006. Propagating native Salicaceae for riparian restoration on the Hopi Reservation in Arizona. Native Plants Journal 7:52-60.

Leclerc, C.R.; Chong, C. 1984. Influence of willow and poplar extracts on rooting cuttings. The International Plant Propagators' Society, Combined Proceedings 33: 528-536.

Scagel, C.F.; Reddy, K.; Armstrong, J.M. 2003. Mycorrhizal fungi in rooting substrate influences the quantity and quality of roots on stem cuttings of hick's yew. HortTechnology 13(1): 62-66.

Landis, T.D.; Dreesen, D.R.; Pinto, J.R.; Dumroese, R.K. 2006. Propagating native Salicaceae for riparian restoration on the Hopi Reservation in Arizona. Native Plants Journal 7:52-60.

APPENDIX 9A. PLANTS MENTIONED IN THIS CHAPTER

Alberta penstemon, *Penstemon albertinus*

antelope bitterbrush, *Purshia tridentata*

ash, *Fraxinus* species

blueberry, *Vaccinium* species

buffaloberry, *Shepherdia* species

camas, *Camassia quamash*

cascade mountain-ash, *Sorbus scopulina*

cottonwood, *Populus* species

cranberry, *Vaccinium macrocarpon*

dogbane, *Apocynum cannabinum*

fourwing saltbush, *Atriplex canescens*

fritillary, *Fritillaria* species

jointfir, *Ephedra* species

juniper, *Juniperus* species

kinnikinnick, *Arctostaphylos uva-ursi*

lily, *Lilium* species

maple, *Acer* species

mariposa lily, *Calochortus* species

Missouri iris, *Iris missouriensis*

Pacific yew, *Taxus brevifolia*

quaking aspen, *Populus tremuloides*

redosier dogwood, *Cornus sericea*

rhododendron, *Rhododendron* species

silverberry, *Elaeagnus commutata*

shooting star, *Dodecatheon* species

snowberry, *Symphoricarpos albus*

trout lily, *Erythronium* species

twinflower, *Linnaea borealis*

wapato, *Sagittaria latifolia*

white-rooted sedge, *Carex barbarae*

wild strawberry, *Fragaria* species

willow, *Salix* species

yellowbells, *Fritillaria pudica*

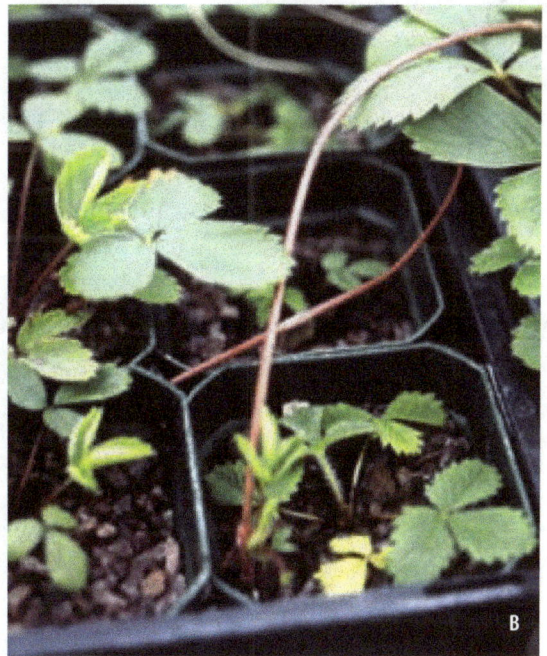

Figure 9.35—*(A) Stolons of twinflower and (B) runners of wild strawberry can be collected and used to root several plants from one plant.* Photos by Tara Luna.

Water Quality and Irrigation

Thomas D. Landis and Kim M. Wilkinson

10

Water is the single most important biological factor affecting plant growth and health. Water is essential for almost every plant process: photosynthesis, nutrient transport, and cell expansion and development. In fact, 80 to 90 percent of a seedling's weight is made up of water. Therefore, irrigation management is the most critical aspect of nursery operations (figure 10.1).

Determining how, when, and how much to irrigate is a crucial part of nursery planning as well as day-to-day operations. One missed watering session can cause serious injury and even death to plants at any stage in their development. Adequate watering is particularly important with container plants, whose roots cannot access water beyond the container walls and therefore are entirely dependent on receiving enough water through irrigation. Excessive watering is also problematic; it is the major cause of root diseases and contributes to other problems with seedling growth. Therefore, good design and operation of an irrigation system is central to managing a nursery successfully.

Every nursery is unique and native plant nurseries typically grow a wide range of species with differing water requirements. In addition, the distinct phases of growth that plants go through (establishment, rapid growth, and hardening) will require differing watering regimes. Designing an effective and efficient irrigation system is not a matter of deciding on one specific system but rather choosing which types of irrigation systems and practices best serve the needs of the plants. The nursery might have various propagation environments and corresponding irrigation zones that provide for the changing needs

Redosier dogwood after irrigation by Tara Luna.

Figure 10.1—*A supply of good-quality water is one of the most critical requirements for a native plant nursery.* Photo by Thomas D. Landis.

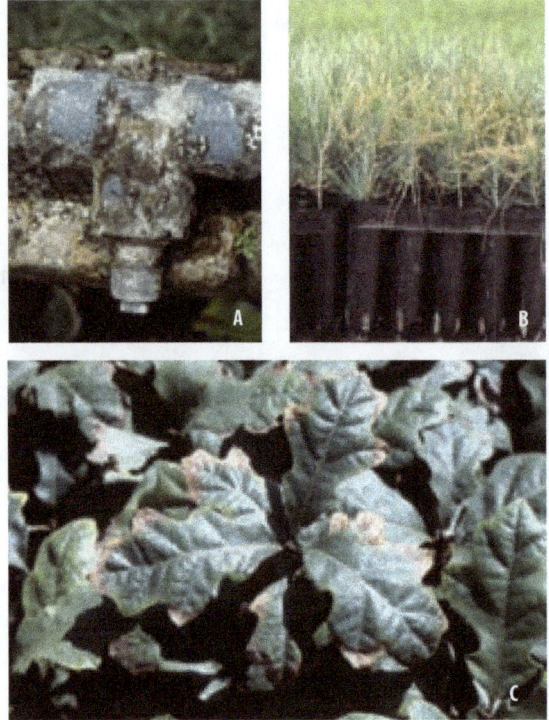

Figure 10.2—*Agricultural water quality is determined by the level of soluble salts because they can (A) build up on irrigation nozzles (B) accumulate in containers, usually around the drainage holes, and (C) eventually "burn" seedling foilage (C).* Photos by Thomas D. Landis.

of plants during all phases of growth. For example, one nursery might have a mist chamber for the germination phase, overhead sprinklers in the growing area, a selection of rare plants that receive daily hand watering, and some large plants under driplines. In addition, irrigation may be needed for other horticultural purposes such as cooling or frost protection.

The best design for any irrigation system will come from understanding the needs of the plants, the factors that affect water availability, and the details of how, when, and why to water. The first step in setting up a nursery and its watering system is to ensure that the nursery has a reliable and high-quality source of water for irrigation.

WATER QUALITY

The quality of irrigation water is a critical factor in the site selection for and management of a container nursery. Improving poor-quality irrigation water is expensive, often prohibitively so. Therefore, water quality should be a primary consideration during nursery site evaluation. A water sample should be collected and sent to a laboratory to test its quality—soluble salts in water can clog nozzles (figure 10.2A), accumulate in containers (figure 10.2B), and eventually harm plants (figure 10.2C).

For irrigation purposes, water quality is determined by two factors: (1) the types and concentrations of dissolved salts (total salinity and individual toxic ions) and (2) the presence of pests (pathogenic fungi, weed seeds, algae, and possible pesticide contamination).

A WELL-DESIGNED IRRIGATION SYSTEM HAS MANY ADVANTAGES

— Better plant quality and health.
— Lower labor costs.
— Improved crop uniformity and reliability.
— Reduced runoff and waste of water.

Water Quality: Salts

For our purposes, a *salt* can be defined as a chemical compound that releases charged particles called ions when dissolved in water. Some salts are fertilizers that can have a beneficial effect, while other salts have a neutral effect. Salts such as sodium chloride (ordinary table salt), however, dissolve into harmful ions that can damage or even kill plant tissue.

An excess of dissolved salts in nursery irrigation water has negative effects, including the following:

- Water availability is reduced resulting in growth loss.
- Some ions (sodium, chloride, and boron) are directly toxic to plants.
- Other ions (calcium) affect mineral nutrient availability.
- Other ions (bicarbonate or iron) cause salt crusts or staining.
- Salt deposits may also accumulate on sprinkler nozzles and reduce their efficiency.

An excess of dissolved salts in the water can be the result of a number of factors. First, the topographic location of a container nursery can have an effect on irrigation water quality because of local climatic or geologic influences. In locations with arid or semiarid climates where evapotranspiration exceeds precipitation, salts naturally accumulate in the soil, and groundwater irrigation sources are often high in salt content. Second, in coastal areas, irrigation water may be contaminated by saltwater intrusion. Third, the high fertilization rates used in container nurseries can lead to a salinity problem. Fourth, improper horticultural practices can compound salt problems: the soluble salt level doubles when the growing medium dries from 50 to 25 percent moisture content.

The symptoms of salt injury vary with species but can include foliar tip burn, scorching or bluish color on leaves, stunting, patchy growth, and eventual mortality. Most native plants are extremely sensitive to salinity damage. The principal damage of high salinity is reduced growth rate, which usually develops before more visible symptoms become evident.

It is expensive to remove salts from irrigation water, so ideally the nursery should be established on a site where water salinity is within acceptable levels. Test results for salinity are traditionally expressed as electrical conductivity (EC); the higher the salt concentration, the higher the EC reading (table 10.1). The EC can be checked at the nursery using a conductivity meter (see figure 11.12), or by sending water samples to a local laboratory. The most commonly used salinity units in irrigation water quality are micromhos per centimeter (abbreviated as umho/cm and pronounced "micro-mows") and the International System of Units of microsiemens per centimeter, which are equivalent. Microsiemens per centimeter (abbreviated as μS/cm)

GOOD WATER MANAGEMENT HAS THE FOLLOWING ATTRIBUTES

- Efficient use of water.
- Reliable source of water.
- High uniformity of water distribution.
- An approach that is flexible and tailored to the changing needs of the species grown and their phases of development.

will be used as the standard EC unit in this handbook. General guidelines for salinity ranges are in table 10.1.

As already mentioned, irrigation water salinity tests should be tested prior to nursery establishment and retested periodically. It is particularly important to do an initial test in areas with high salinity because the addition of fertilizer could raise salinity to unacceptable levels (figure 10.3). In these cases, a nursery would need to be careful to use very dilute liquid fertilizers or controlled-release fertilizers to keep salinity within acceptable ranges. Horticultural practices such as increasing the porosity of the growing medium and leaching more frequently during waterings can help alleviate the effects of saline water.

Water Quality: Pests

Container nurseries that use irrigation water from surface water sources such as ponds, lakes, or rivers may encounter problems with biotic pests; that is, weeds, pathogenic fungi, moss, algae, or liverworts. Surface water that originates from other nurseries or farmland is particularly likely to be contaminated with water-mold fungi, such as *Pythium* and *Phytophthora*, which cause damping-off. Recycled nursery irrigation water should also be suspect and should be analyzed. Many weed seeds and moss and algal spores are small enough to pass through the irrigation system and can cause real problems in container nurseries. Waterborne pests can be killed with chlorination, and some specialized filtration systems can remove many disease organisms from irrigation water. See the following sections for more information on chlorination and filtration.

Irrigation water, especially in agricultural areas, may have become contaminated with residual pesticides. Herbicides applied to adjacent cropland or to control aquatic weeds in reservoirs can affect irrigation water

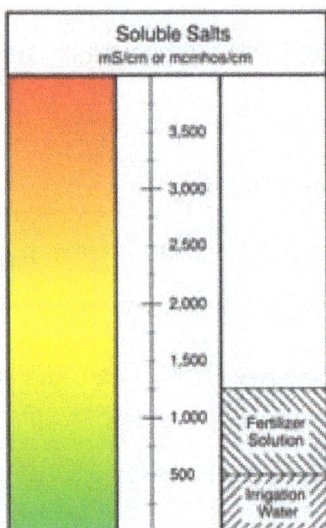

Soluble Salts
mS/cm or mcmhos/cm

- 3,500
- 3,000
- 2,500
- 2,000
- 1,500
- 1,000 — Fertilizer Solution
- 500 — Irrigation Water

Figure 10.3—*When soluble fertilizers are injected into the irrigation system, salinity levels are cumulative. For example, a nursery with a base irrigation salinity of 500 µS/cm has good quality but, after fertigation is added, the total salinity can reach into the zone of caution.* Illustration by Jim Marin.

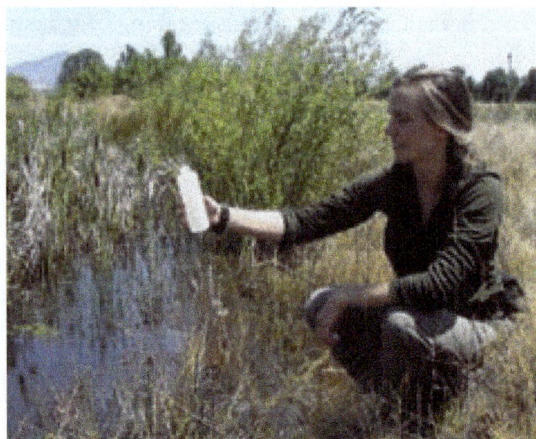

Figure 10.4—*Water quality should be tested before nursery establishment, and then yearly to make certain that quality has not changed.* Photo by Thomas D. Landis.

quality. Potential sources of irrigation water should be tested for pesticide contamination when a nursery site is being evaluated.

Testing Water Quality

Ideally, a water-quality test is done when the nursery is established and then at yearly intervals (figure 10.4). Many existing nurseries, however, have never had a detailed water analysis performed. A complete analysis of irrigation water quality should consist of a salinity evaluation listing the concentrations of eight specific ions that should be reported in parts per million (ppm). For a small additional fee, it is possible to test for the other nutrient ions at the same time. In addition to the ion concentrations, the testing laboratory should report three standard water-quality indices: EC, toxic ion concentrations, and pH.

Irrigation water should also be tested for the presence of pathogenic fungi, preferably during the site selection process but also if a problem is observed at a later date. Most plant pathology laboratories can conduct bioassays of irrigation water. Testing for residual herbicides is also possible but can be expensive because of the sophisticated analytical procedures required. Because of the different chemical structures of various pesticides, a separate analysis for each suspected pesticide is usually required. Therefore, specialized pesticide tests are generally considered only when a definite problem is suspected.

Collecting a sample for irrigation water testing should be done properly. Use a clean plastic bottle with a firm, watertight lid. A 16-fluid-ounce (475-ml) container is ideal for most water tests. To begin, let the water run for several minutes, and then rinse the sample bottle well before collecting the sample. Label the sample bottle properly with a waterproof marker before sending it to the analytical laboratory. The sample should be sent away for testing as quickly as possible but can be stored under refrigeration for short periods, if necessary. Most laboratories charge $25 to $50 and will provide results within a few weeks.

Table 10.1—Water-quality standards for nursery irrigation water (modified from Landis and others 1989)

Quality Index	Optimal	Acceptable	Unacceptable
pH	5.5 to 6.5		
Salinity (µS/cm)	0 to 500	500 to 1,500	1,500
Sodium (ppm)			> 50
Chloride (ppm)			> 70
Boron (ppm)			> 0.75

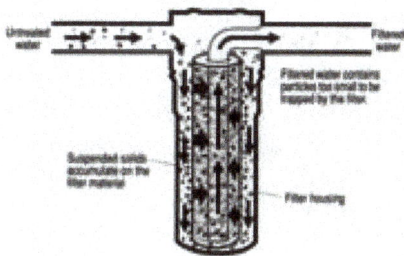

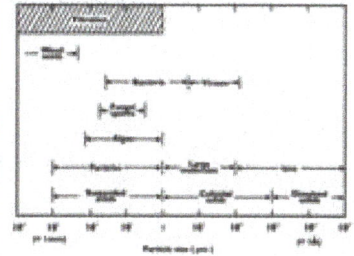

Figure 10.5—*(A) Cartridge filters are an effective and inexpensive way to treat irrigation water; (B) they should be installed before the fertilizer injector; and (C) they can remove sand, silt, fungal and algal spores.* Photo by Thomas D. Landis, illustrations by Jim Marin.

Practical Water Treatments: Filtration, Chlorination, and Temperature Modification

Establishing the nursery on a site with tested, good-quality water is the best way to preclude water-related problems. If existing water quality is poor, methods such as deionization and reverse osmosis can treat and improve irrigation water, but they are often prohibitively expensive and not feasible for most native plant nurseries. To correct or safeguard against minor problems with otherwise good-quality water, however, some treatments are low cost and highly effective for container nurseries. These treatments include filtration, chlorination, and temperature modification.

Filtration is used to remove suspended or colloidal particles such as very fine sand or silt. Filtration is recommended for most container nurseries because it prevents problems such as the plugging of nozzles or damage to the irrigation or fertilization equipment (figure 10.5). Filters also remove unwanted pests such as weed seeds or algae spores. Two types of filters are commonly used in nurseries: granular medium filters and surface filters. A local irrigation supplier may be able to help you select a good-quality filter. Granular medium filters consist of beds of granular particles that trap suspended material in the pores between the particles, whereas surface filters use a porous screen or mesh to strain the suspended material from the irrigation water. Granular medium filters can be used to remove fine sand or organic matter and are constructed so that they can be backflushed for cleaning. Surface filters include screens or cartridges of various mesh sizes to

remove suspended material; screens must be physically removed and cleaned whereas cartridge filters are not reusable and must be regularly replaced.

Filters should be installed before the water passes through the nutrient injector to intercept sand particles that can cause excessive wear of plumbing or plug valves (figure 10.5B). Jones (1983) recommends cartridge filters because they are easy to change. Backflushing screens or granular medium filters is not practical with many nursery irrigation systems. Handreck and Black (1984) recommend using filters small enough to remove particles greater than 5 microns in diameter, which will take care of most suspended materials (figure 10.5C). Specialized filtration systems, such as those manufactured by Millipore Corporation, can remove particles around 1 micron in diameter; such a system is therefore capable of removing some disease organisms and most suspended solids (figure 10.5C). Unfortunately, more sophisticated filtration systems are relatively expensive and require frequent maintenance (Jones 1983). Chlorination can be used to kill fungi, bacteria, algae, or liverworts introduced through the irrigation system and causing problems.

Another aspect of water quality that can be controlled in container seedling nurseries is the temperature of the irrigation water. Cold irrigation water can significantly lower the temperature of the growing medium and has been shown to reduce plant growth in ornamental container nurseries. Water uptake by plants is minimal below 50 °F (10 °C), so cold soil temperatures can definitely reduce seedling growth. Cold irrigation water is most damaging during seed germi-

Figure 10.6—*Black water storage tanks are a low-cost way to warm irrigation water with solar energy before it is applied to crops.* Photo by Thomas D. Landis.

nation and could delay seedling emergence. A low-cost way of obtaining warmer water is to store water in black plastic storage tanks that are exposed to the sun or located inside a heated location (figure 10.6).

WATER QUANTITY

The amount of water necessary to produce a crop of container plants depends on many factors, such as climate, type of growing structure, type of irrigation system, growing medium, and seedling characteristics. Therefore, it is difficult to estimate the amount of water that a native plant nursery will require, but some water use data from conifer nurseries in the Western United States are provided in table 10.2.

Remember that a nursery needs water for operational requirements other than irrigating crops. For example, mixing growing media; cleaning containers, structures, and equipment; and staff personal water needs all increase water use. Also, a nursery that starts small may choose to expand. Therefore, make sure an abundance of water is available to meet present and future needs.

Even in cases in which the nursery has access to a very steady, reliable, and high-quality municipal water source, a backup system is always a good idea in case of emergency. A prudent investment is a backup water storage tank containing sufficient water to meet the nursery's needs for at least a week (figure 10.6). Backup systems may be pumped into the normal irrigation system, but it is advantageous to locate the storage tank upslope so that water can be supplied by gravity in case of power failure.

FACTORS AFFECTING WATER AVAILABILITY TO PLANTS

After a good site with tested, reliable, high-quality water has been obtained, the next step in irrigation design is to understand the factors that affect water availability. Plant water use is affected by environmental conditions such as humidity, temperature, season, and the amount of sunlight the plants receive. The growth phase of the crop will also affect the rate of evaporation and transpiration. During seedling germination and early emergence, evaporation is the primary cause of water loss (figure 10.7A). After the seedling's roots occupy the container, however, transpiration becomes the primary force for water loss (figure 10.7B).

Several factors unique to container nurseries affect water availability. These factors include the types of containers, especially their volume (table 10.2), and the types of growing media. These factors make water management in a container nursery distinct from water management in a bareroot nursery, a garden, or other agricultural setting.

The type of growing medium has a large impact on water availability and use. Common components of artificial media behave very differently than soil. Peat moss and vermiculite have a high water-holding capacity, whereas perlite and pumice do not. Water infiltration and drainage are much higher than with mineral soils. The average pore size of the growing medium is the most significant influence. All things being equal, a finer-textured growing medium with a smaller average pore size holds more water than a coarser textured medium does (figure 10.8). See Chapter 5, *Growing Media*, for more details on this topic.

Container type, volume, and shape also affect water availability. Water in a container behaves differently

Table 10.2—Typical irrigation use in forest and conservation nurseries for a crop of 1,000 conifer seedlings

Nursery and Location	Container Type and Volume	Irrigation Water Use per Week gallons (liters)	
		Establishment Phase	Rapid Growth Phase
University of Idaho, Moscow	Ray Leach Cone-tainers™ 4 in³ (66 ml)	10 (38)	15 (57)
Mt. Sopris, Colorado	Ray Leach Cone-tainers™ 10 in³ (172 ml)	15 (57)	50 (189)
University of Idaho, Moscow	Styroblock™ 20 in³ (340 ml)	60 (227)	125 (473)

than water in unconfined soil because it does not drain completely, resulting in a layer of saturated medium at the bottom (figure 10.7B). The height of this saturated medium is a function of the growing medium, but taller containers will have a smaller proportion of saturated medium than shorter ones (see figure 6.2).

The small top opening in some types of plant containers is important operationally because it is extremely difficult to distribute irrigation evenly between containers, which leads to considerable variation in growing medium water content. This distribution problem becomes even more critical when the plants become larger and their foliage begins to intercept water before it can reach the surface of the container. Foliage interception is particularly serious for broad-leaved species.

Because small containers have a correspondingly small volume of growing medium, they have limited moisture reserves and require frequent irrigation, especially in times of high evapotranspirational losses. These factors should be kept in mind when designing the irrigation system the plants will depend on to grow and survive.

Irrigation as a Cultural Treatment: Determining How Much To Irrigate

The next step is to determine how much water will need to be applied per irrigation event. The most important concept in container irrigation is to apply enough water during each event to more than saturate the medium so that a small amount of leaching occurs. This practice simply means applying enough water that some drips out the bottom of the container

Figure 10.7—The amount and type of water use changes dramatically during the growing season. (A) During the establishment phase, a relatively small amount of water is used by evaporation but, (B) when the seedling fully occupies the container during the rapid growth phase, a much greater amount is used for transpiration with relatively little lost to evaporation. In all containers, a layer of saturated media always exists at the bottom. Illustration by Jim Marin.

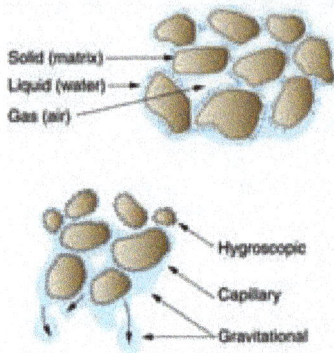

Figure 10.8—*Water is held in the pores between growing media particles by capillarity. For a given type of media, the larger the particles, the less water will be held.* Illustration by Jim Marin.

(although not so much that water streams out the bottom). Because of the unique properties of artificial growing media in containers (as discussed in Chapter 5, *Growing Media*), enough water must be applied to the surface to force the air out of the medium pores (figure 10.9A). If the irrigation period is too short, the water will never reach the bottom of the container. The result will be a perched water table with a layer of dry growing medium underneath (figure 10.9B). To avoid this result, it is important to always fully saturate the medium profile; otherwise, only the top part of the medium will be wetted. If the growing medium throughout the container is not completely saturated after irrigation, the seedling will never develop roots in the dry medium at the bottom of the container, resulting in a poorly formed plug. Another hazard is that fertilizer salts will accumulate in the medium and cause salinity damage or "fertilizer burn."

So, the general rule of thumb for sprinkler irrigation is to apply approximately 10 percent more water than is needed to completely saturate the entire growing medium profile during irrigation. The best procedure is to actually check to make sure that drainage is occurring during or immediately after irrigation by direct inspection.

IRRIGATION AND SEEDLING GROWTH PHASES

The amount of irrigation to apply varies during the growing season in a native plant seedling nursery because of the stages of seedling development and the horticultural objectives of the nursery manager. Because water is so essential to plant growth, the irrigation regime can be manipulated to control seedling growth. As discussed in Chapter 3, *Planning Crops and Developing Propagation Protocols*, plants go through three phases of development: establishment, rapid growth, and hardening. Irrigation is an important strategy for both manipulating and supporting plant growth and health as the crop moves through these phases.

Irrigating During the Establishment Phase

Immediately after the sown containers are placed in the growing area, the growing medium should be completely saturated. Thereafter, watering needs during establishment should be monitored carefully and tailored to the needs of the species. Some species such as quaking aspen, willow, and cottonwood require nearly continuous misting and may even benefit from a fog chamber. Other species may require less water, but, nevertheless, inadequate or too-infrequent irrigations will cause the seeds to dry out, which will decrease germination success or even cause total crop loss. Conversely, excessive irrigation may create overly wet conditions around the seeds for some species, promoting damping-off and/or delaying germination. For these reasons, irrigation should be applied on an as-needed basis as determined by the germinant requirements and by observation of the crop (figure 10.10). Remember that until the seeds germinate and begin to grow, the major water loss is from evaporation from the top of the container, not from water use by the germinating plants. Irrigation during this period, therefore, must be applied with the goal of replenishing the moisture in the thin surface layer of the medium. This practice is usually best accomplished by periodic mistings or light irrigation as necessary with a very fine spray nozzle, which also protects germinating seeds from being moved or damaged by the force of the

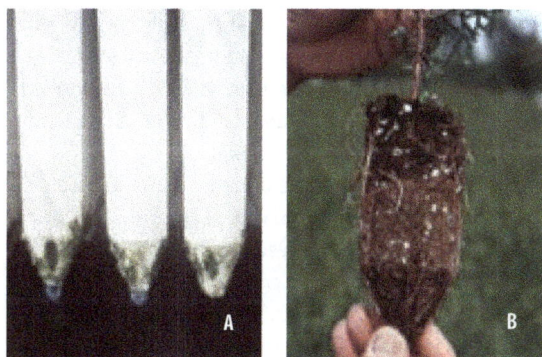

Figure 10.9—*(A) Apply enough water during irrigation to fully saturate the growing medium profile and permit some leaching out the bottom of the container; (B) insufficient irrigation will result in a "perched water table" within the growing media.* Photos by Thomas D. Landis.

water. In some cases, if seeds are mulched, irrigation may not be necessary for the first week or so of the germination phase. The water status of the medium can be checked by scratching through the surface of the mulch or grit and making sure the medium is moist enough.

Irrigation can also be used to control the temperature around germinating seeds. Germinants, particularly those covered by dark-colored mulches, can be injured by high temperatures on the surface of the growing medium. If temperatures exceed 86 °F (30 °C), light mistings will help keep germinants cool. This practice is sometimes called "water shading." These waterings should not add much water to the medium but are merely used to dissipate the heat from the mulch. After plants develop thicker stems, they become more resistant to heat injury. Water shading should not be used in nurseries with saline water or salts can build up in surface layers of the growing medium.

Irrigating During the Rapid Growth Phase

After the seedling's roots have expanded throughout the container, the amount of water lost through transpiration increases greatly and so irrigations must be longer and more frequent. As seen in table 10.2, water use will double or even triple during the rapid growth phase. A tremendous amount of variability occurs between different native species. Full saturation is best for some species whereas others might benefit from periods of slight moisture stress between waterings. No plants should ever be allowed to dry out completely (figure 10.11). Nursery managers should be

aware of the varying water requirements for the species grown and adjust their irrigation practices accordingly. Grouping species together by their water requirements ("wet," "moderate," or "dry") makes this practice much easier.

Another factor that comes into play during the rapid growth phase is the effect of foliage interception of water. During the rapid growth phase, the leaves of plants begin to form a tight canopy that causes a significant reduction in the amount of irrigation that can reach the growing medium surface. This "umbrella effect" is particularly serious with broad-leaved species. Water will often drip through the foliage irregularly so that one container is fully saturated whereas the one right next to it may receive almost no water. To compensate for foliage interception, growers tend to irrigate more frequently and longer. This practice still results in uneven irrigation and wasted water, because the water runs off the leaves and onto the floor. The types of irrigation systems discussed below will help address this issue for broad-leaved species, particularly through subirrigation and/or hand watering practices.

The rapid growth phase is also the time when liquid fertilizers are most concentrated and water loss through transpiration is high, so growers must be concerned about the accumulation of salts. Remember that fertilizers are salts, too, and that injecting liquid fertilizers adds to the base salinity level of the irrigation water. Increases in salinity are a concern with controlled-release fertilizers as well because their release rate is controlled by water and temperature and salts can accumulate quickly in small containers. Frequent leachings with regular irrigation water ("clearwater flush") are needed to push excess salts out the bottom of the container. One of the first signs of a salinity problem is salt crust around the drainage holes (see figure 10.2B).

Irrigating During the Hardening Phase

The manipulation of irrigation practices is an effective way to initiate the hardening of plants prior to storage or shipment. Because seedling growth is so critically tied to moisture stress levels, growers can

---

The key concept to irrigating during the establishment phases is to keep the growing medium "moist, but not saturated."

Figure 10.10—*During the establishment phase, watering should be tailored to the requirements of the species during germination and early growth. For many species, light irrigations as needed using a fine spray ("misting") provide enough water for germination and early growth and also protect the young germinants from heat injury. Overwatering must be avoided.* Photo by Thomas D. Landis.

Figure 10.11—*These quaking aspen seedlings are suffering from severe water stress due to improper irrigation. Seedlings use a lot of water during the rapid growth phase and frequent irrigation is needed to prevent growth loss.* Photo by Thomas D. Landis.

affect shoot growth (slow or stop it by inducing bud set), increase general resistance to stress (especially water stress), and/or initiate the development of cold hardiness in many species of container plants by horticulturally inducing moderate water stress. This "drought stressing" procedure consists of withholding irrigation for short periods of time until the plants can be seen to wilt or until some predetermined moisture stress is reached. After this stress treatment, the crop is returned to a maintenance irrigation schedule.

Implementing moisture stress, however, can be challenging with native plants because (1) dormancy and hardiness are affected by other environmental conditions, such as day length, (2) considerable varia-

tion in growing medium moisture content can exist between adjacent containers, so it is hard to achieve a uniform level of moisture stress, and (3) if the growing medium is allowed to dry too far, it can become hydrophobic and difficult to rewet.

Water stressing must be done correctly and conscientiously, and there is no substitute for experience. Most of the water stress research has been done with commercial conifers and good guidelines have been published (for example, Landis and others 1989) for monitoring container weights. Unfortunately, little is known about the response of most native plants. Inducing moisture stress, therefore, can be risky if the plant's tolerance is unknown. Drought stressing simply does not work for some species and in some environments. Growers should conduct their own trials of operational moisture stressing to determine the effect on their own species in their respective growing environments. Careful scheduling and communication with other nursery workers is essential (figure 10.12). Be sure to keep good records of how the crops respond.

In spite of these caveats, the induction of mild moisture stresses should be considered as a horticultural technique to manipulate seedling physiology and morphology. A further discussion of the hardening process, including moisture stress, is provided in Chapter 12, *Hardening*.

Irrigating for Frost Protection

In cold climates, irrigation can also be used for the frost protection of plants in open growing compounds. Container plants that are raised in outdoor growing areas or stored in sheltered storage may require protection against freezing temperatures in autumn or spring. Proper hardening procedures will help protect shoots against frost injury, but unusually cold weather can sometimes occur suddenly before the plants have had time to harden sufficiently. Roots do not achieve a high degree of cold hardiness and should always be insulated if plants are to be stored under exposed conditions.

Sprinkler irrigation protects against cold injury because heat is released when water freezes on the seedling foliage, and the ice layer provides some degree of insulation (figure 10.13). The main protection comes from the heat released from the freezing water, however, so this protective effect lasts only as long as irrigation continues to be applied. Irrigation should begin as soon as the temperature drops below freezing and continue until the ice is melted. Water is not effective in all freezing weather, especially in areas with advective frosts where dry, cold winds can actually drive the temperature down if water is applied. Therefore, growers must carefully monitor conditions before taking action. Some nurseries test their plants for frost hardiness and base their determinations of when frost protection should begin on these tests. Frost protection with sprinkler irrigation cannot protect against severe "hard" freezes, but agricultural crops have been saved in temperatures as low as 17 °F (-8 °C).

Detrimental Results of Irrigation on Plant Growth and Health

When not applied properly, irrigation can cause serious problems. Excessive irrigation leads to root diseases and fungus gnats and during hardening can delay the normal development of frost hardiness. On the other hand, applying a too-severe moisture stress treatment may actually inhibit the development of frost hardiness. The most common mistake is not adjusting irrigation for different growth phases.

FERTIGATION

Irrigation is critical to the proper application of fertilizers, especially when injecting liquid fertilizer solution into the irrigation system—a practice called "fertigation." Fertigation can be used with many different types of irrigation systems, from hand-watering to automated sprinkler or drip systems. Fertilizer injectors range from simple, low-cost siphons for hand-watering to sophisticated pumps for automated sprinklers. Because it can be designed to apply to proper mineral nutrients at the proper concentration and at the proper time, fertigation has several advantages over other types of fertilization. See Chapter 11, *Fertilization*, for more information on fertigation and how it can be applied in native plant nurseries.

TYPES OF IRRIGATION SYSTEMS

The best method of applying irrigation water in container nurseries depends on the size and complexity of the operation and on the water requirements of the plants being grown. Small nurseries and those growing a variety of species may prefer hand-watering for irri-

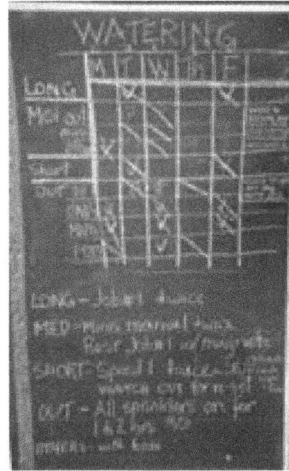

Figure 10.12—*Because so little is known about the response of most native plants to water stress, good scheduling, recordkeeping, and communication is critical in order to determine what the plants need.*
Photo by Thomas D. Landis.

Figure 10.13—*Water can be used to protect succulent plant tissue during unseasonal frosts in spring or autumn.* Photo by Thomas D. Landis.

Figure 10.14—*Hand-watering is often the best way to irrigate small lots of native plants in various types of containers and with different water requirements.* Photo by Tara Luna.

gation, whereas large nurseries growing only a few species of commercial conifers may use some sort of mechanical irrigation system, such as mobile overhead booms. Most native plant nurseries use a combination of systems in different watering zones to fulfill their irrigation needs, and these will be discussed one by one in the following sections.

Hand-Watering

Hand-watering is often the most practical irrigation strategy for small native plant nurseries or for nurseries producing a diversity of species with radically different water requirements (figure 10.14). Often hand-watering is also the best strategy for a nursery in the startup phase. After the different watering requirements are understood for each nursery crop, then, if desired, an investment can be made in appropriate irrigation systems to meet the plant's needs.

Hand-watering requires simple and inexpensive equipment; a hose, a couple of different nozzle types, and a long-handled spray wand are all that are absolutely necessary. The watering job will be more pleasant and efficient with a few additional small investments, such as overhead wires to guide the hoses and rubber boots for the staff (Biernbaum 1995). Although the task may appear easy, good technique and the application of the proper amount of water to diverse species of native plants in a variety of containers and at various growth stages is challenging. Nursery managers should make sure that irrigators have a conscientious attitude and are properly trained to work effectively with water application.

Good hand-watering practices include the following:

— Direct water to the roots of the plants.
— Avoid spraying the foliage to conserve water and preclude foliar diseases.
— Angle the watering nozzle straight down (not sideways) so the water does not wash out seeds, medium, or mulch.
— Use an appropriate nozzle type and water volume for the crops: a very fine, gentle spray for young germinants and a larger volume nozzle for larger plants.
— Adjust the flow, volume, and pace of watering to irrigate efficiently without wasting water or compacting or washing out root medium (Biernbaum 1995).
— Achieve uniformity of water distribution so all plants are well irrigated; account for microclimate differences in nursery (for example, plants on the outer edge of a south-facing wall might need more water).
— Attend to the individual needs of each crop and its development phase so none are overwatered or underwatered; develop a "feel" for the watering needs of crops over time.

Overhead Irrigation Systems

Overhead sprinkler systems are a common type of irrigation system and the kind that most people imagine when they think of irrigation. Many types of overhead irrigation systems exist, ranging from fixed sprinklers to moving boom systems. Fixed irrigation systems consist of a grid of regularly spaced irrigation nozzles and are popular because they are less expensive than mobile systems. Mobile systems have a traveling boom that distributes water to the crops, a system that works well but is too costly for most small-scale nurseries. The following section discusses two types of fixed irrigation systems: fixed overhead sprinklers and fixed basal sprinklers.

Fixed Overhead Sprinklers

Fixed overhead sprinkler systems consist of a series of parallel irrigation lines, usually constructed of plastic polyvinyl chloride pipe, with sprinklers spaced at uniform intervals to form a regular grid pattern. Overhead sprinklers apply water at a fairly rapid rate and will do an acceptable job if properly designed and maintained.

Generally, the propagation environment is divided into irrigation "bays" or "zones" depending on the num-

WATER QUALITY AND IRRIGATION

HAND-WATERING

Advantages

— Requires inexpensive equipment that is simple to install.
— Is flexible and can adjust for different species and container sizes.
— Irrigators have a daily connection to the crop and can scout out diseases or other potential problems.
— Allows water to be directed under plant foliage, reducing risk of diseases.

Disadvantages

— Is time consuming and labor intensive.
— Involves a daily responsibility; weekends and holidays do not exist for plants.
— Requires skill, experience, and presence of mind to do properly (the task should not be assigned to inexperienced staff).
— Presents a risk of washing out or compacting medium from the containers.
— Presents a risk of runoff and waste.

Figure 10.15—(A) Both overhead and basal sprinkler systems feature rotating "spinners" or (B) stationary nozzles which, in addition to full circles, are (C) available in quarter, half, and three-quarters coverage. (D) Because of their greater coverage, rotating-impact (Rain Bird®-type) sprinklers (E) are commonly used in outdoor growing areas. Photos by Thomas D. Landis, illustration by Jim Marin.

ber of nozzles that the pump can operate at one time at the desired water pressure. Ideal operating pressures vary with the type of sprinkler, and specifications are available from the manufacturer. Some sprinklers come in different coverages such as full-circle, half-circle, and quarter-circle, so that full overlap coverage can be obtained by placing irrigation lines around the perimeter of the irrigation bay. Each bay should be able to be separately controlled with a solenoid valve, which can be connected to an irrigation timer so that the duration and sequence of irrigation can be programmed. The size of each irrigation bay can be designed so that species with differing water requirements can be grown within a larger growing structure. When designing a new irrigation system, it is a good idea to obtain the help of an irrigation specialist to ensure that the system is balanced in terms of coverage and water pressure.

Several types of spray nozzles are used for fixed overhead irrigation systems. Spinner sprinklers, which have offset nozzles at the end of a rotating arm, spin in a circle when water pressure is applied (figure 10.15A). Stationary nozzles (figure 10.15B) have no moving parts but distribute water in a circular pattern; these nozzles also come in half-circle and quarter-circle pat-

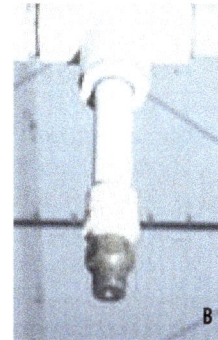

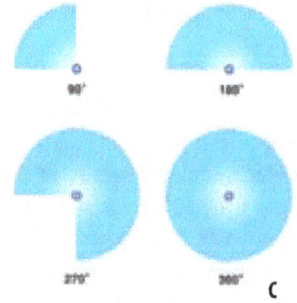

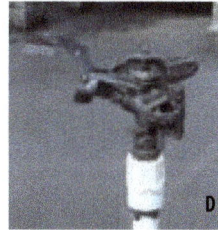

terns (figure 10.15C). Mist nozzles are also sometimes installed on overhead irrigation lines and are primarily used during the germination period and for cooling and humidity control.

Fixed Basal Sprinklers

Basal irrigation systems are commonly used in large outdoor growing or holding areas. They are similar to overhead systems in design and operation in that they use a regular grid of permanent or movable irrigation lines with regularly spaced sprinklers. Both stationary sprinklers (figures 10.15A and B) and rotating-impact nozzles (figure 10.15D) are commonly used. These sprinklers rotate slowly due to the impact of a spring-loaded arm that moves in and out of the nozzle stream (figure 10.15E). Rotating-impact sprinklers are available from several manufacturers in a variety of nozzle sizes and coverages. Because the impact arm is driven by the water pressure out of the nozzle jet, the water distribution pattern of these sprinklers is particularly dependent on proper water pressure. One advantage of basal irrigation systems is that impact sprinklers have relatively large coverage areas, which means that fewer nozzles and less irrigation pipe are required.

Moveable Boom Irrigation Systems

The most efficient but most expensive type of sprinkler irrigation is the moveable boom (figure 10.16A), which applies water in a linear pattern (figure 10.16B) and only to the crop. Moveable booms are generally considered too expensive for most small native plant nurseries but should be considered whenever possible. For more information, see Landis and others (1989).

Designing and Monitoring Fixed Sprinkler Systems

The efficiency of an irrigation system is primarily dependent on its original design. Uniform irrigation is a function of five factors: (1) nozzle design, (2) size of the nozzle orifice, (3) water pressure and application rate at the nozzle, (4) spacing and pattern of the nozzles, and (5) effects of wind. Few operational procedures can improve a poorly designed system. Therefore, it is important to consult an irrigation engineer during the planning stages. Basic engineering considerations, such as friction loss in pipes or fittings and the effect of water pressure on sprinkler function, must be incorporated into the irrigation system design.

The spacing and pattern of the sprinklers in fixed irrigation systems are related to sprinkler function and the effect of wind. The size of the sprinkler nozzle and its resultant coverage pattern can be determined by consulting the performance specifications provided by the sprinkler manufacturer. Container nursery managers should select a nozzle size that is coarse enough to penetrate the plant's foliage and minimize wind drift but not large enough to create splash problems.

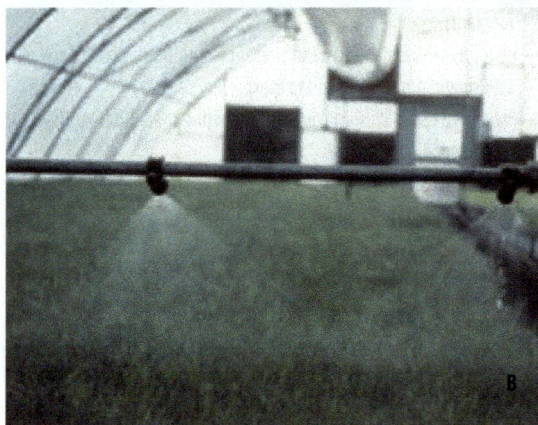

Figure 10.16—(A) Moveable boom irrigation systems (B) apply water in a very efficient linear pattern but may be considered too expensive for many small native plant nurseries. Photos by Thomas D. Landis

All types of stationary sprinklers throw water in a circular distribution pattern (figure 10.17A), so irrigation systems should be designed to provide adequate overlap between sprinklers. This consideration is especially important in shadehouses or outdoor growing areas where wind drift can be a problem (figure 10.17B). Too often, sprinklers are spaced at greater intervals in a cost-saving effort, but this practice is economically shortsighted considering the profound effect of water and injected nutrients on plant growth.

Because water pressure has such an effect on sprinkler function and efficiency, it should be checked regularly. Performance specifications for sprinklers at standard water pressures can be obtained from the manufacturer. The water pressure should be regularly monitored with a gauge permanently mounted near the nozzles (figure 10.17C) or with a pressure gauge equipped with a pitot tube directly from the sprinkler nozzle orifice. The pressure should be checked at several different nozzles including the nozzle farthest from the pump. The importance of regular water pressure checks cannot be overemphasized because many factors can cause a change in nozzle pressure. Water pressure that is either too high or too low can cause erratic stripe or doughnut-shaped distribution patterns (figure 10.17D).

Subirrigation

Overhead irrigation systems have always been the choice of container nurseries because the systems are relatively cheap and easy to install. The inherent inefficiency of overhead systems, however, becomes a very serious problem with native plants, especially those with broad leaves. Wide leaf blades combined with the close spacing of most containers create a canopy that intercepts most of the water applied through overhead irrigation systems, reducing water use efficiency and creating variable water distribution among individual containers (figure 10.18A). These problems can be precluded by subirrigation systems, which offer a promising alternative for native plant nurseries.

Subirrigation is a relatively new irrigation option. Subirrigation has been used to grow several species of wetland plants, but its applications are being expanded for many types of plants, including forbs (Pinto and others 2008), conifers (Dumroese and others 2006), and hardwood trees (Davis and others 2008). In subirrigation systems, the bottoms of containers are temporarily immersed in water on a periodic basis (for example, for a few minutes once a day). The water then drains away, leaving the growing medium thoroughly wet while the leaves remain dry. Subirrigation thereby bypasses the problem of large leaves intercepting overhead water and precludes other problems inherent in overhead irrigation.

All subirrigation systems rely on capillary action to move water up through the growing medium against gravity. Capillarity is the result of the attraction of water molecules for each other and other surfaces. Once the subirrigation tray is flooded, water will move up through the growing medium in the containers (figure 10.18B), with the extent of this movement depend-

SPRINKLER IRRIGATION SYSTEMS

Advantages

— They are relatively simple and inexpensive to design and install.
— A variety of nozzle patterns and application rates are available.
— Water distribution patterns can be measured with a "cup test."

Disadvantages
— Foliar interception makes overhead watering ineffective for large-leaved crops.
— Irrigation water can be wasted due to inefficient circular patterns.
— An increased risk of foliar diseases is possible from excessive water on leaves.
— For overhead sprinklers, nozzle drip from residual water in lines can harm germinants and young plants.
— For basal sprinklers, irrigation lines must run along the floor, creating obstacles for workers and equipment.

ent on the characteristics of the container and the growing medium, mainly the latter. The smaller the pores between the growing medium particles, the higher water will climb. Once the root systems are saturated (usually a few minutes), the water drains away. Thus, subirrigation is the practice of periodically recharging the moisture in the growing medium by providing water from the bottom.

Several different subirrigation systems have been developed but some, such as capillary beds and mats, will not work with the narrow-bottomed containers often used in native plant nurseries. A couple of others, however, have promise. For example, with ebb-and-flow or ebb-and-flood systems, containers sit on the floor in a shallow structure constructed from pond liner material surrounded by a raised border of wood or masonry.

Subirrigation trays, troughs, and bench liners are filled with water and drained after the growing medium in the containers has been saturated (figure 10.18B).

Either of these subirrigation systems should work for a variety of native plants. Although prefabricated subirrigation systems are available commercially, nurseries on a limited budget may consider designing their own systems using available materials. A trough system can be made out of concrete blocks and pond liner or out of prefabricated drainable plastic ponds.

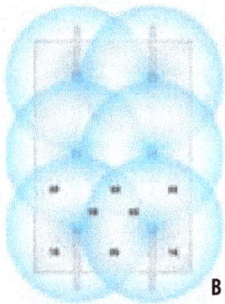

Figure 10.17—(A) Because sprinklers produce a circular irrigation pattern, (B) proper spacing is critical to produce enough overlap. (C) The water pressure of irrigation systems should also be checked annually to make sure that nozzles are operating efficiently (D) before irrigation problems become apparent. Photos by Thomas D. Landis, illustration by Jim Marin.

(Note: some materials, such as galvanized metal, are inappropriate due to zinc toxicity.) There are some important design considerations for subirrigation systems. The holes of the containers must have good contact with the water in order for the water to enter the container. Subirrigation may be less effective during the establishment phase, when medium in the upper portions of containers needs to be moist to promote germination and early growth; therefore, supplemental hand-watering or sprinkler irrigation may be necessary at first. Air root pruning is usually reduced with this system, resulting in a need for hand-pruning of roots and making this system inadvisable for use with very sensitive plants. Of the four irrigation systems mentioned in this chapter, subirrigation may require the most upfront planning and design work. Nevertheless, we think that it has very good potential for native plant nurseries.

Microirrigation

For nurseries that grow plants in 1-gal (4-L) or larger containers, microirrigation can be a very efficient method for water delivery. Microirrigation usually involves poly pipe fitted with microsprayers (sometimes called "spitters" or "spray stakes") (figure 10.19A) or drippers (figure 10.19B) inserted individually into each container, sometimes with the use of small tubing to extend the emitters beyond the poly tube. Microsprayers are often preferred to drippers because they wet more surface area and distribute water more evenly throughout the container. It is also easier to visually verify the operation of a sprayer than a dripper. Filtration is a necessity for microirrigation systems in order to prevent emitters from clogging. Because of the slow infiltration rate of microirrigation systems, each irrigation station will need to run a long time in order to deliver adequate water to plants. Also, if containers are allowed to dry out, hand-watering may be necessary to rewet the growing medium before drip irrigation will work.

Automating Irrigation Systems

Several types of automatic controllers are available, some using time clocks and one using container weight, so that irrigation can be automatically applied. This equipment allows the nursery manager to preprogram periods of irrigation and saves time and labor. The prudent grower, however, will never become com-

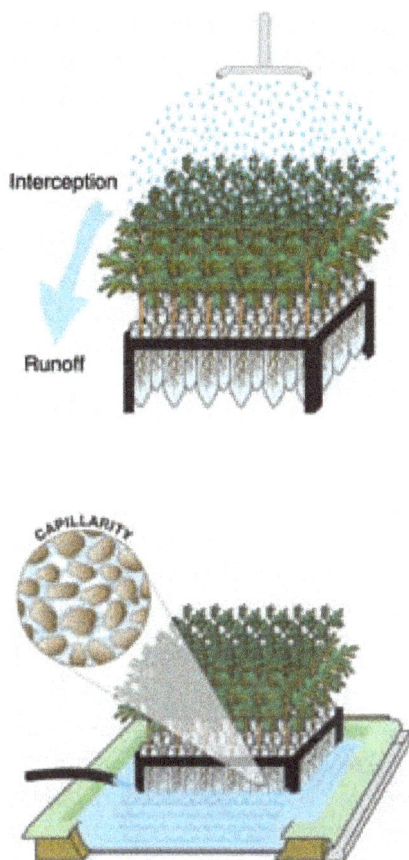

Figure 10.18—*(A) Overhead irrigation is ineffective for broad-leaved plants because so much water is intercepted by the foliage, called "the umbrella effect." (B) Subirrigation works because water is drawn upward into the containers by capillarity.* Illustrations by Jim Marin.

SUBIRRIGATION SYSTEMS

Periodic saturation with water supplied from below the containers followed by drainage of growing medium.

Advantages

—Although commercial products are available, subirrigation systems can be constructed from affordable, local materials.

—Foliage remains dry, reducing the risk of foliar diseases.

—Water use (up to 80 percent less than overhead watering systems) is efficient.

—Application among plants is very uniform.

—Lower fertilizer rates are possible.

—Reduced leaching of mineral nutrients is possible.

—Drainage water can be recycled or reused.

—No soil splashing disrupts or displaces mulch, germinants, or medium.

—Provides the ability to irrigate different size containers and different age plants concurrently.

—Is efficient in terms of time and labor requirements following installation.

Disadvantages

—Overhead or hand watering may be required to ensure sufficient surface moisture until seeds germinate.

—No leaching occurs, so it cannot be used with poor-quality water because salt buildup would occur.

—Less air pruning of roots occurs.

—Risk of spreading waterborne diseases is greater.

—High humidity within plant canopy is possible.

—Almost nothing is known about the response of most native plant crops to subirrigation practices.

pletely reliant on automatic systems and will continue to directly monitor irrigation efficiency and its effect on plant growth on a regular basis.

Monitoring Water in Containers: Determining When To Irrigate

Determining the moisture status of the growing medium in most of the containers used in native plant nurseries is a challenge because it is difficult to observe or sample the medium in small containers. Some containers, such as the "book" type, can be opened up to allow direct observation of the moisture content of the medium. In other cases, it is difficult to ascertain whether plants are getting adequate water saturating throughout the root system. In spite of these difficulties, it is absolutely necessary to regularly monitor the moisture status of container growing

media. The limited volume of moisture reserves in small containers means that critical moisture stresses can develop quickly.

Many types of equipment can be used to test and assess the effectiveness of water application. These tools include tensiometers, electrometric instruments, balances for measuring container weight, commercial moisture meters, or pressure chambers. Some of these tools are described in Landis and others (1989). Currently, no inexpensive yet accurate instrument is available to measure growing media moisture content in containers. Any method must be supported by actual observation

CHARACTERISTICS OF MICROIRRIGATION SYSTEMS

Advantages
—Water is delivered directly to the root zone of plants (not to foliage, where it may cause disease).
—Use of water is very efficient; less than 10 percent of applied water is wasted.
—Delivery is uniform; an even amount of water is applied to each container.
—Infiltration rate is good (due to slow delivery).
—The amount of leachate is also reduced.

Disadvantages
—Designing the system and installing individual emitters for each plant is difficult and time consuming.
—It is not generally efficient to install for plants grown in containers smaller than 1 gallon in size.
—Each irrigation station must run a long time due to slow water delivery.
—Emitters can plug easily (water filtration and frequent irrigation system maintenance is required).
—It is difficult to verify water delivery visually; often, problems are not detected until it is too late.

Figure 10.19—*(A) Spray stakes are effective only for larger containers and work well because you can see them functioning and they have more even distribution. (B) Drip emitters can also be used for larger containers such as these quaking aspen.* Photo by Thomas D. Landis, illustration by Jim Marin.

and the grower's experience; indeed, visual and tactile assessments are the most common method of monitoring irrigation effectiveness. This monitoring sometimes includes formal or informal assessments of container weight. Visual and tactile monitoring and monitoring with container weights are discussed in the following sections.

Visual and Tactile Assessment
Most nurseries successfully monitor the effectiveness of irrigation based on the feel and appearance of the plants and the growing medium (figure 10.20A). The best technique is to observe the relative ease with which water can be squeezed from the medium and attempt to correlate this moisture condition with plant appearance and container weight (figure 10.20B). This process requires a lot of experience and is very subjective. In spite of its obvious limitations, the visual and tactile technique is still widely used and can be very effective when used by a knowledgeable, experienced nursery manager.

Looking at the root systems or the growing media may involve damage to the plants that are examined, especially if they must be pulled from their containers. This practice may be necessary during the learning phase of growing a new crop. With time and experience, however, nondestructive indicators such as the appearance of the plant, the look and feel of the growing medium, and the weight of the containers will be practiced most of the time, and the need for destructive sampling is reduced or eliminated.

Monitoring Irrigation with Container Weights
Developing a container weight scale requires a significant amount of effort and recordkeeping, but container weight is one of the few objective, non-destructive, and repeatable techniques for monitoring irrigation in container nurseries. Container weight is also the best way to determine irrigation needs early in the growing season before plants are large enough to show moisture stress or use in the pressure chamber. The weight of the container decreases between irrigations because the water in the growing medium is lost through evaporation and transpiration, and the crop is irrigated when the container weight reaches some predetermined level. Workers can develop an intuitive sense of this level based on picking up a few randomly spaced trays (figure 10.20). It

Figure 10.20—(A) Joanne Bigcrane monitors the need for irrigation by careful observation of plant condition so that (B) water can be applied before plants are seriously wilted. (C) Feeling the weight of containers can also be an effective way to determine when to irrigate. (D) Monitoring container weights is a standard method to induce moisture stress in conifer crops.. Photos A–C by Tara Luna, D by Thomas D. Landis.

can also be done objectively, weighing containers on a simple household bathroom scale (figure 10.20D). Container irrigation weights will vary significantly between species due to the physiological response of different species to moisture stress.

In general, however, container weight will be used subjectively, based on experience. The person in charge of irrigation will develop a feel for the proper weight of containers or trays based on experience. As the person is learning this technique, he or she will need to verify his or her conclusions by examining the appearance of the plant's root system, growing medium, and leaves. Eventually, the visual and tactile method of directly observing the amount of moisture in the growing medium can be used to estimate available moisture levels, and the wilting point can be established by observing the turgidity of the leaves.

Assessing the Evenness of Irrigation Systems

A simple test called a "cup test" can be carried out periodically to evaluate the evenness of irrigation distribution. Both new and existing irrigation systems should be assessed periodically to see if they are performing properly. Existing irrigation systems need to be checked every few months because nozzles can become plugged or wear down to the point that they are no longer operating properly. The cup test measures the irrigation water caught in a series of cups laid out on a regular grid system throughout the growing

VISUAL AND TACTILE CLUES FOR MONITORING IRRIGATION

—Leaves should look and feel firm, not wilted.
—Potting medium should be moist throughout the plug; moisture should come out when squeezed.
—Containers should feel relatively heavy when lifted.

Figure 10.21—Periodic checks of water distribution can easily be done with a "cup test" in which cups are arranged in a grid pattern. The depth of water in them is measured after a standard watering period. Photo by Kim M. Wilkinson.

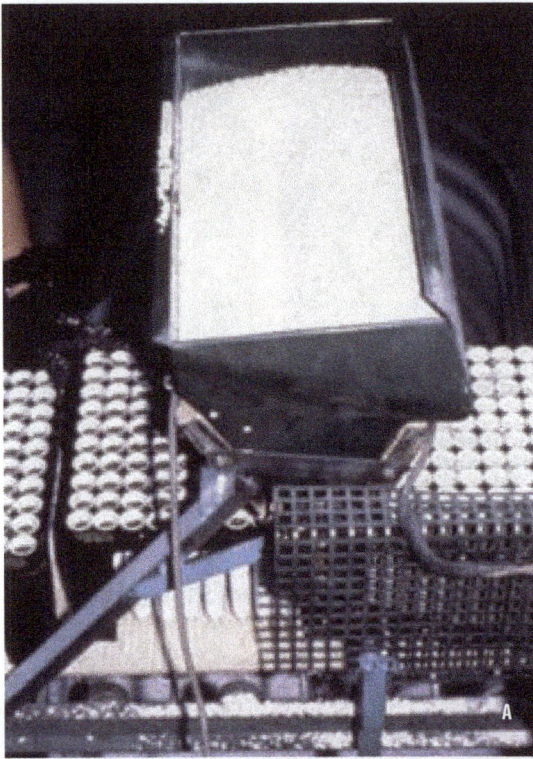

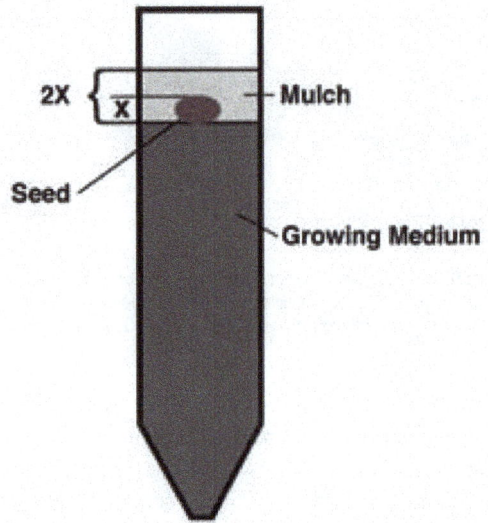

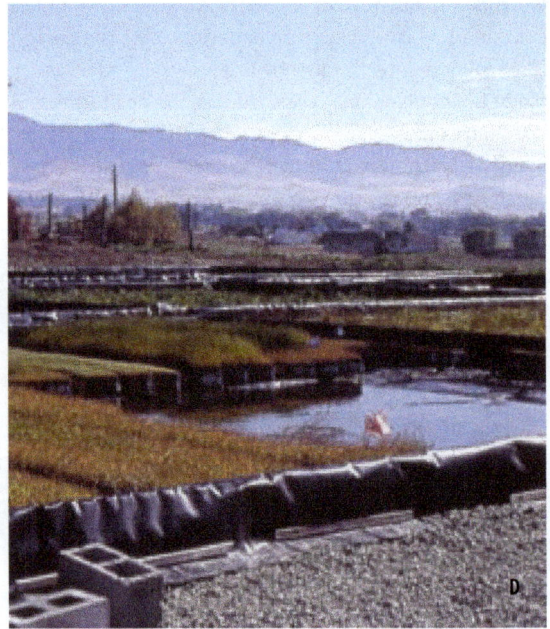

Figure 10.22—*(A and B) Water can be conserved by many horticultural practices such as covering sown containers with a mulch or (C) reducing water loss with shade and straw bale insulation around the perimeter of the growing area. (D) Recycled runoff water can be used to propagate riparian and wetland plants or to irrigate surrounding landscaping or orchards.*
Photos A and D by Thomas D. Landis, C by Tara Luna, illustration by Jim Marin.

area (figure 10.21). Containers for cup tests should have circular openings that have narrow rims; the shape of the container below the opening is not important as long as the cups are stable and 2 to 4 in (5 to 10 cm) deep to hold water without any splashing out.

Distribute empty cups evenly throughout the nursery irrigation station and run the irrigation as usual for a standard time period. Turn off the system and measure the depth of the water in the cups, which should be relatively even. If not, check pressure in the line and also check for clogs or problems with individual nozzles. If the test is done in a hand-watering system, the person applying the water can learn where the application is uneven.

WATER CONSERVATION AND MANAGING NURSERY WASTEWATER

Depending on the efficiency of the irrigation system, nursery runoff and wastewater may be important factors to consider. Overhead sprinkler irrigation is very inefficient. Microirrigation or subirrigation systems are much more efficient but are impractical for some types of containers and plants.

The problem of poor irrigation efficiency involves more than simply wasted water because many container nurseries apply some or all of their fertilizer and pesticides through the irrigation system. Liquid fertilizer is usually applied in excess of the actual amount needed to saturate the growing medium to stimulate the leaching of excess salts. Most pesticides are applied in a water-based carrier through the irrigation system, and some of these chemicals inevitably end up in the wastewater runoff; growing medium drenches are particularly serious in this regard.

Originally, it was thought that the soil filtered out or absorbed fertilizer salts and pesticides, but this belief has recently been refuted. Leaching tests in conifer nurseries have shown that excess fertilizer nutrients and pesticides drain out of containers and may contaminate groundwater. Maximizing the efficiency of irrigation systems and implementing water conservation strategies, such as the use of mulch, is the most effective way to handle the problem. When runoff is created, it can be collected for treatment and/or redirected to landscaping or other crops for absorption by other plants.

Table 10.3—Native plant nurseries should incorporate a variety of horticultural practices to use irrigation water effectively and efficiently

Nursery Practice	Conservation Effect
Mulches (figure 10.22A and B)	Reduce evaporation from the surface of the growing medium; the larger the container the greater the savings
Windbreaks (see figure 1.12)	Reduce water loss and seedling stress due to wind
Remove cull plants and minimize space between containers, including aisles	Reduce wasted water and resultant runoff
Shadecloth and shadehouses (figure 10.22C)	Reduce water use for species that don't require full sunlight
Catch runoff and recycle water (figure 10.22D)	Can save considerable amounts of water, and reduce fertilizer use as well

Good irrigation design and application minimizes the amount of water used while providing for the needs of the plants. Several horticultural practices can help conserve water and reduce water use and runoff in the nursery (table 10.3).

The direct recycling of used nursery water is generally not done on a small scale because of the expense of water treatment and the risks of reintroducing excess salts or pests. However, these kinds of capture-and-recycle systems for nursery runoff water may be economically viable in very water-limited areas. Less high-tech options for water reuse can make use of an impermeable nursery floor (pond liner, for example) to collect water runoff from the nursery. This water can be stored in a tank or run directly to other crops. Crops that are more tolerant of salts, such as rushes, may benefit from using runoff water, and these crops will even clean and filter the water. Nurseries growing aquatic or semi-aquatic plants may be able to direct runoff to these plants and thereby increase the water-use efficiency of the nursery operation. Crops in the field, such as seed orchards, wetland crops, surrounding landscaping, or tree crops, can all benefit from receiving nursery runoff water for irrigation. Generally, in these cases, an additional sandbed or filter is used to strain the particulates out of the water before running it through an irrigation system to these crops. If the crops are located downhill from the nursery itself, the system can be gravity fed. Otherwise, a pump will be necessary to apply the water to the crops.

SUMMARY

Because of the overriding importance of water to plant growth, managing irrigation is the most critical cultural operation in native plant nurseries. Water must be managed differently in container nurseries growing systems (in comparison with agricultural crops, gardens, or bareroot nurseries) because of the restrictive effects of the containers. Growing media composed of materials such as peat moss and vermiculite have different properties than soil, including a higher water-holding capacity. The container also has an effect on the water properties of the growing medium because water does not drain completely out of the container, which results in a layer of saturated media at the bottom. The depth of this layer is a function of container height and the properties of the growing medium.

The quantity and quality of the irrigation water is probably the most important consideration in site selection for a nursery. Sufficient quantities of water must be available throughout the year to supply all the various uses at the nursery. The quality of the nursery irrigation water is primarily a function of the concentration and composition of dissolved salts, although the presence of pathogenic fungi, weed seeds, algae, and pesticides must also be considered. Because water treatment is impractical and costly in most instances, irrigation water sources should be thoroughly tested during nursery site selection. Most plants are very sensitive to soluble salts, so water should be tested at all stages of the irrigation process at regular intervals during the growing season.

In general, four types of irrigation systems are appropriate for native plant nurseries: hand-watering, overhead sprinklers, subirrigation, and microirrigation. Optimal irrigation system design usually creates several irrigation zones to meet the unique needs of the diverse species of plants and to cater to their changing needs as they pass through each phase in their growth and development. Each type of irrigation system has advantages and disadvantages as well as important design considerations. For new native plant nurseries, hand-watering may be the most practical strategy during the start-up phase until the plant's needs are thoroughly understood. Every irrigation system must be tested periodically to ensure that it is working properly.

Determining both when and how much to irrigate is one of the most important day-to-day decisions of the nursery manager. Because of the physical limitations of small containers used in nurseries, there is currently no way to directly monitor the water potential of the growing medium within the container. Experienced growers develop an intuitive skill for determining when irrigation is required, using the appearance and feel of the growing medium and the relative weight of the container. Because of the restrictive drainage characteristics of containers, growers must apply enough water during each irrigation event to completely saturate the entire volume of growing medium and flush excess salts out the bottom of the container. The amount of water supplied at each irrigation period is a function of the growth stage of the plants and the environmental conditions. In addition to promoting rapid germination and growth, water can be used as a cultural tool to help harden the plants and induce dormancy. In cold climates, irrigation can be used for frost protection of plants in open growing compounds.

Because of the excess amounts of irrigation required and the poor efficiency of some irrigation systems, the disposal of wastewater is an important consideration in nursery management. Injected fertilizer nutrients, such as nitrate nitrogen and phosphorus, and pesticides applied through the irrigation system may affect groundwater quality and could become a serious problem. In addition to the importance of collecting and managing runoff, practices to conserve water, such as mulching and efficient nursery design, are important for ecologically sound nursery management.

LITERATURE CITED

Biernbaum, J. 1995. How to hand water. Greenhouse Grower 13(14): 39, 24, 44.

Davis, A.S.; Jacobs, D.F.; Overton, R.P.; Dumroese, R.K. 2008. Influence of irrigation method and container type on growth of *Quercus rubra* seedlings and media electrical conductivity. Native Plants Journal 9(1):4–13.

Dumroese, R.K.; Pinto, J.R.; Jacobs, D.F.; Davis, A.S.; Horiuchi, B. 2006. Subirrigation reduces water use, nitrogen loss, and moss growth in a container nursery. Native Plants Journal 7(3):253–261.

Handreck, K.A.; Black, N.D. 1984. Growing media for ornamental plants and turf. Kensington, Australia: New South Wales University Press. 401 p.

Jones, J.B., Jr. 1983. A guide for the hydroponic and soilless culture grower. Portland, OR: Timber Press. 124 p.

Landis, T.D.; Tinus, R.W.; McDonald, S.E.; Barnett, J.P. 1989. The container tree nursery manual: volume 4, seedling nutrition and irrigation. Agriculture Handbook 674. Washington, DC: U.S. Department of Agriculture, Forest Service. 119 p.

Pinto, J.R.; Chandler, R.; Dumroese, R.K. 2008. Growth, nitrogen use efficiency, and leachate comparison of subirrigated and overhead irrigated pale purple coneflower seedlings. HortScience 42:897–901.

APPENDIX 10.A. PLANTS MENTIONED IN THIS CHAPTER

cottonwood, *Populus* species
quaking aspen, *Populus tremuloides*
redosier dogwood, *Cornus sericea*
rushes, *Juncus* species
willow, *Salix* species

Fertilization

Douglass F. Jacobs and Thomas D. Landis

11

Fertilization is one of the most critical components of producing high-quality nursery stock. Seedlings rapidly deplete mineral nutrients stored within seeds, and cuttings have limited nutrient reserves. Therefore, to achieve desired growth rates, nursery plants must rely on root uptake of nutrients from the growing medium. Plants require adequate quantities of mineral nutrients in the proper balance for basic physiological processes, such as photosynthesis, and to promote rapid growth and development. Without a good supply of mineral nutrients, growth is slowed and plant vigor reduced. Proper fertilization can promote growth rates three to five times greater than normal.

In this chapter, the importance of fertilization to plant growth development is briefly described and typical fertilization practices for producing native plants in small tribal nurseries are detailed.

BASIC PRINCIPLES OF PLANT NUTRITION

A common misconception is that fertilizer is "plant food" (figure 11.1A), but the basic nutrition of plants is very different from that of animals. Using the green chlorophyll in their leaves, plants make their own food, called "carbohydrates," from sunlight, water, and carbon dioxide in a process called "photosynthesis" (figure 11.1B). These carbohydrates provide energy to the plant, and when combined with mineral nutrients absorbed from the soil or growing medium, carbohydrates are used to synthesize proteins and other compounds necessary for basic metabolism and growth.

Adding controlled-release fertilizer to kinnikinnick by Tara Luna.

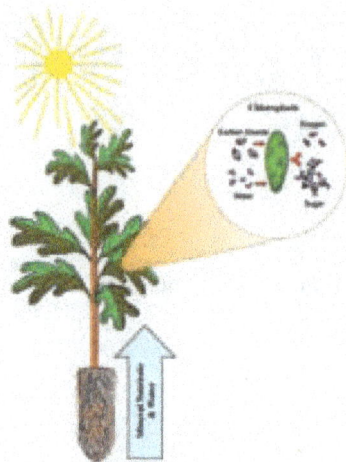

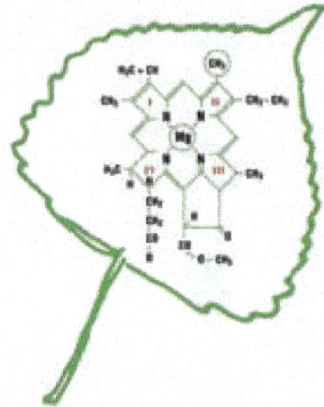

Figure 11.1—(A) Although some fertilizers are advertised as "plant food," (B) plants create their own food through the process of photosynthesis in their green leaves. Photo by Thomas D. Landis., illustration by Jim Marin.

Figure 11.2—Mineral nutrients such as nitrogen and magnesium are important components of chlorophyll molecules, which give leaves their green color and are essential for photosynthesis. Illustration by Jim Marin.

Although many different factors influence plant growth, the growth rate and quality of nursery stock is largely dependent on mineral nutrient availability. When nutrients are supplied in proper amounts, in the proper ratio, and at the proper time, nursery plants can achieve growth rates many times faster than in nature. High-quality nursery stock can even be fortified with surplus nutrients that can accelerate growth after outplanting.

Thirteen mineral nutrients are considered essential to plant growth and development and are divided into macronutrients and micronutrients based on the amounts found in plant tissue (table 11.1). Mineral nutrients can have a structural function. For example, nitrogen is found in all proteins, and nitrogen and magnesium are structural components of chlorophyll molecules needed for photosynthesis (figure 11.2). Having knowledge of these functions is practical because a deficiency of either nutrient causes plants to be chlorotic (that is, yellowish in color). Other mineral nutrients, have no structural role, but potassium, for example, is critically important in the chemical reaction that causes stomata in leaves to open and close.

Nitrogen is almost always limiting to plant growth in nature, which is the reason why nitrogen fertilizer is applied frequently in nurseries. Nitrogen fertilization is one of the main reasons for the rapid growth and short production schedules of nurseries, and a nitrogen deficiency often shows up as stunted growth.

Many container nurseries that grow native plants use artificial growing media such as peat moss and vermiculite. Because media are essentially infertile, nurseries either incorporate a starter dose of fertilizer or start liquid fertigation (irrigation water containing liquid fertilizer) soon after germination.

An important concept to understand in regard to fertilization is Liebig's Law of the Minimum. This law states that plant growth is controlled by the mineral nutrient in shortest supply, even when sufficient quantities of other nutrients exist. See Chapter 4, *Propagation Environments*, for more discussion about limiting factors. Thus, a single nutrient element may be the only factor limiting to plant growth even if all other elements are supplied in sufficient quantity. A good way to visualize the concept of limiting factors is a wooden bucket with staves of different lengths. If water is poured into the bucket, it can be filled only to the height of the shortest stave—the limiting factor. As mentioned previously, nitrogen is almost always limiting in natural soils (figure 11.3).

Just as important as the absolute quantities of nutrients in the growing media is the balance of one nutrient to another. The proper balance of nutrients to one another seems to be relatively consistent among plant species. A common reference is Ingestad's Ratios, which suggests a ratio of 100 parts nitrogen to 50 phosphorus, to 15 potassium, to 5 magnesium, to 5 sulfur.

Table 11.1—The 13 essential plant nutrients (divided into macronutrients and micronutrients), common percentages of these nutrients within plant tissues, and examples of physiological functions necessary to promote healthy plant development

Element	Percentage of Plant Tissue (ovendry weight)	Examples of Structural or Physiological Functions
Macronutrients		
Nitrogen (N)	1.5	Amino acid and protein formation
Phosphorus (P)	0.2	Energy transfer
Potassium (K)	1.0	Osmotic adjustment
Calcium (Ca)	0.5	Formation of cell walls
Magnesium (Mg)	0.2	Enzyme activation, constituent of chlorophyll
Sulfur (S)	0.1	Amino acid formation, protein synthesis
Micronutrients		
Iron (Fe)	0.01	Component of chloroplasts, RNA synthesis
Manganese (Mn)	0.005	Enzyme activation
Zinc (Zn)	0.002	Enzyme activation, component of chloroplasts
Copper (Cu)	0.0006	Component of chloroplasts, protein synthesis
Boron (B)	0.002	Transport of assimilates and cell growth
Chlorine (Cl)	0.01	Maintenance of cell turgor
Molybdenum (Mo)	0.00001	Component of enzymes

On a practical basis, most native plant nurseries use complete fertilizers that contain a balance of most mineral nutrients. Some nutrients, notably calcium and magnesium, are very insoluble in water and can even cause solubility problems in concentrated fertilizer solutions. So, if they are not present naturally in the irrigation water, they must be added to the growing medium as dolomite or in a separate fertilizer solution.

ACQUISITION OF MINERAL NUTRIENTS

Plants produced in container nurseries may acquire nutrients from several different sources, including the growing medium, irrigation water, beneficial microorganisms, and fertilizers. The inherent infertility of most commercial growing media was mentioned earlier and is discussed in detail in Chapter 5, *Growing Media*. Levels of mineral nutrients in peat-vermiculite media are generally very low, but native soils, including composts, may contain significantly higher nutrient concentrations than do commercial growing media. When using a good soil or compost-based medium, the substrate may contain enough nutrients for sufficient plant growth. These mixes, however, often tend to be nutrient deficient in some way and deficiencies of any

Figure 11.3— *The concept of limiting factors can be illustrated by a wooden bucket that can be filled only to the shortest stave.* Illustrated by Steve Morrison.

single nutrient can significantly limit plant growth. Thus, if homemade growing media will be used, a soil test will reveal which nutrients may be lacking (see details on testing in a following section).

Another potential source of mineral nutrients in container nurseries is irrigation water. Usually only a few mineral macronutrients (sulfur, calcium, magne-

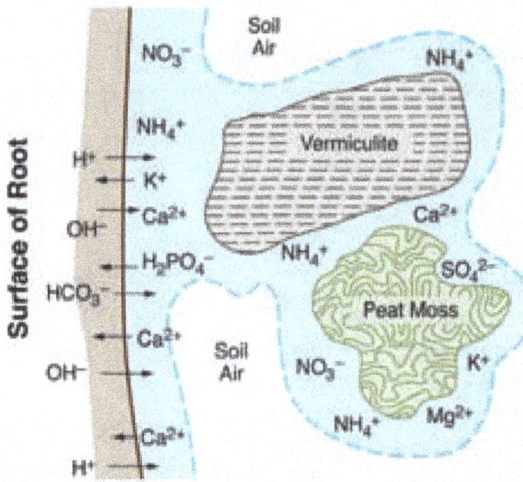

Figure 11.4—*Nutrients are extracted by plant roots through an exchange process with soil, compost, or artificial growing media.* Illustration by Jim Marin.

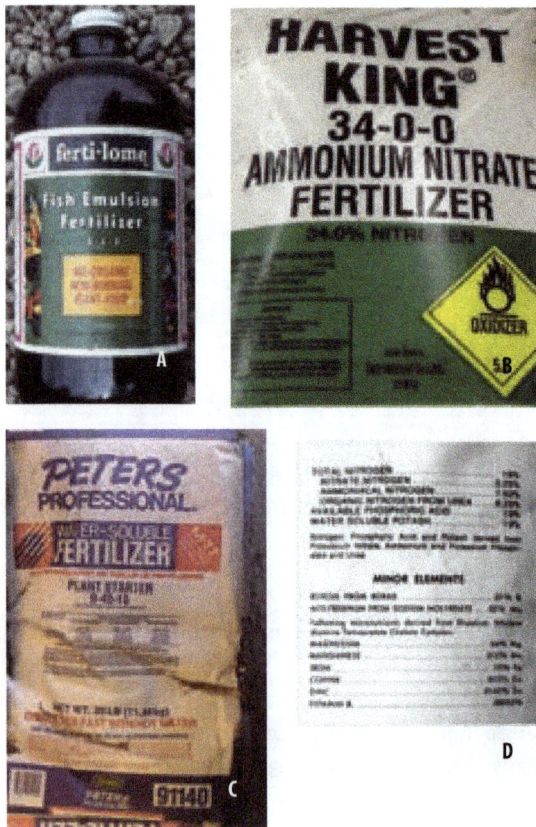

Figure 11.5—*Several different types of fertilizer products, including (A) organic fish emulsion, (B) inorganic granular ammonium nitrate, (C) and an inorganic water-soluble fertilizer. (D) A fertilizer label must show required description of fertilizer nutrient concentrations.* Photos by Thomas D. Landis.

sium) are found in nursery water supplies. Beneficial microorganisms may provide an important source of nitrogen for some species, such as legumes, as detailed in Chapter 14, *Beneficial Microorganisms*. To achieve the desired high growth rates, however, fertilizers are the most common source of mineral nutrients in native plant nurseries. Although fertilizers are powerful horticultural tools, nutrient interactions and the possibility of salt injury from excessive fertilization often occur. In the remainder of this chapter, some of the important considerations when applying fertilizers are discussed.

Mineral nutrients are absorbed by root hairs as two types of ions: cations and anions (figure 11.4). Cations have an electrically positive charge, while anions are negatively charged. Particles of soil, compost, and artificial growing medium are also charged, so nutrient ions can attach to organic matter or clay particles of an opposite charge. Ions may also be chemically bound to minerals, particularly under conditions of high or low media pH. Typically, roots exchange a cation (often H^+) or an anion (for instance, HCO_3^-) for a nutrient ion from the soil or growing media; see Chapter 5, *Growing Media*, for more information on this topic.

Soluble fertilizers are chemical salts that dissolve in water into mineral nutrients. Plant roots are very sensitive to high salinity, so growers must be careful not to apply too much fertilizer. Because all salts are electrically charged, growers may easily monitor levels of fertilizer nutrients in irrigation water or in the growing medium by measuring electrical conductivity (EC). This process is discussed in detail in Chapter 10, *Water Quality and Irrigation,* and in the monitoring and testing section later in this chapter.

TYPES OF COMMERCIAL FERTILIZERS

Many different types of fertilizers are available for use in native plant nurseries (figures 11.5A–C) and vary according to their source materials, nutrient quantities, and mechanisms of nutrient release. All commercial fertilizers are required by law to show the ratio of nitrogen to phosphorus to potassium (actually to the oxides of phosphorus and potassium; $N:P_2O_5:K_2O$) on the package (figure 11.5B). In addition, most show a complete nutrient analysis on the label (figure 11.5D).

Some fertilizers contain only one mineral nutrient whereas others contain several. Examples of single-nutrient fertilizers include ammonium nitrate (34-0-0)

Figure 11.6—*Many tribal nurseries use organic fertilizers to reduce environmental impacts.* Photo by Thomas D. Landis.

(figure 11.5B), urea (46-0-0), and concentrated superphosphate (0-45-0). Multiple-nutrient fertilizers may be blended or reacted to provide two or more essential nutrients. An example of a blended fertilizer is a 12-10-8 + 4 percent magnesium + 8 percent sulfur + 2 percent calcium product, which was formed by adding triple superphosphate (0-46-0), potassium magnesium sulfate (0-0-22 plus Mg and S), and ammonium nitrate (34-0-0). An example of a reacted multiple nutrient fertilizer is potassium nitrate (13-0-44).

Organic Fertilizers

Historically, all fertilizers were organic and applied manually. Organic fertilizers are often not well balanced and the release of nutrients can be unpredictable—either too fast or too slow. Some tribes, however, are using organic materials with great success in native plant nurseries (figure 11.6), showing that experience may lead to the successful identification of organic fertilizer options. Examples of organic fertilizers include animal manure, sewage sludge, compost, fish emulsion (figure 11.5A), and other animal wastes. Because plants take up all their mineral nutrients as ions, it does not matter whether the ions came from an organic or an inorganic

(synthetic) source. Many people are concerned about the higher energy input to create inorganic fertilizers and therefore prefer to use organic sources.

An obvious advantage of inorganic fertilizers is that they are widely available and relatively inexpensive. A disadvantage is that many are bulky, heavy, or unpleasant to handle. In many cases, however, organics such as manure may be free if the nursery provides labor and transport. Most organic fertilizers have relatively low nutrient analyses (table 11.2), although actual concentrations may vary considerably depending on the type of material and stage of decomposition. On the one hand, an advantage of the low-nutrient concentrations of organic fertilizers, such as compost, is that it is more difficult to apply excessive amounts of fertilizer, to "overfertilize," but other organic fertilizers, such as fresh chicken manure, could damage plants. On the other hand, the low levels of nutrients provided by organic materials may be insufficient to achieve the rapid plant growth expected of nursery stock and to have plants reach an acceptable size for outplanting within a desired time frame. Anyone considering the use of bulk organic fertilizers should have them tested first to establish the need for composting, determine

Table 11.2—Mineral nutrients supplied by a variety of organic materials (from Jaenicke 1999)

Source	Nitrogen (% N)	Phosphorus (% P$_2$O$_5$)	Potassium (% K$_2$O)
Manures			
Cow	0.35	0.2	0.1 – 0.5
Goat/sheep	0.5 – 0.8	0.2 – 0.6	0.3 – 0.7
Pig	0.55	0.4 – 0.75	0.1 – 0.5
Chicken	1.7	1.6	0.6 – 1.0
Horse	0.3 – 0.6	0.3	0.5
Compost	0.2 – 3.5	0.2 – 1.0	0.2 – 2.0
Fish emulsion	5.0	2.0	2.0
Kelp	1.0	0.2	2.0

Table 11.3—Advantages and disadvantages of controlled-release fertilizers compared with immediately available fertilizers

Advantages

— Better suited to the longer term nutrient requirements of perennial species

— Extended nutrient availability with single applications

— Potential improvement in efficiency of fertilizer use, leading to decreased leaching of nitrogen and other nutrients from the nursery

— Potential reduction in salt damage to root systems

—More conducive to beneficial microorganisms

Disadvantages

— Cost is higher (this may be alleviated somewhat by labor and machine application savings)

— Availability may be limited in some locales

— Nutrient release rates can be unpredictable in nurseries

— More difficult to adjust nutrient inputs to match needs based on growing cycle

proper application rates, and identify potential nutrient toxicities or deficiencies.

Inorganic Fertilizers

With the advent of inorganic (synthetic) fertilizers, the use of organic fertilizers has declined over the years. Nutrients in synthetic materials are derived either from mine extraction or by chemical reaction to capture nitrogen from the atmosphere. These products are readily available at most garden supply shops and through horticultural dealers.

Although these fertilizers work consistently well, it should be noted that some organic options can provide significant nutrition to plants, as described previously. Sometimes, growers of native plants tend to favor organic over inorganic fertilizers due to simple preference. Experimentation and experience will help to determine the best fertilizer type for your nursery.

Immediately Available versus Controlled-Release Inorganic Fertilizers

Two general categories of inorganic fertilizer materials are those with immediately available forms of nutrients and those that release nutrients slowly over time ("slow-release" or "controlled-release" fertilizers). These two forms have several notable advantages and disadvantages (table 11.3).

Fertilizers immediately available to plants include water-soluble fertilizers commonly used in container nurseries (figure 11.5C). Other immediately available fertilizers, such as urea, are seldom used in native plant nurseries. Soluble fertilizers are typically injected into the irrigation system, a process known as fertiga-

tion (see the discussion in the following section). Their popularity stems from the fact that the application rates can be easily calculated, distribution is as uniform as the irrigation system, the nutrients are readily available, and, if properly formulated and applied, the chance of fertilizer burn is very low.

All controlled-release fertilizers are synthetic, including plastic-coated fertilizers and those manufactured from nitrogen reactions. Coated fertilizers consist of a water-soluble fertilizer core covered with a less-insoluble barrier, which affects the nutrient release rate. Coatings must be thin and free of imperfections, which is challenging because fertilizer granules are relatively porous, rough, and irregularly surfaced.

The most common coatings for controlled-release fertilizers are sulfur or a polymer material. With sulfur-coated products, nutrients are released by water penetration through micropores or incomplete sulfur coverage. These materials are typically less expensive than polymer-coated fertilizers. Nutrient release rates, however, are less consistent than those with polymer-coated fertilizers. For example, with sulfur-coated urea, a rapid initial release of nutrients is followed by a rapidly decreasing release rate.

Polymer-coated fertilizers (figure 11.7) are considered the "state of the art" controlled-release fertilizer for horticultural plant production and are widely used in native plant nurseries. The round, polymer-coated "prills" have a more uniform nutrient release than sulfur-coated products. In addition, the prills can be formulated to contain the proper balance of both macronutrients and micronutrients, whereas sulfur-coated products generally release only nitrogen. Note that calcium is the only nutrient missing from a popular controlled-release fertilizer (table 11.4).

Nutrient release from polymer-coated fertilizer is a multistep process. During the first irrigation, water vapor is absorbed through microscopic pores in the coating. This process creates an osmotic pressure gradient within the prill, causing the flexible polymer coating to expand. This expansion enlarges the tiny pores and the mineral nutrients are released into the soil or growing medium (figure 11.8). Besides water, temperature is the primary factor affecting the speed of this process, so nutrient release generally increases with rising temperature. Release rates of polymer-coated products are adjusted by the manufacturer by altering the thickness and nature of the polymer material, and longevities vary from about 3 to 16 months. Popular brands of polymer-coated fertilizers include Osmocote®, Nutricote®, and Polyon®.

Another category of controlled-release fertilizers are the nitrogen-reaction products, such as ureaform and IBDU Micro Grade Fertilizer. These fertilizers are created through a chemical reaction of water-soluble nitrogen compounds, which results in a more complex molecular structure with very limited water solubility. The rate of nutrient release of ureaform is controlled by many factors, including soil temperature, moisture, pH, and aeration, while IBDU becomes available primarily through hydrolysis. These materials are rarely used in native plant container production but are more commonly applied at outplanting.

FERTILIZER APPLICATION

Fertigation

Most forest and conservation nurseries apply soluble fertilizers through their irrigation systems, a process known as fertigation. The fertigation method varies depending on the type of irrigation and the size and sophistication of the nursery. The simplest method is to combine soluble fertilizers (figure 11.9A) in a watering container or use a hose injector (figure 11.9B), and water plants by hand. This method can be tedious and time consuming, however, when fertigat-

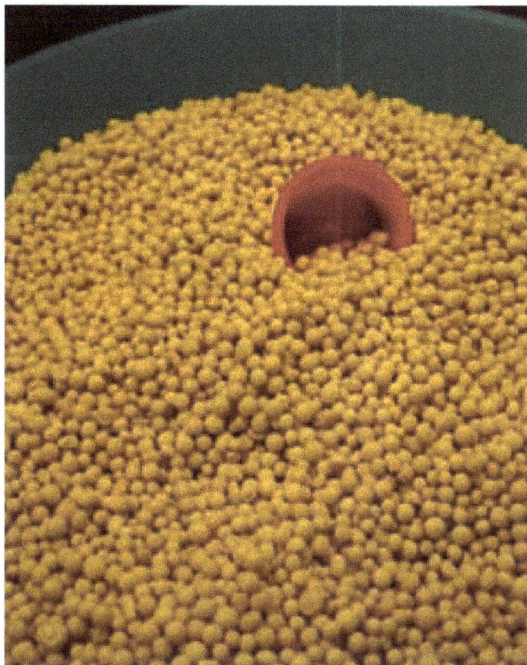

Figure 11.7—Polymer-coated fertilizers occur as round "prills" in which nutrients are encapsulated by a plastic coating that controls the rate of nutrient release. Photo by Douglass F. Jacobs.

Table 11.4—Nutrient analysis of 15-9-12 Osmocote® Plus controlled release fertilizer (from The Scotts Company 2006)

Mineral Nutrient	Percentage
Macronutrients	
Nitrogen (7% ammonium; 8% nitrate)	15.0
Phosphorus (P_2O_5)	9.0
Potassium (K_2O)	12.0
Calcium	0.0
Magnesium	1.0
Sulfur	2.3
Micronutrients	
Iron (chelated 0.23%)	0.45
Manganese	0.06
Zinc	0.05
Copper	0.05
Boron	0.02
Molybdenum	0.02

CRF Release Rate Varies With:
1) Coating Type & Thickness
2) Water
3) Increasing Temperature

Figure 11.8—*Nutrient release from polymer-coated fertilizers occurs after water is absorbed through the prill membrane, creating an osmotic pressure gradient that expands the pores within the coating and allows fertilizer nutrients to pass through to the growing medium.* Illustration by Jim Marin.

ing a large quantity of plants. Nonetheless, this method may be best when growing a variety of species with different fertilizer needs in small areas.

Fertilizer injectors are used when growing large numbers of plants with the same fertilizer requirements. The simplest injectors are called siphon mixers and the HOZ Hozon™ and EZ-FLO® are common brands. Siphon injectors are attached to the water faucet and have a piece of rubber tubing that is inserted into a concentrated fertilizer solution (figure 11.9C). When an irrigation hose is attached to the other end and the water is turned on, the flow through the hose causes suction that pulls the fertilizer stock solution up and mixes it with the water at a fixed ratio. For example, the Hozon™ injects 1 part of soluble fertilizer to 16 parts of water, which is a 1:16 injection ratio. Note that this injector requires a water pressure of at least 30 pounds per square inch (psi) to work properly whereas the EZ-FLO® functions at water pressures as low as 5 psi.

More complicated but more accurate fertilizer injectors cost from around $300 to more than $3,000. For example, the Dosatron® is a water pump type of injector that installs directly into the irrigation line and pumps the fertilizer solution into the irrigation pipe at a range of injection ratios (figure 11.9D).

Any injector must be calibrated after it is installed to verify the fertilizer injection ratio and then must be

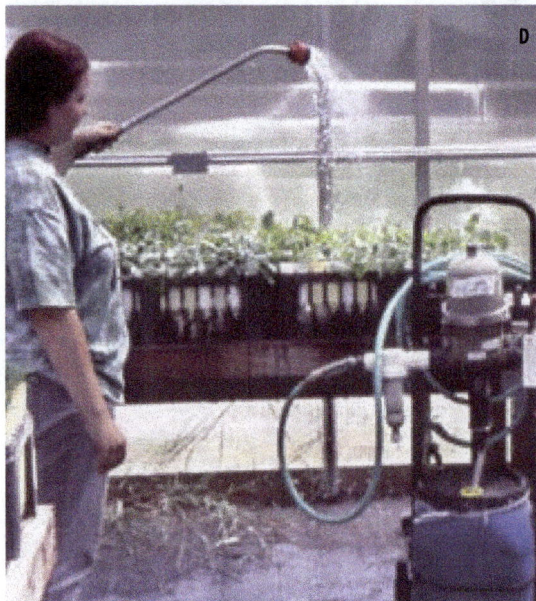

Figure 11.9—*(A) Soluble fertilizers can be mixed with water and applied to the crop, a process called "fertigation." (B) The liquid fertilizer solution can be applied in a watering can or through a hose-end sprayer. (C) A siphon injector sucks up concentrated soluble fertilizer solutions and mixes it with irrigation water that can be applied with a hose. (D) The Dosatron® injector allows more precise control of injection ratios.* Photos A and B by Thomas D. Landis, D by Tara Luna, illustration from Dumroese and others (1998).

checked monthly to make sure that it is still working properly. Among the most technologically advanced fertigation systems is the automated hydraulic boom, which provides very consistent and uniform coverage of water and fertilizer to the crop. The relatively high price of irrigation booms makes them cost prohibitive for many small native plant nurseries.

Special types of water-soluble fertilizers are sometimes applied directly to the crop to stimulate uptake through the foliage ("foliar feeding"). Although this method seems to be a good way to fertilize, remember that all leaves are covered with a water-repelling cuticle, so foliar feeding is very inefficient. More nutrient uptake actually occurs through the roots as a result of the fertilizer solution washing down into the soil or growing medium rather than through the foliage itself. Special care must be taken to prevent salt damage to foliage, so we do not recommend foliar feeding for smaller nurseries.

Applying Granular or Controlled-Release Fertilizers

Applying dry fertilizers directly to the tops of the containers ("topdressing"), should never be attempted with granular dry fertilizers because of the possibility of "burning" plants with succulent tissue. Controlled-release fertilizers can be topdressed, however, if care is taken to make sure that each container or cell receives an equal number of prills (figure 11.10A). A special drop-type application wand can be used to topdress larger (> 1-gal [4-L]) containers because a measured dose of fertilizer can be applied to the base of each plant. This method avoids the potential of fertilizer granules being lodged in foliage and burning it as soon as the crop is watered.

A better option for applying dry granular fertilizers or controlled-release prills is to incorporate them into the growing medium (figure 11.10B). The only time dry granular fertilizers are used is when a "starter dose" is incorporated into the growing medium by the manufacturer. The incorporation of controlled-release fertilizers, however, is the method of choice. If nurseries decide to mix their own growing medium, controlled-release fertilizers can be incorporated, but special care must be taken to ensure even distribution (figure 11.10C) and prevent damage to the prill coating. If the coating is fractured, then the soluble fertilizer releases immediately, which may cause severe salt injury. We recommend that nurs-

> **! ●** **Caution:** Every fertilizer injector must be installed with a backflow preventer to eliminate the possibility that soluble fertilizer could be sucked back into the water line and contaminate drinking water.

eries purchase their growing media with controlled-release fertilizers already incorporated with more accurate commercial mixing equipment.

Calculating Fertilizer Application Rates

Fertilizer application rates depend on the growing environment and other cultural factors such as container volume, type of growing media, and irrigation frequency. In particular, the size of the growth container has a profound effect on the best application rate and timing. Very small containers require lower rates, but more frequent application, whereas larger containers can tolerate higher application rates applied less frequently.

Soluble Fertilizers

For most fertilizer products, manufacturers provide general recommended application rates for container nursery plants on package labels. Few recommendations can be found, however, for most native plants. Experimentation, consultation with other growers, and propagation protocols (see Chapter 3, *Crop Planning and Developing Propagation Protocols*) will help develop better fertilizer application rates for native plants (table 11.5). Note that species with wide geographic distribution, such as Douglas-fir, have different nutrient requirements based on their source. Seedlings of coastal sources can be produced with 100 parts per million (ppm) nitrogen, whereas high-elevation sources and those from the Intermountain Region require much more nitrogen (200 ppm) to grow plants of the same size (Thompson 1995).

Detailed descriptions of liquid fertilizer calculations for commercial conifer crops and adjustments for fertilizer injector ratios can be found in Landis and others (1989). These calculations can be confusing at first but get easier with practice and growing experience.

Controlled-Release Fertilizers

Much less has been published about how much controlled-release fertilizer to apply to native plants in containers. Growers can use the general recommenda-

CRF Prills

Figure 11.10—(A) Controlled-release fertilizers can be applied directly to containers ("top-dressing") if care is taken to achieve uniform application. (B) Incorporating controlled-release fertilizers when the growing medium is mixed (C) is a better to achieve even distribution of prills in small containers. Photos A and C by Tara Luna, B by Thomas D. Landis.

tions if they classify their crops by relative nutrient uses: low, medium, or high (table 11.6). Of course, these applications rates should be used conservatively until their effect on individual plant growth and performance can be evaluated.

DETERMINING WHEN TO FERTILIZE

How do you determine when to fertilize? Because artificial growing media such as peat-vermiculite media are infertile, fertilization should begin as soon as the seedlings or cuttings become established. Some brands of growing media contain a starter dose of fertilizer, so fertilization can be delayed. Native soil mixes that have been amended with compost or other organic fertilizers may not need fertilization right away.

In nurseries, plant growth rates can be controlled by fertilization rates, especially nitrogen rates. As plants take up more nutrients, growth rate increases rapidly until it reaches the critical point (A in figure 11.11A). After this point, adding more fertilizer does not increase plant growth but can be used to "load" nursery stock with extra nutrients for use after outplanting. Overfertilization can cause plant growth to decrease (B in figure 11.11A) and eventually results in toxicity.

Much depends, however, on the species of plant. Some natives require very little fertilizer but others must be "pushed" with nitrogen to achieve good growth rates and reach target specifications. Small-seeded species (for example, quaking aspen) expend their stored nutrients soon after germination whereas those with large seeds (for example, oak) contain greater nutrient reserves and do not need to be fertilized right away. Some native plants require minimal fertilization whereas others need relatively large fertilizer inputs to sustain rapid growth. Experience in growing a particular species is the best course of action to develop species-specific fertilizer prescriptions.

Native plant growers should never wait for their crops to show deficiency symptoms before fertilizing. Plant growth rate will slow down first and, even after fertilization, it can take weeks before growth will resume. Evaluating symptoms of nutrient deficiencies based on foliar characteristics can be challenging even for experts. Many different nutrient deficiencies may result in similar characteristic symptoms and considerable variation in these symptoms may occur among species. In addition, typical foliar symptoms such as chlorosis may

Table 11.5—Examples of liquid fertilizer application rates for a variety of native plants, applied once or twice per week during the rapid growth phase

Low Rate: 25–50 ppm N	buffaloberry[a]
	ceanothus[a]
	dogbane
	fourwing saltbush
	hawthorn
	sagebrush
Medium Rate: 50–100 ppm N	chokecherry
	cottonwood
	cow parsnip
	elderberry
	redoiser dogwood
	serviceberry
	wild hollyhock
	willow
High Rate: 100–200 ppm N	blue spruce
	Douglas-fir [b]
	limber pine
	western white pine
	whitebark pine

ppm = parts per million
[a] = Plants that fix nitrogen (see Chapter 14, *Beneficial Microorganisms*).
[b] = Coastal sources require low N levels but high elevation and interior sources require very high N levels.

Table 11.6—Manufacturer's recommendations for applying 5-9-12 Osmocote® Plus controlled-release fertilizer (from The Scotts Company 2006)

Nutrient Release Rate (months)	Incorporation: Ounces per Cubic Foot of Growing Medium[a]		
	Low	Medium	High
3–4	1.8	3.6	7.1
5–6	2.4	4.7	7.1
8–9	4.1	5.9	8.3
12–14	2.0 – 4.0	5.0 – 7.0	8.0 – 12.0
14–16	8.0	12.0	16.0
	Topdressing: Ounces per 5-inch Diameter Container[b]		
3–4	0.07	0.10	0.25
5–6	0.07	0.18	0.25
8–9	0.14	0.21	0.28
12–14	0.07–0.14	0.18–0.25	0.25–0.39
14–16	0.25	0.39	0.53

[a] Multiply ounces by 1,000 to obtain grams per cubic meter or multiply ounces by 1 to obtain grams per cubic liter.
[b] Multiply ounces by 28.35 to obtain grams. Multiply inches by 2.54 to obtain centimeters.

sometimes be a result of an environmental response unrelated to nutrient stress, such as heat damage or root disease. The position of the symptomatic foliage can also be diagnostic. For instance, nitrogen is very mobile within the plant and will be translocated to new foliage when nitrogen is limiting (figure 11.11B). Therefore, nitrogen deficient plants will show yellowing in the older rather than newer foliage. Compare this condition to the symptoms of iron deficiency; iron is very immobile in plants, so deficiency symptoms first appear in newer rather than older foliage (figure 11.11C). *Deficiency symptoms are visible only after a severe nutrient deficiency has developed, so they should never be used as a guide to fertilization.* Keep in mind that excessive fertilization can cause toxicity symptoms (figure 11.11D).

MONITORING AND TESTING

What is the best way to monitor fertilization during the growing season? As previously discussed, by the time deficiency symptoms appear, plant growth has already seriously slowed. Instead, the EC of fertilizer solutions and chemical analysis of plant foliage can determine if fertilization is sufficient and prevent problems from developing.

EC Testing

Remember that all fertilizers are taken up as electrically charged ions, so the ability of a water solution to conduct electricity is an indication of how much fertilizer is present. Growers who fertigate should periodically check the EC of the applied fertigation water and the growing medium solution. Note that this practice is much more useful for artificial growing media than native soil mixes.

Simple handheld EC meters (figure 11.12) are fairly inexpensive and are very useful for monitoring fertigation. Measuring the EC of fertigation water as it is applied to the crop can confirm that the fertilizer solution has been correctly calculated and that the injector is functioning. Remember that the total EC reading is a combination of fertilizer salts and natural salts present in the water source. Normal readings in applied ferti-

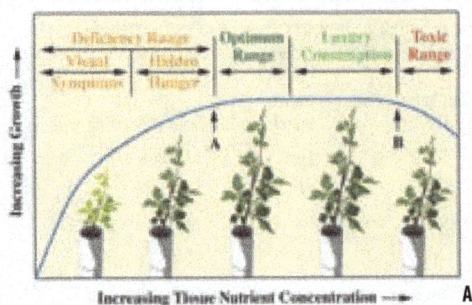

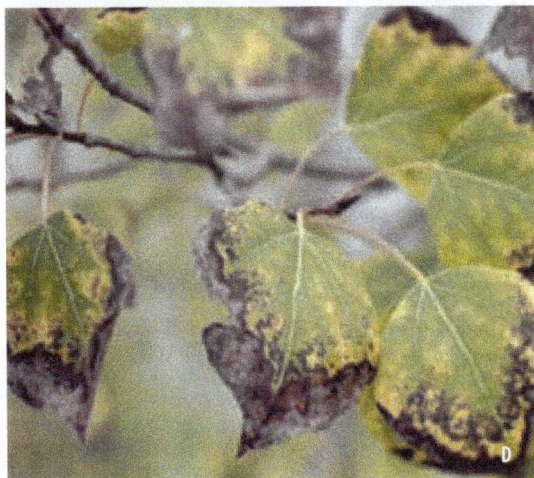

Figure 11.11—*(A) Fertilization is one of the major ways to increase plant growth, which follows a characteristic pattern. Deficiency symptoms such as chlorosis (yellowing) are common but can be caused by several different nutrients. (B) Nitrogen chlorosis is seen first in newer foliage, whereas (C) iron chlorosis occurs in older foliage. (D) Excessive fertilization can cause toxicity symptoms, such as the chlorosis and leaf margin scorch of boron toxicity.* Photo B from Erdmann and others (1979), C and D by Thomas D. Landis, illustration by Jim Marin.

gation should range from 0.75 to 2.0 μS/cm. See Chapter 10, *Water Quality and Irrigation*, for information about units of measure for EC. Measuring the EC of water leached from containers can also help pinpoint problems of improper leaching and salt buildup within the growing medium. Measurements of the EC of the solution in the growing medium, however, provide the best estimate of how much fertilizer is available to plant roots. In the growing medium, the typical range of acceptable EC values for most native plant species is about 1.2 to 2.5 μS/cm. If the EC is much over 2.5, it is probably a good idea to leach out the salts with clean irrigation water.

Foliar Testing

The best way to monitor fertilization is to test plant foliage. This test shows the exact level of nutrients that the plant has acquired. By examining tissue nutrient concentrations and simultaneously monitoring plant growth, it is possible to identify when nutrients are limiting to growth, if they are in optimal supply, or if they are creating growth toxicities (figure 11.11A).

Foliar samples must be collected in a systematic manner and be sent to a reputable laboratory for processing (recommendations are often available through local county extension agents). The analyzed nutrient concentration values are then compared to some known set of adequate nutrient values to determine which specific elements are deficient. The cost to analyze these samples is relatively inexpensive considering the potential improvement in crop quality that may result from conducting the tests.

Growth Trials

Small growth trials are another good way to monitor fertilization. This concept is especially true for native plants because so little published information is available. Detailed documentation of growing conditions, fertilizer inputs, and resulting plant response can help to formulate future fertilizer prescriptions for a specific species within a nursery. See Chapter 17, *Discovering Ways to Improve Crop Production and Plant Quality*, for more information on how to make these discoveries through trials and experiments.

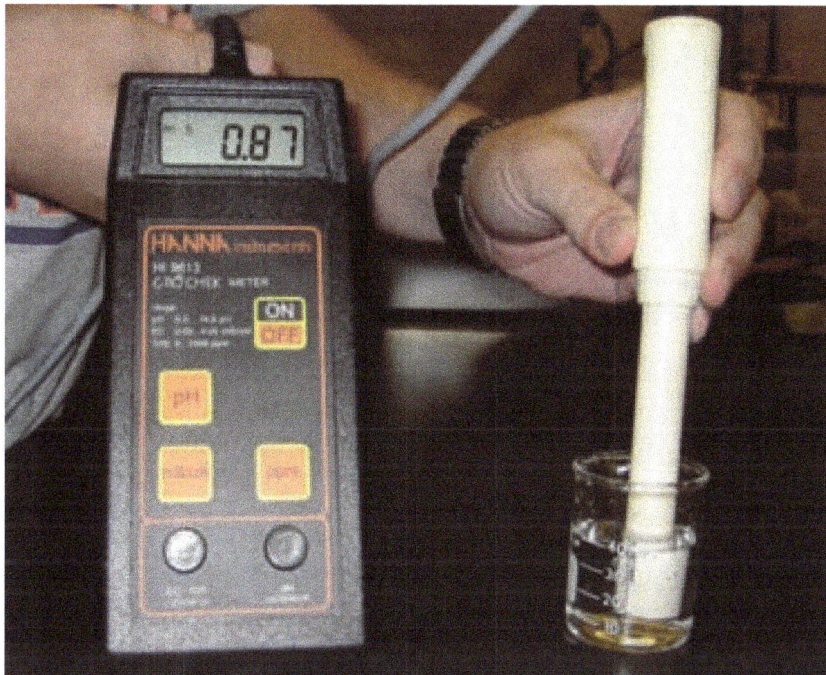

Figure 11.12—*An electrical conductivity meter used to estimate fertilizer salt concentrations in irrigation water, fertigation water, or saturated media extract.* Photo by Douglass F. Jacobs.

FERTILIZATION DURING PLANT GROWTH PHASES

As discussed in Chapter 3, *Crop Planning and Developing Propagation Protocols*, plants go through three growth phases during their nursery tenure. Growers should be aware of the different nutrient requirements during each of these phases and adjust fertilizer prescriptions accordingly (table 11.7). These adjustments are particularly important for nitrogen (especially the ammonium form of nitrogen), which tends to be a primary driver of plant growth and development (figure 11.13).

Establishment Phase

Soon after germination, we want to stimulate plant growth. Small plants, however, are particularly vulnerable to root damage from high salt concentrations. In addition, high nitrogen at this time results in succulent tissue at the root collar that can cause plants to be vulnerable to damping-off fungi and other pest problems. Thus, it is recommended to use moderate nitrogen applications during this period.

Rapid Growth Phase

This phase is the period when plants attain most of their shoot development; high levels of nitrogen tend to

promote this growth. Growers must closely monitor plant growth and development, however, to ensure that shoots do not become excessively large. A good rule of thumb for conifer seedlings (and a good starting point for many native plants) is that hardening should begin and nitrogen fertilization should be reduced when shoots have reached 75 to 80 percent of the target size. Plants take several weeks to respond to this change in fertilization and will continue to grow even though nitrogen fertilization has been reduced. Leaching the growing medium with several irrigations of plain water is a good way to make certain that all excess nitrogen is eliminated.

Hardening Phase

Adjusting fertilizer inputs during the hardening phase is perhaps one of the most critical procedures to follow in plant fertilization. See Chapter 12, *Hardening*, for a complete description of this topic. The objective of the hardening phase is to prepare plants for the stresses of shipping, storage, and outplanting by slowing shoot growth while simultaneously promoting stem and root growth. It is recommended to use low nitrogen applications during this period. Typically, a lower ratio of nitrogen to that of phosphorus and potassium is helpful (table 11.7). In addition, changing

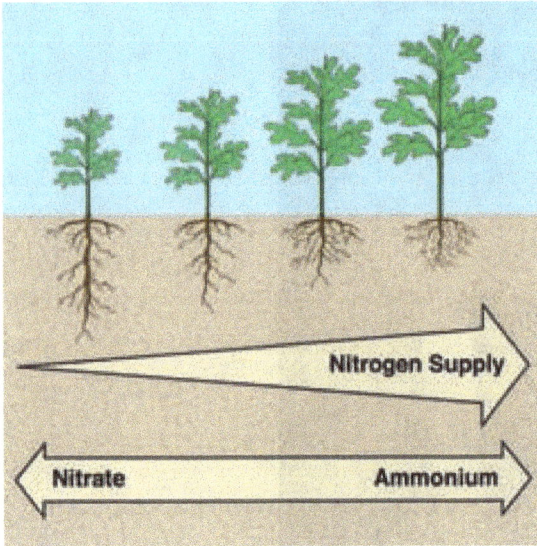

Figure 11.13—*Because nitrogen is so critical ot seedling physiology, nitrogen fertilization can be used to speed up or slow down plant growth and to control the shoot-to-root ratio.* Illustration by Jim Marin.

Table 11.7—Examples of fertilization regimes adjusted for plant growth phases

Growth phase	Nitrogen Inputs[a]	Proportion of		
		Nitrogen	Phosphorus	Potassium
Establishment	Half-strength	Medium	High	Low
Rapid growth	Full-strength	High	Medium	Medium
Hardening	Quarter-strength	Low	Low	High

[a] An example of a "full-strength" solution might be 100 ppm nitrogen

to fertilizers containing the nitrate form of nitrogen (as opposed to ammonium) is helpful because nitrate-nitrogen does not promote shoot growth. Calcium nitrate is an ideal fertilizer for hardening because it provides the only soluble form of calcium, which has the added benefit of helping to promote strong cell wall development. It is important to distinguish granular calcium nitrate from liquid calcium ammonium nitrate, which does contain ammonium.

ENVIRONMENTAL IMPACT OF FERTILIZATION

Regardless of the method of fertilizer application, a major concern is the impact that the application of fertilizers has on the water quality of the natural environment outside of the nursery. Nutrient ions, such as nitrate and phosphate, which easily leach from container nurseries, may potentially enter adjacent water supplies and degrade water quality. Thus, as growers of native plants for ecological restoration, do everything you can to minimize environmental impacts associated with fertilization. From an economic standpoint as well, as much applied fertilizer as possible should be taken up by plants (as opposed to leached away from roots). Fertilizer applications should be carefully calculated and applied only as necessary in pursuit of high-quality plant production. Continued investigation of options such as organic materials, controlled-release fertilizer, and self-contained systems (for example, subirrigation) that may minimize resulting nutrient leaching would be a logical future direction to reduce the impact of fertilization in native plant nurseries on the environment.

LITERATURE CITED

Dumroese, R.K.; Landis, T.D.; Wenny, D.L. 1998. Raising forest tree seedlings at home: simple methods for growing conifers of the Pacific Northwest from seeds. Moscow, ID: Idaho Forest, Wildlife and Range Experiment Station. Contribution 860. 56 p.

Erdmann, G.G.; Metzger, F.T.; Oberg, R.R. 1979. Macronutrient deficiency symptoms in plants of four northern hardwoods. General Technical Report. NC-53. Washington, DC: U.S. Department of Agriculture, Forest Service. 36 p.

Jaenicke, H. 1999. Good tree nursery practices: practical guidelines for research nurseries. International Centre for Research in Agroforestry. Nairobi, Kenya: Majestic Printing Works. 93 p.

Landis, T.D.; Tinus, R.W.; McDonald, S.E.; Barnett, J.P. 1989. The container tree nursery manual: volume 4, plant nutrition and irrigation. Agriculture Handbook 674. Washington, DC: U.S. Department of Agriculture, Forest Service. 119.

The Scotts Company. 2006. Scotts fertilizer tech sheet. http://www.scottsprohort.com/products/fertilizers/osmocote_plus.cfm (19 Jan 2006).

Thompson, G. 1995. Nitrogen fertilization requirements of Douglas-fir container seedlings vary by seed source. Tree Planters' Notes 46(1): 15-18.

ADDITIONAL READINGS

Bunt, A.C. 1988. Media and mixes for container grown plants. London: Unwin Hyman, Ltd. 309 p.

Marschner, H. 1995. Mineral nutrition of higher plants. 2nd ed. London: Academic Press. 889 p.

APPENDIX 11.A. PLANTS MENTIONED IN THIS CHAPTER

blue spruce, *Picea pungens*

buffaloberry, *Shepherdia* species

ceanothus, *Ceanothus* species

chokecherry, *Prunus virginiana*

cottonwood, *Populus* species

cowparsnip, *Heracleum* species

dogbane, *Apocynum* species

Douglas-fir, *Pseudotsuga menziesii*

elderberry, *Sambucus* species

fourwing saltbush, *Atriplex canescens*

hawthorn, *Crataegus* species

kinnikinnick, *Arctostaphylos* species

limber pine, *Pinus flexilis*

oak, *Quercus* species

quaking aspen, *Populus tremuloides*

redoiser dogwood, *Cornus sericea*

sagebrush, *Artemisia* species

serviceberry, *Amelanchier* species

western white pine, *Pinus monticola*

whitebark pine, *Pinus albicaulis*

wild hollyhock, *Iliamna* species

willow, *Salix* species

Hardening

Douglass F. Jacobs and Thomas D. Landis

12

To promote survival and growth following outplanting, nursery stock must undergo proper hardening. Without proper hardening, plants do not store well over winter and are likely to grow poorly or die on the outplanting site. It is important to understand that native plant nurseries are different from traditional horticultural systems in that native plants must endure an outplanting environment in which little or no aftercare is provided.

Hardening refers to a series of horticultural practices during the nursery cycle that increase plant durability and resistance to stresses. Plant hardiness primarily develops internally, although certain external characteristics such as thickening stems, a tougher feel to the foliage, and leaf abscission of deciduous species are indicators of the hardening process. Promoting hardiness is critical to prepare plants for the stresses they will endure after leaving the nursery. This process takes time and a common mistake of nursery growers, particularly with novice or inexperienced growers, is not to schedule adequate time to harden their crops.

To properly harden plants to withstand stresses of outplanting, it is important to consider the Target Plant Concept presented in chapter 2. Using knowledge of the expected conditions of a given outplanting site, nursery cultivation may be adjusted to acclimatize plants for site conditions by promoting specific traits. For instance, on sites where drought is anticipated, a larger proportion of roots relative to shoots may be desirable to improve plant resistance to moisture stress.

Cascade mountain-ash showing fall colors by R. Kasten Dumroese.

Figure 12.1—*Succulent shoot tissue can be damaged by freezing temperatures during early spring, late fall, or during overwinter storage. Note that the dormant buds on the plant in the middle are not injured.* Photo by Thomas D. Landis.

Figure 12.2—*(A) Nursery plants always undergo some degree of "transplant shock" as soon as they are outplanted. (B) This shock is primarily due to moisture stress and lasts until the roots are able to grow out into the surrounding soil.* Photo by Thomas D. Landis, illustration by Jim Marin.

In this chapter, we illustrate the importance of proper hardiness in promoting plant performance following outplanting, discuss how hardiness naturally changes through the course of the nursery growing cycle, describe how plants may be conditioned to prepare them for the characteristics of a particular outplanting site, and suggest horticultural treatments that may be used in small native plant nurseries to help promote hardiness.

EXPOSURE TO STRESSES FOLLOWING NURSERY CULTURE

During the nursery growing cycle, the objective is to promote ideal plant growth and development. This is largely accomplished by providing optimal levels of all potentially limiting factors and minimizing environmental stresses. Following nursery culture, however, plants must be hardened prior to outplanting.

Nursery plants are exposed to a series of stresses starting with harvesting. The harvesting process requires the moving and handling of plants, which creates potential for physical and internal damage. Following harvesting, nursery stock is usually overwintered outdoors or sometimes stored under refrigeration for several months while awaiting transport to the outplanting site. To withstand cold temperatures, plants must be sufficiently dormant and hardy or else injury may occur (figure 12.1). After nursery plants are transported to the outplanting site, they are often exposed to unfavorable environmental conditions until actually outplanted. For instance, sunny and windy conditions on the outplanting site can result in overheating or desiccation damage.

After nursery stock is outplanted, the plants must tolerate a period of "transplant shock" (figure 12.2A). This *shock* occurs as a result of moving plants that have grown under a favorable nursery environment, in which they receive plenty of water and fertilizer, to the outplanting site, where these factors are always limiting. Recently outplanted nursery stock must rapidly develop new roots that can grow out into the surrounding soil to access water and nutrients (figure 12.2B), compete with other plants, resist animal browse damage, and endure extreme high or low temperatures. All these stresses create the potential for physical or physiological injury that may limit outplanting success. Thus, it is important to understand how the capacity of plants to resist these stresses changes over the growing cycle and how nursery horticultural practices can increase hardiness.

PLANT GROWTH STAGES, DORMANCY, AND STRESS RESISTANCE

It is very important for growers to understand the relationship between plant growth stages and their ability to tolerate stresses. When plant shoots are actively growing, their ability to resist stress is relatively low. This condition is particularly true during the rapid growth phase of nursery production. See Chapter 3, *Crop Planning and Developing Propagation Protocols*, for descriptions of the phases. Toward the end of the rapid growth period, growth slows as plants begin to physiologically prepare to endure the stresses of winter by entering a state of dormancy (figure 12.3).

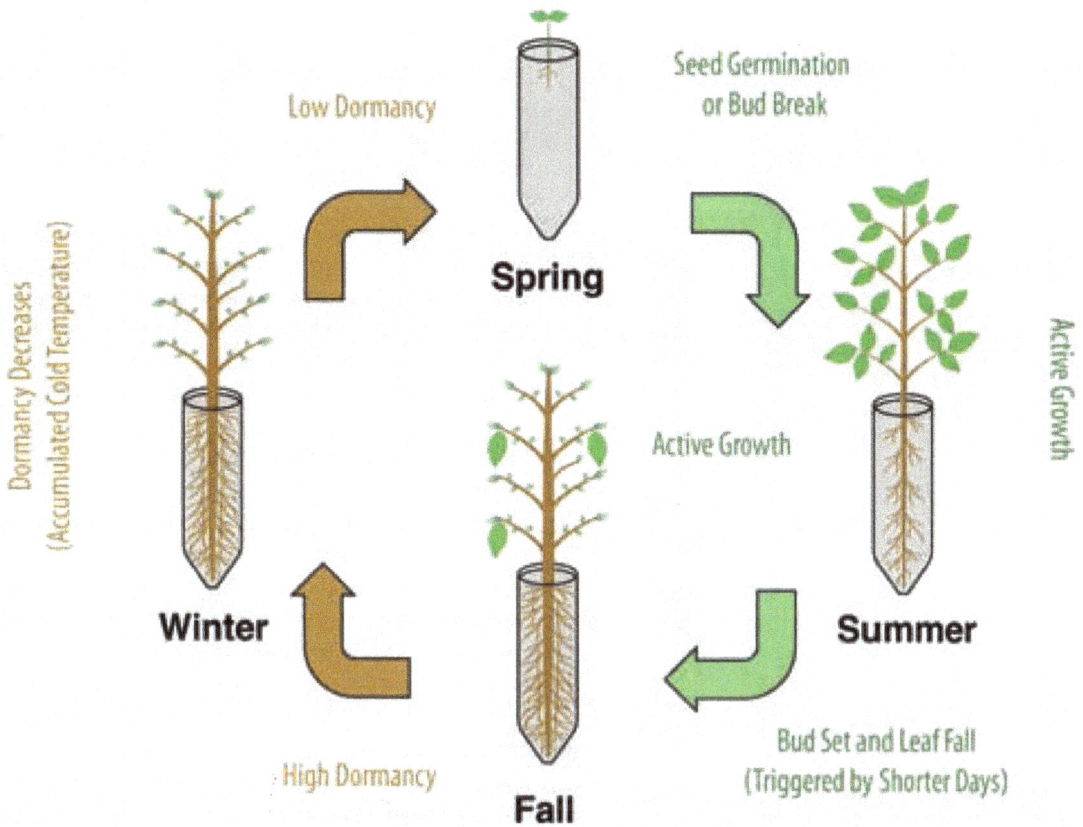

Figure 12.3—*Plants go through an annual cycle of active gowth and dormancy. As they become more dormant in autumn, their resistance to stress increases and is greatest during mid-winter.* Ilustration by Jim Marin.

A seed or plant is dormant when it will not grow even when all environmental conditions are ideal. It is important to realize that dormancy refers only to the growing points of a plant, which are known as "meristems." Plants have three meristems: foliar buds, a lateral meristem just inside the bark of the stem, and the root tips (figure 12.4A). Dormancy refers only to foliar buds because the lateral meristem and the roots never undergo true dormancy and will grow whenever conditions are favorable (figure 12.4B). Plants rely on environmental cues, especially shortening daylength, to trigger the onset of dormancy (figure 12.4C). Dormancy deepens through late autumn to early winter, when it is at its greatest (see figure 12.3). Exposure to temperatures just above freezing is an environmental cue to increase dormancy. After reaching full dormancy, the accumulated exposure to cold temperatures ("chilling hours") gradually releases dormancy (figure 12.3). This release continues until late winter to early spring, when all dormancy has been lost and buds are ready to grow again. Buds "break" under a combination of warm temperatures, moisture, and longer days, which initiates the growth cycle again.

Dormancy is related to stress resistance because plants that have stopped growing are more hardy than those that are still growing. Although the stress resist-

TWO STEPS IN THE HARDENING PHASE

Plants must first be cultured into reducing shoot growth and setting buds, and then conditioned to withstand stress

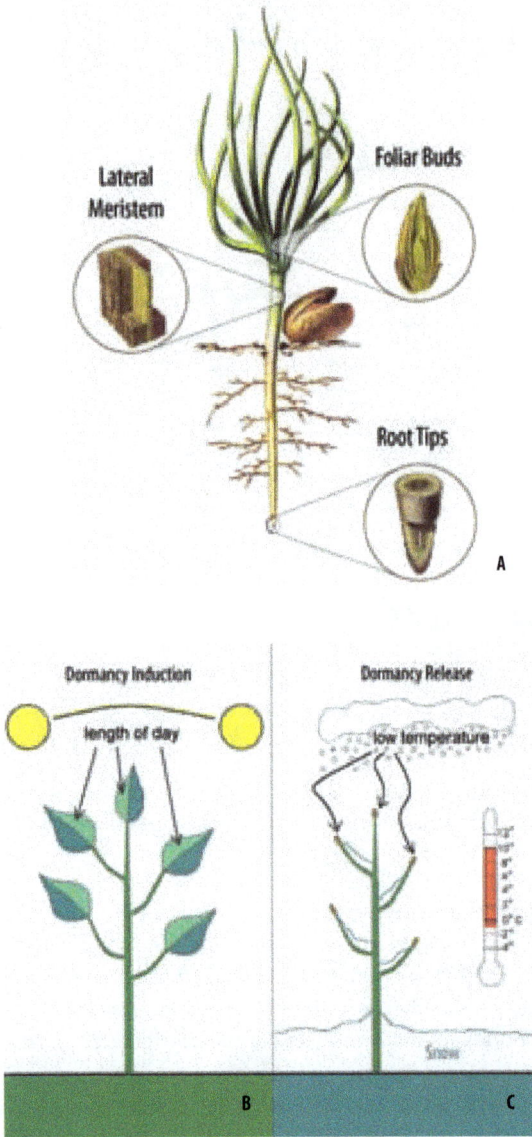

Figure 12.4—(A) Of the three growing points in plants, only foliar buds undergo true dormancy. The lateral meristem in the stem and especially the root meristem will grow whenever temperatures permit. (B) In late summer, plant leaves receive the cue of shortening days to begin the dormancy process. (C) As dormant buds accumulate more "chilling hours" through the winter, they gradually lose dormancy. Illustration by Jim Marin.

ance of plants cannot be measured, it is closely related to their cold hardiness. Thus, during winter, when plants must endure exposure to cold temperatures, they are also most able to resist other stresses. This midwinter dormant period is the best time to harvest, ship, and outplant nursery stock. Because many native plant nurseries are located in milder climates than their clients, plants must be harvested and stored until conditions on the outplanting sites are optimal. For more information on the "lifting window" and "outplanting window," see Chapter 13, *Harvesting, Storing, and Shipping*.

CREATING HARDY PLANTS IN NURSERIES

In nature, plants harden gradually as summer changes into autumn, but it is possible to achieve a greater level of stress resistance in a shorter amount of time through horticultural treatments in the nursery. These treatments must not be too severe, however, because overly stressed plants will actually be less hardy. In particular, plants with low levels of photosynthetic reserves cannot acclimate properly. In the following sections we describe how specific nursery horticultural treatments may be adjusted to induce hardiness and properly condition plants to resist stresses.

Scheduling the Hardening Phase

Scheduling enough time for the hardening phase is one of the most critical concepts in growing native plants, but it is not appreciated by inexperienced nursery managers growing their first crops. Proper hardening takes time, and it is a common mistake to try to rush the process. This mistake often happens when growing more than one crop per season or when growers try to force a little extra height growth with crops that grow more slowly than expected.

Many growers do not realize that root-collar diameter and root growth require a steady supply of photosynthate, so the hardening phase must be scheduled when there is still enough solar energy to fuel this growth.

Environmental Factors Affecting Hardening

To better understand how horticultural practices affect hardening, growers need to know the role that environmental conditions play on dormancy induction. The four main factors that affect plant dormancy and hardiness are intensity and duration of light, temperature, soil moisture, and fertility (figure 12.5). When nursery plants have reached their optimum ("target") height, nurseries horticulturally manipulate these four factors to stop shoot growth and induce hardiness.

Light

In temperate regions, daylength begins to slowly decrease following the summer solstice. Plants have adapted to recognize this change in daylength and use this environmental cue to start their preparation to resist the stresses of winter. Horticultural lighting should be discontinued at the beginning of the hardening period to ensure that the shortening daylength is recognized by plants. This practice is very effective for plants from high elevations and northern environments, in which they are particularly attuned to daylength.

as along ocean coastlines, moisture stress is more effective than other treatments. Typically, watering frequency should be gradually reduced but it is important that plants do not permanently wilt or come under severe water stress (figure 12.7). As you can imagine, inducing water stress is one of the trickiest parts of growing native plants and requires close observation and experience. The best way to quickly and accurately evaluate the water status of container plants is to weigh the growth container. With experience in monitoring container weights, a grower can gain a feel for when watering is necessary. See Chapter 10, *Water Quality and Irrigation*, for a discussion of irrigation monitoring with container weight.

After nursery plants reach their target size (usually expressed in terms of target height) at the end of the rapid growth phase, they need about 2 months to continue growing stem tissue and roots and then to harden enough to tolerate the stresses of harvesting, storage, shipping, and outplanting.

Temperature

Air temperature also has a large effect on dormancy induction in temperate regions. Exposure to gradually lower temperatures during autumn provides another important cue that winter is approaching and increasing cold hardiness is needed to resist freezing temperatures. For nursery plants in greenhouses, the temperature settings are lowered in stages. Attempting to induce hardening in an enclosed greenhouse is difficult, however, because the intense sunlight keeps temperatures warm and humidity levels high. It is more effective to expose crops to a moderate level of temperature shock to help facilitate the hardening process. Therefore, moving nursery plants from the greenhouse to ambient conditions in a shadehouse or open compound is a good strategy (figure 12.6). See Chapter 4, *Propagation Environments*, for a description of these structures.

Water Stress

A reduction in water availability to create a mild moisture stress also slows shoot growth and helps induce hardiness. For plants from mild climates, such

Mineral Nutrition

Reducing fertilizer also acts to slow growth and harden plants to help prepare them to withstand outplanting stresses. Among the mineral nutrients, nitrogen, particularly in the ammonium form, is the primary driver of shoot growth. During hardening, it can be helpful to reduce or stop nitrogen fertilization for several weeks to induce a mild nutrient stress. The use of controlled-release fertilizers with more than a 6- to 8-month release period can prevent hardiness from developing and potentially make plants susceptible to frost injury in autumn.

Some fertilizers have been specifically developed to aid in plant hardening, often containing a low nitrogen– high potassium formulation. Calcium nitrate is also a useful hardening fertilizer because it contains the nitrate form of nitrogen, which does not promote shoot growth. Calcium also helps develop strong cell walls and leaf waxes to protect plants during overwinter storage. Be sure not to use a similar product known as calcium ammonium nitrate because this fertilizer can stimulate shoot growth.

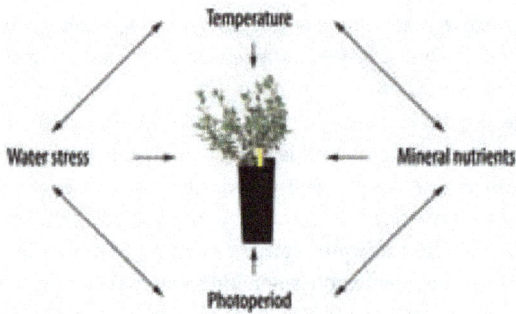

Figure 12.5—*Nurseries manipulate four environmental factors to stop shoot growth and induce hardiness.* Illustration by Jim Marin.

Figure 12.6—*Proper hardening requires that plants be exposed to the natural (ambient) environment, in which they receive environmental cues such as decreased daylength and temperatures. Moving plants from an enclosed greenhouse to a shadehouse or open compound is an effective method to begin hardening.* Photo by Thomas D. Landis.

Figure 12.7—*Reducing irrigation to induce a mild moisture stress helps induce dormancy and begin hardening but severe stress, such as in these quaking aspen plants, can be harmful.* Photo by Thomas D. Landis.

CONDITIONING PLANTS FOR OUTPLANTING

Other horticultural techniques can be used to properly condition nursery plants for outplanting. In determining how to properly condition plants for the intended outplanting site, it is important to consider several factors. First, we must understand the ecological characteristics of the species being grown. For instance, is this a light-demanding species or a shade-loving species? Next, we must be aware of the potential stresses on the outplanting site. Is this site an open field or a riparian zone or will the stock be outplanted underneath an existing canopy of trees? Will the site be prone to extended dry periods? Understanding the character of the site is best accomplished by interacting closely with the client ordering the plants.

These factors all reflect the main principles of Chapter 2, *The Target Plant Concept,* which suggests that the characteristics of nursery stock must be matched to those of the intended outplanting site. The goal of these treatments is to acclimatize plants to conditions on the outplanting site. Some of the major considerations regarding plant conditioning are shoot-to-root balance, shade, water stress conditioning, and root or shoot pruning.

Shoot-to-Root Balance

Shoot-to-shoot balance is the relative ratio of shoot biomass to root biomass, not shoot length to root length. It is one important way to describe plant size. Growing nursery plants to the appropriate size for a specific container is critical, and container volume and plant spacing are important. Plants in small containers and those grown close to one another grow tall and spindly and do not have enough stem strength to resist physical stresses after outplanting (figure 12.8A). These "topheavy" plants do not have enough roots to provide moisture to the foliage, so water stress can develop after outplanting. Roots in containers that are too small often begin to spiral and become compacted (figure 12.8B). In these "rootbound" plants, most roots become woody and less effective in water uptake and, after outplanting, do not grow out from the compacted root mass to promote structural stability.

The key to developing a plant with a sturdy shoot and well-balanced root system (a good shoot-to-root balance) is to select a container that is appropriate for the species and conditions on the outplanting site. Plants should be moved from the greenhouse as soon as they

Table 12.1—Examples of native plant species commonly grown under full sun versus shaded conditions, along with some species which often grow under either condition

Species	Common Name	Sun Requiring	Shade Requiring	Sun or Shade
Artemisia tridentata	Big sagebrush	▪		
Carex aquatilis	Water sedge	▪		
Juniperus virginiana	Eastern redcedar	▪		
Pinus edulis	Pinyon	▪		
Prunus virginiana	Chokecherry	▪		
Dryopteris filis-mas	Male fern		▪	
Chimaphila umbellata	Pipsissewa		▪	
Gymnocarpium dryopteris	Oakfern		▪	
Rubus pedatus	Strawberry leaf raspberry		▪	
Abies bifolia	Subalpine fir			▪
Ceanothus sanguineus	Redstem ceanothus			▪
Rubus parviflorus	Thimbleberry			▪
Streptopus amplexifolius	Twisted stalk			▪
Pteridium aquilinum	Bracken fern			▪

have reached their target height. Experienced growers know that moving plants from the greenhouse to a shadehouse or open compound is an easy and effective way to keep them in proper shoot-to-root balance.

Shade

The use of shading as a conditioning treatment depends on the shade tolerance of the species (table 12.1) and the conditions on the outplanting site. The amount of light a plant receives can be reduced by installing shadecloth or moving the crop to a shadehouse (figure 12.9). Shading is probably an overused treatment in nurseries, however, because most species (even those classified as shade tolerant) tend to grow best in full sunlight. In addition, many native plants tend to grow excessively in height ("stretch") under excessive shade, which may create a shoot-to-root imbalance. Nonetheless, if the species is shade loving and will be planted onto a site underneath an existing canopy, then shading may be a useful treatment. Plants that will be planted into full sun conditions should receive minimal or no shading at any point during nursery cultivation including during the hardening phase.

Figure 12.8—*(A) Maintaining a proper shoot-to-root balance can be difficult with fast growing species in small containers. (B) Often, these plants become "rootbound" when they are held too long* . Photos by Thomas D Landis.

Proper conditioning requires that we think like a plant!

Figure 12.9—*A shadehouse is sometimes used to help condition shade-loving plants or those for outplanting sites with shady conditions, such as underplantings in existing forests.* Photo by Thomas D. Landis.

Irrigation

Reducing irrigation duration or frequency can help condition nursery stock to withstand droughty conditions on dry outplanting sites. How would this characteristic occur? Giving plants less water slows shoot growth, reducing the possibility of producing top-heavy plants, but nursery stock can also physiologically adjust to mild water stress. Less irrigation also encourages the formation of smaller leaves with thicker cuticles that transpire less after outplanting. Because moisture stress is the primary cause of transplant shock, it makes sense to precondition plants only before shipping them.

Root Culturing

Native plants that are grown in containers with root-controlling features encourage the formation of a healthy, fibrous root system that is not damaged during harvesting, is easily planted, and is able to rapidly proliferate after outplanting to access water and nutrients. Containers should always have vertical ribs to limit root spiraling and should be designed to promote good air pruning at the drainage hole (figure 12.10A). Other root culturing features such as sideslit air pruning and copper pruning are effective, especially with very vigorous rooted species. See Chapter 6, *Containers*, for more information on these features. Whether root

Figure 12.10—*(A) Containers and benches must be designed to promote "air pruning" of roots at the drainage hole. (B) Plants in an open-growing area should be placed on benches designed to facilitate air pruning and prevent roots from growing into the ground.* Photos by Thomas D. Landis.

culturing features are cost worthy for all species of native plants needs to be determined by nursery trials, as explained in Chapter 17, *Discovering Ways to Improve Crop Production and Plant Quality*.

After plants are moved to a shadehouse or an outdoor compound, it is important not to place the containers directly on the ground. Instead, plants should be placed on benches or pallets to facilitate air pruning of roots (figure 12.10B). Otherwise, roots may grow directly into the ground, which will require the added expense of root pruning during harvest. This severe pruning immediately before storage or outplanting makes the plants more vulnerable to pathogenic fungi and may delay quick root outgrowth after outplanting.

Shoot Pruning

Pruning shoots ("top pruning") is sometimes required if the top is growing too large for the root system. Shoot pruning can maintain a proper shoot-to-root balance and reduce water stress resulting from an excessively high transpirational demand. In addition, the shock of pruning stimulates more stem and root growth and allows all plants to receive more irrigation and fertigation. One of the most important reasons to prune shoots is to make the height of the entire crop more uniform. When done properly, pruning occurs at the level just above the height of the smaller plants that have been overtopped (figure 12.11A). This practice releases smaller plants and the additional light helps them re-establish a growth rate that is consistent with the rest of the crop (figure 12.11B).

It is critical that shoot pruning treatments not be too severe; a rule of thumb is never to remove more than one-third of the total shoot. Plants to be pruned should also be in general good health and have enough stored energy to rapidly grow new tissue. It is best to prune succulent tissue because woody stem tissue tends to split and has less regenerative ability (figure 12.11C). Some native plants respond better than others, however, so a small trial is always recommended. Generally, grasses, forbs, and shrubs respond well to pruning and their shoots may be pruned several times during the growing season.

Other Conditioning Practices

The horticultural techniques described previously prepare plants to endure the stresses that occur during the processes of lifting, handling, transport, and outplanting. Experience is the best teacher—experiment on a few plants and discover which treatment or combination of practices work best in your circumstances.

One such treatment is known as "brushing." This practice developed from the observation that plants that are repeatedly handled during crop monitoring tend to develop greater root-collar diameter. Growers tried to replicate this effect by moving a pole through the crowns of the plants in both directions (figure 12.12A). Of course, this practice must be done gently, especially

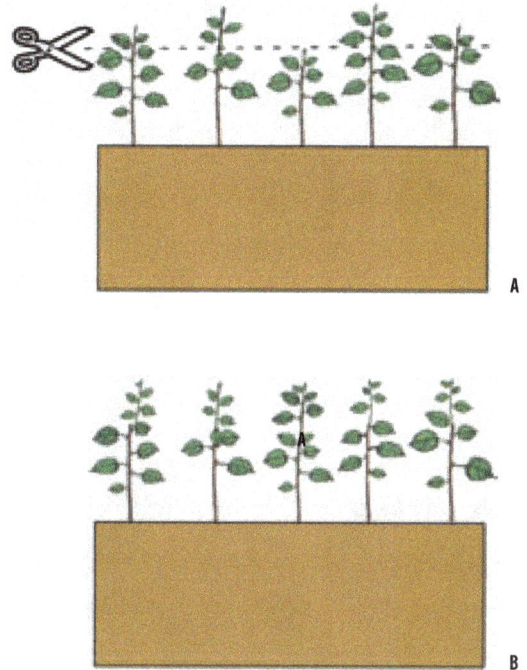

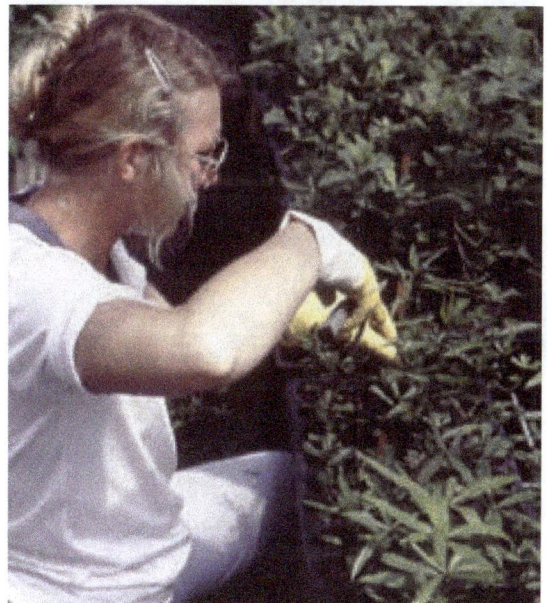

Figure 12.11—*(A) The objective of pruning shoots is to reduce height of taller plants and (B) "release" smaller ones. It is best to prune non-woody stem tissue so new buds and shoots can form. (C) It is best to prune shoots while they are still succulent.* Photo by Thomas D. Landis, illustration by Jim Marin.

when the foliage is still very succulent. Nurseries with traveling irrigation booms have mechanized the process by attaching a polyvinyl chloride pipe to the boom. A good time to brush plants is right after overhead irrigation because the rod also shakes excess water from the foliage and reduces the potential for foliar diseases such as *Botrytis* later in the season.

Increased distance between individual cells or containers allows more sunlight to reach lower leaves, improves air circulation, and promotes hardening. Increased spacing encourages the development of shorter plants with more root-collar diameter and also promotes thickening of the leaf cuticle. One real advantage of containers comprising individual, removable cells is that individual containers can be moved to every other slot to increase spacing within the trays during the hardening period (figure 12.12B).

SUMMARY

Remember two key aspects of hardening: (1) the goal of these treatments is to acclimatize plants to the harsher conditions of the outplanting site and (2) nursery managers must plan sufficient time in the crop schedule to allow for sufficient hardening. Plants can be hardened by adjusting the type and amount of nitrogen fertilizer, reducing irrigation frequency, moving plants from inside greenhouses to outdoor areas, increasing exposure to colder temperatures, and manipulating the intensity and duration of light. The hardening principle is connected directly to the target plant concept, which emphasizes the need for good communication between nursery managers and their clients. Because the "all-purpose" plant does not exist, hardening regimes will need to be developed for specific species and seed sources.

Figure 12.12—*(A) "Brushing" plants promotes greater stem diameter and (B) increasing the distance between plants promotes air circulation and the development of a sturdier plant.* Photos by Thomas D. Landis.

ADDITIONAL READINGS

Landis, T.D.; Tinus, R.W.; Barnett, J.P. 1999. The container tree nursery manual: volume 6, seedling propagation. Agriculture Handbook 674. Washington, DC: U.S. Department of Agriculture, Forest Service. 167 p.

APPENDIX 12.A. PLANTS MENTIONED IN THIS CHAPTER

big sagebrush, *Artemisia tridentata*
bracken fern, *Pteridium aquilinum*
Cascade mountain-ash, *Sorbus scopulina*
chokecherry, *Prunus virginiana*
eastern redcedar, *Juniperus virginiana*
male fern, *Dryopteris filix-mas*
oakfern, *Gymnocarpium dryopteris*
oaks, *Quercus* species
pinyon pine, *Pinus edulis*
pipsissewa, *Chimaphila umbellata*
quaking aspen, *Populus tremuloides*
redstem ceanothus, *Ceanothus sanguineus*
strawberry leaf raspberry, *Rubus pedatus*
subalpine fir, *Abies lasiocarpa*
thimbleberry, *Rubus parviflorus*
twisted stalk, *Streptopus amplexifolius*
water sedge, *Carex aquatilis*

13

Harvesting, Storing, and Shipping

Thomas D. Landis and Tara Luna

Plants are ready for harvest and delivery to clients after they have reached target specifications (see Chapter 2, *The Target Plant Concept*) and have been properly hardened (see Chapter 12, *Hardening*). Originally, nursery stock was grown in soil in fields; nursery managers would "lift" those seedlings out of the ground to harvest them. That traditional nursery term is still used today, and we refer to the traditional "lifting window" (usually late autumn to very early spring) as the time period during which plants are at maximum hardiness, most tolerant to stress, and therefore in the best condition for harvesting.

SCHEDULING HARVESTING: THE "LIFTING WINDOW"

Just 50 years ago, the process of harvesting, storing, and shipping native plants was much simpler than it is today. Conifer trees were the main native plants used for restoration after fire or logging. Seedlings were grown bareroot and the traditional lifting window described above allowed foresters to have plants in time for the traditional "outplanting window," which was always springtime. Now, container stock allows a much wider planting window so plants can be outplanted almost year-round if site conditions are favorable (table 13.1). For example, in northern Idaho, native plants can be outplanted starting in February at the lowest elevations through July at the highest elevations, and, if autumn rains are sufficient, again in September and October. Still, most container stock is outplanted in the spring, when soil moisture and temperature are most favorable for survival and growth.

Packaging seedlings for storage by R. Kasten Dumroese.

This potential variety in outplanting times makes defining the lifting window and scheduling harvesting more difficult. Our primary focus in this chapter will be on lifting plants during the more traditional season (autumn through early spring) because storage, shipping, and outplanting require special techniques. Summer and fall lifting is discussed in the *Special Outplanting Windows* section found near the end of this chapter.

In native plant nurseries, four different methods of scheduling seedling harvesting have been used: calendar and experience, foliar characteristics, time and temperature, and seedling quality tests.

1. Calendar and Experience

Scheduling harvesting according to the calendar is the most traditional technique, and, when based on the combined experience of the nursery staff, can be quite effective. The procedure is simple: if it takes 4 weeks to harvest the plants, then that amount of time is scheduled on the calendar. The dates are selected based on past weather records and how well plants harvested on those dates have survived and grown after outplanting.

A good rule of thumb is to use the frost date (Mathers 2004). To estimate the autumn frost date, take the average date of the first frost in autumn and add 30 to 45 days before that. The spring frost date, which is calculated as 30 to 45 days before the last average frost, can be used to determine when to uncover plants in spring.

2. Foliar Characteristics

Native plant growers use several morphological indicators to help them determine when plants are

Figure 13.1—*(A and B) When plants are ready to harvest, the foliage changes color and becomes more hardy. (C and D) Many woody plants form firm dormant buds with overlapping scales.* Photos by Thomas D. Landis.

Table 13.1—If outplanting site conditions are favorable, container plants can be planted year-round

Region	Potential Outplanting Windows	Outplanting Conditions
Northeast	April and May	Typical spring outplanting
Rocky Mountains (high elevations)	June and July	Good soil moisture and warmer soil temperatures; spring access prohibited by snow
Southwest	July and August	Coincides with summer rains
Northern California	September and October	Adequate soil moisture exists; poor spring access
Southeast	November through February	Outplanting conditions favorable throughout winter
Pacific Northwest	February and March	Typical late winter outplanting

becoming hardy, including changes in foliage, buds, and roots.

Foliar Changes

All plants give visual cues when they are dormant and hardy enough to harvest. With grasses and sedges, the chlorophyll dies, so foliage becomes straw colored (figure 13.1A). The leaves of deciduous woody shrubs and trees also change color, from yellows to reds, depending on the species (figure 13.1B). Even evergreen plants show signs when they are becoming dormant. For example, the cuticle of leaves or needles becomes thicker and waxier so that the seedling can tolerate desiccation during winter. Experienced growers can feel when plants are becoming hardy and the needles of some species even show a slight change in color. In spruces, the actively growing foliage is bright green whereas dormant foliage becomes bluer in color because of the waxy cuticle that develops on the surface (figure 13.1C).

Buds

Plants of many native plant species form a bud at the end of the growing season and many people look for large buds with firm bud scales as an indication of shoot dormancy (figure 13.1D). However, some species with indeterminate growth patterns (junipers, cedars) do not form a dormant bud.

Presence of White Root Tips

Roots never truly go dormant and will grow whenever soil temperatures are favorable, so the presence or absence of white roots should not be used as an indication of when harvesting should begin.

3. Time and Temperature

The technique of accumulating "chilling hours" is intuitive; plants need a certain amount of exposure to cold before they can be harvested. Therefore, the cumulative exposure of plants to cold temperatures should help indicate when they are becoming dormant and hardy. The chilling hours technique is relatively simple: record the daily temperatures, calculate the chilling hours, and then correlate this numerical index to some measure of seedling quality, such as outplanting performance. Inexpensive temperature loggers are available that will calculate chilling hours directly.

Plants are harvested over the duration of the potential lifting season and outplanted to determine survival and growth. Because chilling hours will vary from year to year, data should be gathered for at least 3 to 5 years. Seedling performance data are then plotted against the accumulated chilling hours, and the resulting graph shows when it is safe to begin harvesting the plants.

4. Seedling Quality Tests

Larger forest nurseries use cold hardiness tests as an indication of storability. Plants are placed in a programmable freezer and taken down to the predetermined temperature threshold of 0 °F (-18 °C). After a period of exposure, the plants are placed in a warm greenhouse and evaluated for cold injury to the foliage or cambium. A modification of this technique could be used for other native species.

HARVESTING OR LIFTING

In container nurseries, the process of harvesting consists of two contiguous operations: grading and packing.

Grading

This operation consists of evaluating plant size and quality and removing plants ("culls") that are outside the size specifications or are damaged or deformed (figure 13.2A). Culls that are damaged or diseased are discarded or, better yet, incorporated into the compost pile. Plants that are just too small can be held over for additional growth but usually must be transplanted into larger containers.

Typical grading criteria include size measurements such as shoot height and stem diameter at the root collar, which is known as "caliper" (figure 13.2B). Plants that meet size and quality standards are called "shippable," and they are counted to establish an accurate inventory; the inventory should be shared with the client or person in charge of outplanting.

Packing

This next step in the harvesting process depends on how plants will be stored and shipped. The two options for packing include (1) plants remain in their containers or (2) plants are extracted and placed in bags or boxes and refrigerated or frozen. Large container stock (> 1 gal) is typically stored and shipped in their growth containers but smaller plants can be removed from their containers. The type of processing depends on the type of container. With single-cell containers, the culls are removed from the racks and replaced with shippable plants, a process known as "consolidation." Containers are then stored on the ground in a shade-house until they are shipped to the field. With block containers, however, cull plants are difficult to remove, so larger nurseries extract the shippable plants, wrap or bag them, and place them in bags or boxes for refrigerated storage, a practice known as "pull and wrap" (figure 13.2C).

PLANT STORAGE

Unlike many other products that can be stored for extended periods without a decrease in quality, nursery crops are living and have a very limited shelf life. Therefore, well-designed seedling storage facilities are needed at all native plant nurseries. Seedling storage is *an operational necessity, not a physiological requirement* because of the following conditions:

1. Traditionally, forest nurseries were located at great distances from outplanting sites, and refrigerated storage was needed to preserve plant quality and to facilitate shipping. Tribal nurseries, however, are more local and closer to where stock will be outplanted, so sophisticated storage facilities may not be needed.

2. The lifting window at the nursery may not coincide with outplanting windows on all the various sites. This scenario is particularly true in mountainous areas where nurseries are located in valleys with much different climates than what project sites at higher elevations have. If a client wants to outplant during summer or fall, then short-term storage is all that is necessary. Often, however, the best conditions for outplanting occur the following spring, so it is necessary to protect plants throughout the winter.

3. The large number of plants being produced at today's nurseries means that it is physically impossible to lift, grade, process, and ship plants all at the same time. Storage facilities help to smooth out the lift-pack-ship process.

4. Refrigerated storage is a cultural tool that can be used to control plant physiology and morphology. For example, plants can be harvested at

Figure 13.2—*(A) Grading consists of separating "shippable" plants from "culls" based on criteria such as (B) root-collar diameter. (C) Larger nurseries "pull and wrap" their stock and pack it into bags or boxes.* Photos by Thomas D. Landis.

the peak of dormancy in autumn and then the chilling requirement can be met by temperatures in refrigerated storage.

Short-Term Storage for Summer or Fall Outplanting

Often, native plants are simply held in a shadehouse (figure 13.3A) or open compound (figure 13.3B) until they are shipped. Both structures are typically equipped with sprinklers, so irrigation and fertigation are possible. Larger containers are stored in wire racks to keep them upright and to stop roots from growing into the ground (figure 13.3C). To aid in drainage, prevent seedling roots from growing into the soil, and retard weeds, plants can be placed on a layer of pea-sized gravel covered with landscape fabric. Fabric impregnated with copper can also be purchased that chemically prunes roots as they emerge from the bottom of the containers (figure 13.3D).

Overwinter Storage

The importance of properly overwintering stock is often overlooked by novice native plant growers because their emphasis is on growing the plants. Many plants have been damaged and some crops have been completely lost as a result of poorly designed or managed overwinter storage. In northern climates, it is common for native plant nurseries to lose 10 percent or more of their nursery stock during winter, and growers who fail to provide adequate protection may lose half or more of their plants.

Causes of Overwinter Damage

Before we talk about types of storage systems, let us first discuss the main causes of injury to stored plants (table 13.2).

COLD INJURY

Cold injury can develop from a single frost or during an extended period of cold weather. Damage is most common in late autumn or early spring, when plants are entering or coming out of dormancy (figure 13.4A). Cold injury is directly related to seedling dormancy or cold hardiness. Properly hardened shoots of northern native plants can tolerate temperatures to –40 °F (–40 °C) or lower, but cold hardiness and dormancy are lost as winter progresses.

Root systems are injured at much higher temperatures than shoots, so roots need special protection.

ROOTS ARE NOT AS HARDY AS SHOOTS

12 °F (–11 °C) is the lowest temperature that young seedling roots should have to endure.

–20 °F (–29 °C) is the lowest temperature that mature roots can handle, but this critical temperature varies greatly by species.

34 to 41 °F (1 to 5 °C) is an ideal temperature for overwintering newly rooted cuttings.

Figure 13.3—(A) For short-term storage, plants may be held in shadehouses, or (B) open compounds. (C) Large single containers require metal racks to keep them upright. (D) Copper-treated fabrics are ideal for ground storage because they chemically prevent plant roots from growing into the ground. Photos A–C by Thomas D. Landis, D by Stuewe & Sons, Inc.

Furthermore, mature roots are hardier and will tolerate colder temperatures than will younger, less hardy roots. Rooted cuttings are particularly vulnerable to injury because their roots have not developed protective layers yet. Young roots are typically located on the outer portion of the root plug and are the first to be injured by cold temperatures (figure 13.4B). In areas where freezing temperatures occur, cold injury to roots is the most common type of overwinter damage. Because shoots do not show symptoms immediately, root injury often goes unnoticed and the damage becomes evident after outplanting. Therefore, growers should design their overwintering systems to protect roots from damaging temperatures during overwinter storage.

DESICCATION

Winter drying is actually desiccation injury and occurs whenever plants are exposed to extreme moisture stress caused by wind and/or direct sunlight (figure 13.4C). Damage is most severe when the growing medium and roots remain frozen for extended periods. Plants can even become desiccated when stored under frost-free refrigeration without proper packaging. Winter drying is not directly related to seedling dormancy or cold hardiness; even the most dormant and hardy stock can be damaged. Plants stored near the perimeter of sheltered storage are most susceptible (figure 13.4D), but any plant can be damaged if its shoot becomes exposed. This type of desiccation can be prevented if plants can be irrigated during the winter storage period, and perimeter insulation is effective.

FROST HEAVING

Repeated freezing and thawing can cause young seedlings or new transplants to be physically lifted out of the growing medium. Frost heaving is much more common in bareroot nurseries but can still occur when small container plants are exposed to freezing and thawing. Mulches are effective in preventing damage.

LOSS OF DORMANCY

This type of injury is most common when container stock is overwintered in greenhouses. Often, periods of clear, sunny weather during the winter warm the greenhouse and can cause plants to lose dormancy. This condition is particularly true of root systems because roots grow whenever temperatures permit.

Loss of dormancy becomes progressively more serious during late winter and early spring when plants have fulfilled their chilling requirements and cold temperatures are the only factor keeping them from growing (figure 13.4E). Use white or reflective coverings to reflect sunlight and reduce heat buildup, and ventilate greenhouses frequently.

STORAGE MOLDS

The type of storage conditions will determine the types of disease problems that will be encountered. Although fungal diseases can be a problem in open storage or shadehouses, they are most serious when plants are overwintered under refrigeration (table 13.2). Some fungi, such as *Botrytis cinerea*, actually prefer the cold, dark conditions in storage bags and boxes and will continue to grow and damage plants whenever free moisture is available (figure 13.4F). Some nurseries apply fungicides before overwinter storage but careful grading to remove injured or infected plants is the best prevention. Freezer storage has become popular because it prevents the further development of storage molds.

DESIGNING AND LOCATING A STORAGE FACILITY

The type and design of a storage system depends on the general climate of your nursery location, the characteristics of the nursery stock, and the distance to the outplanting sites. Most people would think that overwintering would be most difficult the farther north or higher in elevation you go, but that is not the case. Nurseries in the Midwest or the Great Plains are the most challenging because of the extreme fluctuations in temperature that occur during winter. The east slope of the Cascade Mountains or Rocky Mountains can be just as challenging because of the number of clear, windy days and temperatures that can fluctuate 40 °F (22 °C) within a 24-hour period.

Some plants are easier to store than others. Native species that tend to overwinter well are those that achieve deep dormancy and can withstand low or fluctuating temperatures. Therefore, storage systems must be matched to the plant species being grown and the local climate. In tropical or semitropical climates, plants never undergo a true dormancy and can be outplanted almost any time of the year. Plants from coastal areas are never exposed to freezing and tend to be less hardy than those from inland areas.

Figure 13.4—*(A) Overwinter storage is a time of considerable risk for nursery stock. Cold temperatures can damage nonhardy tissue, even buds. (B) Roots are particularly susceptible because they will grow whenever temperatures permit. (C) Drought injury ("winterburn") is actually desiccation and (D) is particularly severe around the perimeter of storage areas. (E) Overwintered plants gradually lose dormancy and can resume shoot growth during late winter or early spring. (F) Storage molds are most serious in cooler storage whereas (G) animal damage can be a real problem in sheltered storage.* Photos A–D and F by Thomas D. Landis, E by Tara Luna, G by R. Kasten Dumroese.

Table 13.2—Plants can be injured by several types of stress during overwinter storage

Damage	Cause	Preventative Measure for Each Storage Type		
		Open	**Sheltered**	**Refrigerated**
Cold injury (figures 13.4A and B)	Temperature is below seedling cold hardiness level. (Roots are much more susceptible than shoots.)	Properly harden seedlings to tolerate maximum expected cold temperatures		
Drought injury (winter desiccation) (figures 13.4C and D)	Exposed to intense sunlight or drying wind	Bring media to field capacity before storing		
		Shade plants and protect from wind	Cover stock with moisture-retaining film	Use moisture-retentive packaging
Frost heaving	Repeated freezing and thawing of growing medium	Don't overwinter small plants; insulate containers and apply mulch to tops of containers	Not a problem	Not a problem
Loss of dormancy (figure 13.4E)	Temperatures above 40 °F (4 °C)	Not possible to control	Not possible to control	Maintain cold in-box temperatures.
Storage molds (figure 13.4F)	Temperature is warm; latent infections of *Botrytis*	Prevent injury to seedling tissue, cull damaged plants		
		Keep foliage cool and dry	Remove dead foliage before storing	Use freezer storage if longer than 2 months
Animal damage (figure 13.4G)	Small rodents and even rabbits can girdle stored nursery stock	Exclude larger animals with fencing; use poison bait for rodents	Exclude larger animals with fencing; use poison baits for rodents	Not a problem

All temperate and arctic plants go through an annual cycle of growth and dormancy. In nurseries, plants are cultured into an accelerated period of growth, which must be terminated (during the hardening period) before they can be outplanted. See Chapter 12, *Hardening*, for a complete discussion on this topic. Plants that are fully dormant and hardy are in the ideal physiological state for overwinter storage. Dormant, hardy plants can be thought of as being in a state of "suspended animation." They are still respiring and some cell division occurs in the roots and stems; evergreen species can even photosynthesize during favorable periods during winter. The challenge to nursery managers is to design and manage a storage system to keep their stored plants dormant while protecting them from the many stresses discussed in the last section.

THE IDEAL OVERWINTER STORAGE SYSTEM DEPENDS ON:

— Location and climate of the nursery.
— Location and climate of outplanting sites.
— Characteristics of the plant species.
— Number of plants to be stored.

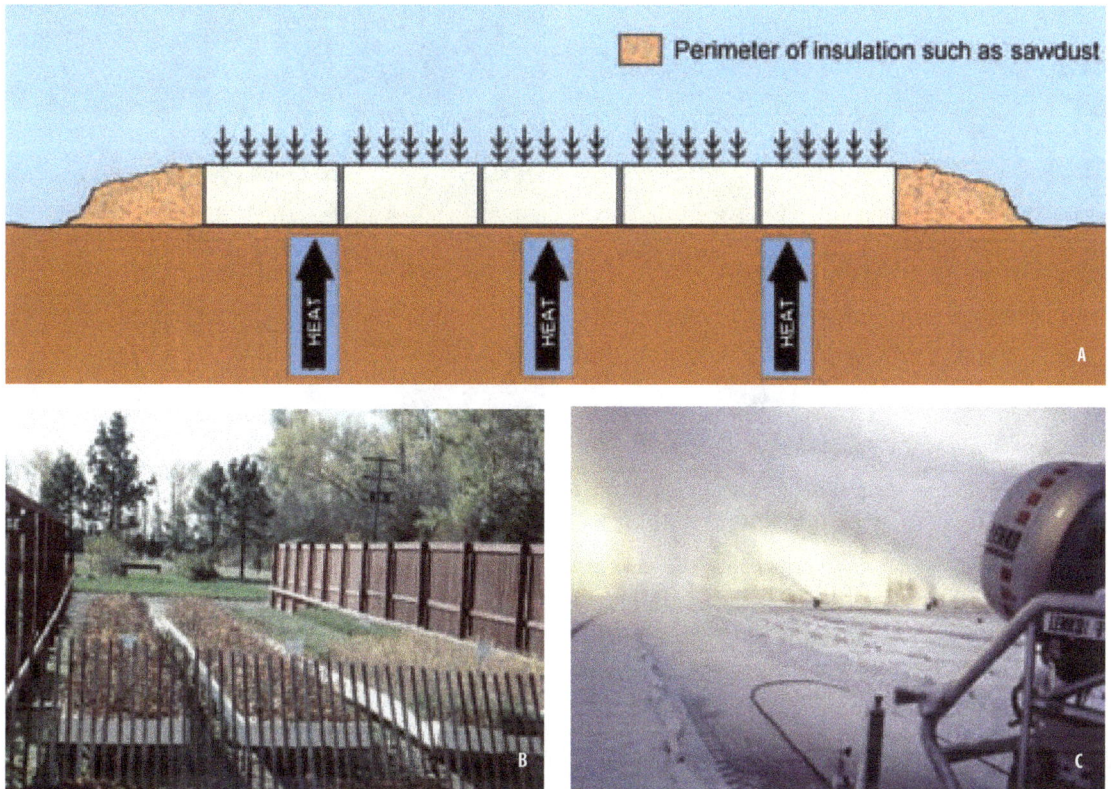

Figure 13.5—(A) Open storage can be effective when plants are blocked on the ground and surrounded by insulation. (B) Plants should be protected from direct sun and wind by natural or artificial snowfences. (C) Snow is an excellent insulator, and some northern nurseries have augmented natural snowfall with snow-making equipment. Photos by Thomas D. Landis, illustration by Jim Marin.

Common Systems for Storing Container Plants

The storage system that you select will depend on the species being grown and the severity of winter conditions where the nursery is located. Generally, four systems are used: open storage, tarp storage, storage structures, and refrigerated storage. In many nurseries, more than one overwintering system is used to accommodate the requirements of different native plant species.

Open Storage

In areas with freezing temperatures, open storage is the least expensive but most risky overwintering option. Select an area of the nursery that has some protection from wind and where cold air will drain away. Use gravel and/or drainage tile to promote the free drainage of rain or snowmelt in the spring. Pack all containers together tightly and insulate the perimeter with straw bales or a berm of sawdust. With this perimeter insulation, the roots of stored plants will be protected by heat stored in the ground (figure 13.5A).

Open storage is most successful in forested areas of northern climates where adjacent trees create both shade and a windbreak and continuous snow cover can be expected. If tree cover is not available, plants can be stored in narrow, east–west oriented bays between vertical snowfence (figure 13.5B). Snow is an ideal natural insulation for overwintering container plants but complete and continuous snow cover is not always reliable. Some northern nurseries have had success with generating snow cover with snow-making equipment (figure 13.5C).

Tarp Storage

These are the simplest and least expensive ways to overwinter container stock. In tarp storage, plants are enclosed with a protective covering that is not mechanically supported. Many different coverings

Figure 13.6—*(A) The simplest and least expensive way to store container stock is to bunch them on the ground and cover with a sheet of white plastic. (B) In colder climates, nurseries use Styrofoam™ sheets and sheeting to provide more insulation.* Photo A by Thomas D. Landis, B by Tara Luna.

have been used but the basic principle is the same: to provide a protective, insulating layer over stored plants. Clear plastic should never be used because it transmits sunlight and temperatures within the storage area can reach damaging levels. Because they reflect most sunlight, white or reflective coverings are preferred. Because the tarp materials are lightweight, they need to be secured against wind by burying the edges with concrete blocks, wooden planks, sand, or soil. All tarps are eventually decomposed by direct sunlight, so store them in a dry, dark location (Green and Fuchigami 1985).

Single layer, white, 4-mil copolymer plastic films are the most common type of storage tarp (figures 13.6A and B). White is preferred because it reflects sunlight and keeps temperatures from building up under the covering. Another tarp, Microfoam™ is a breathable, Styrofoam™-like material that is lightweight, reusable, and easily removed and stored. It can be placed directly on plants or supported by a structure. It prevents wide fluctuations in temperature, minimizes heat

TIMING IS CRITICAL

Any tarp storage system is effective only if applied after plants have developed sufficient hardiness and, most important, is removed before plants lose dormancy in the spring.

buildup, and protects roots even if air temperatures drop to 10 °F (–12 °C). Although one layer of Microfoam™ is probably insufficient in harsh climates lacking reliable snow cover, such as those in northern Minnesota and North Dakota (Mathers 2004), it should not be used in conjunction with plastic sheeting because excessive moisture will develop. Rigid sheets of Styrofoam™ can be used to protect plants, too (Whaley and Buse 1994).

Other tarping material includes plastic Bubble Wrap® sheeting, which has better insulation value than regular plastic sheeting and is reported to be cheaper and more durable than Microfoam™ sheets (Barnes 1990), and frost fabrics. Woven and nonwoven landscape fabrics are usually white in color to retard solar heating but allow the infiltration of air and precipitation. Frost fabrics are available in a range of weights and thicknesses providing 4 to 8 °F (2 to 4 °C) of thermal insulation.

In harsh, northern climates without reliable snow cover, some nurseries cover their plants with a "sandwich" of straw or other insulating material between two layers of clear plastic sheeting. The clear plastic and straw create additional heat on clear, frigid days and provide more constant temperatures during the overwintering period when compared with other systems (Mathers 2003). In general, the sandwich covers provide better insulation but cannot be temporarily removed or vented during periods of sunny, warm, winter weather (Iles and others 1993).

    HARVESTING, STORING, AND SHIPPING

Storage Structures

COLD FRAMES

A variety of cold frames have been used for overwinter storage. In northern Alberta and Alaska, cold frames constructed of sideboards lined and topped with rigid Styrofoam™ sheets have proven effective (figure 13.7A). During warm periods in the winter or as soon as weather conditions permit in the spring, the top layer of insulation is removed so that plants can be irrigated (figure 13.7B). Cold frames constructed of wooden pallets supported by cement blocks and covered with white plastic polyethylene (poly) sheeting are considered the most effective overwintering system for conifer plants at a nursery in eastern Canada (White 2004).

POLYHUTS AND POLYHOUSES

Polyhuts are simple, inexpensive structures, generally less than 4 ft (1.2 m) high, with a wooden, pipe, or cable frame covered by white plastic sheeting (figure 13.8A) or a "sandwich" of Microfoam™ between two layers of plastic (figure 13.8B). Polyhouses are similar in structure but taller (6 to 7 ft [1.8 to 2.1 m] high) to allow better access. The ends of these structures are opened for cooling during sunny, warm periods during winter (figure 13.8C). Although a single layer of white poly sheeting is adequate protection in milder climates, a double layer of white plastic inflated by a small fan provides better insulation in colder locations. If frigid temperatures occur (0 °F [–18 °C] or colder), however, plants would need additional protection such as a sheet of white poly film or a Microfoam™ blanket (Perry 1990). Some nurseries position a separate plastic sheet within the structure so that it can be temporarily pulled over the plants to provide additional protection during unusually cold conditions. Some growers supply just enough heat in their polyhouses to keep the ambient temperatures around 38 °F (3 °C), an effective technique for a wide variety of native plants in Colorado (Mandel 2004).

Any closed structure needs to be monitored carefully. If possible, orient structures with the longest axis north–south to equalize solar heating, which avoids problems with plants drying out unequally. Ventilation is necessary on warm winter and early spring days (figure 13.8C) and can be as complex as thermostat-activated fans and louvers or as simple as opening end doors. To

CHARACTERISTICS TO CONSIDER IN CHOOSING A STORAGE SYSTEM

— Does your system provide adequate protection to plant roots?

— Does it reduce moisture loss thus avoiding tissue burn?

— Is it easily handled and managed?

— Is it economical?

— Can it be adapted to the growth habit of the plant and size of container stock?

Figure 13.7—*(A) Cold frames of wood and rigid Styrofoam™ sheet insulation are used to overwinter container plants in northern climates. (B) When weather conditions permit, the top layer of insulation is removed so that plants can be irrigated.* Photos by Thomas D. Landis.

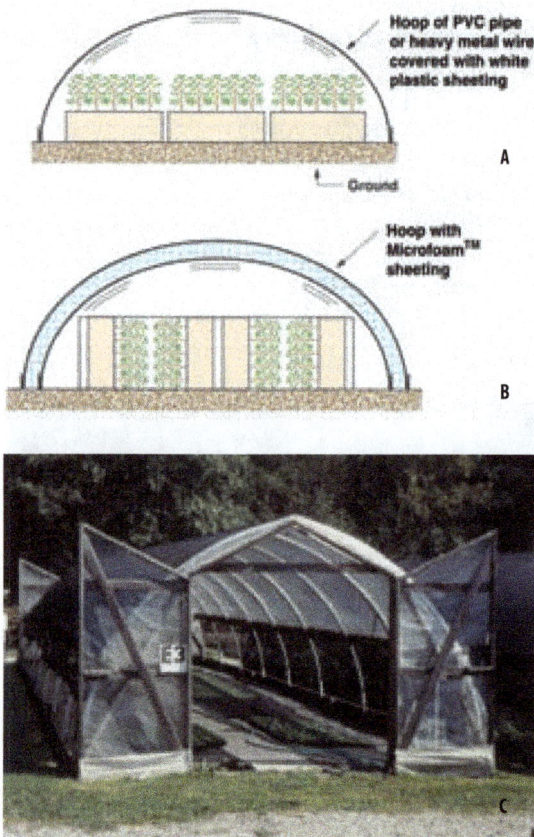

Figure 13.8—(A) Polyhuts are low structures covered with white plastic or (B) a "sandwich" of poly film and Microfoam™. (C) The ends can be opened for ventilation during warm and sunny winter weather. Photo by Thomas D. Landis, illustration by Jim Marin.

prevent desiccation, deflect air movement above the plants toward the roof, where the warmest air will accumulate. Even less-hardy nursery stock, such as newly rooted cuttings, can be overwintered in cold frames provided containers are packed tightly together and perimeter containers and the tops of all containers are insulated.

SHADEHOUSES

Shadehouses are traditional overwintering structures that provide protection from direct sun and wind, and their design will vary with nursery climate and location.

A typical, shadehouse for overwinter storage has shading on both the roof and the sides to protect plants from adverse weather (figure 13.9A), including high winds, intense rains, hail, and heavy snow. Its exact design varies with nursery climate and location. In areas where prolonged cold temperatures are not usual, plants can be overwintered under shadecloth. In wet climates, a waterproof roof is desirable to prevent overwatering by rain and the leaching of nutrients from containers. In areas that receive heavy, wet snowfall, shadehouses must be constructed to bear the load or the covering must be removed to allow the snow to fall onto the crop and insulate it. Shadehouses reduce sunlight by about 30 to 50 percent, thereby reducing seedling temperature. This shade, combined with reduced exposure to wind, can significantly lower transpirational water losses associated with desiccation.

Figure 13.9—(A) Shadehouses, such as this one at the Colorado River Indian Tribes nursery in Arizona, provide adequate overwinter protection in warmer climates. (B) When freezing is expected, roots must be protected. Photos by Thomas D. Landis.

HARVESTING, STORING, AND SHIPPING

Because it regulates dormancy release, temperature is the most important factor to monitor during overwinter storage.

Sensitive roots can be protected by grouping plants on the ground and surrounding them with an insulating material such as sawdust or Microfoam™ panels (figure 13.9B). Because shadehouses have sides, they protect plants from large animal pests such as deer and rabbits.

GREENHOUSES

Very sensitive plants, such as newly rooted cuttings with tender adventitious roots, can be overwintered in a greenhouse set for minimal heating to keep air temperatures above freezing. It must be emphasized, however, that greenhouses should not be considered for routine overwinter storage because temperatures can increase too much during periods of sunny weather, causing plants to rapidly lose dormancy. In snowy climates, heat must be used to melt heavy, wet snow to avoid structural damage (figure 13.10A). Retractable-roof greenhouses (figure 13.10B) are the best option; roofs can be closed during cold weather, opened on sunny days to allow heat to escape, and opened during snowfall to provide a protective layer to the crop.

Smaller quantities of newly rooted cuttings can be held in hot frames; hot frames are ideal because bottom heat can ensure that soil temperatures remain about 38 °F (3 °C) while shoots remain dormant with exposure to cooler air temperatures. See Chapter 4, *Propagation Environments*, for details on these structures. During warm, sunny days, hot frames must be opened to allow heat to dissipate.

Refrigerated Storage

Refrigerated storage is, by far, the most expensive way to store nursery stock and therefore is practical only for large native plant nurseries. Biologically, however, operational experience at the USDA Forest Service nursery in Coeur d'Alene, Idaho, shows that most native plant species can be stored very effectively under refrigeration.

Figure 13.10—*(A) Fully enclosed greenhouses can be used for overwintering sensitive species but snow removal is necessary in cold climates to prevent damage. (B) Retractable-roof greenhouses are better for overwinter storage because they can be opened to allow snow to cover plants.* Photos by Thomas D. Landis.

PRE-STORAGE CHECKLIST

—Ensure that all plants have been properly hardened and dormant; this can vary considerably among species.

—Cover plants as late in the fall as possible so plants can achieve maximum hardiness.

—Ensure that water drains freely from the ground of the overwintering area.

—Remove any remaining leaves from deciduous species.

—Cut leaves back to crown level on grasses and herbaceous perennials.

—Consolidate and pack all containers together tightly.

—Irrigate nursery stock thoroughly 1 to 2 days before covering.

—Do not cover wet plants, wait until the tops are dry before covering.

—Lay large container plants on their sides, packing containers together as tightly as possible without damaging tops.

Table 13.3—A comparison of types of refrigerated storage

Type	In-box Temperature	Length of Storage	Type of Packaging Around Plants
Cooler	33 to 36 °F (1– 2 °C)	1 – 2 months	Kraft-poly bags Cardboard boxes with thin plastics liners
Freezer	30 to 25 °F (−2 to −4°C)	3 – 6 months	Boxes with moisture retentive liner

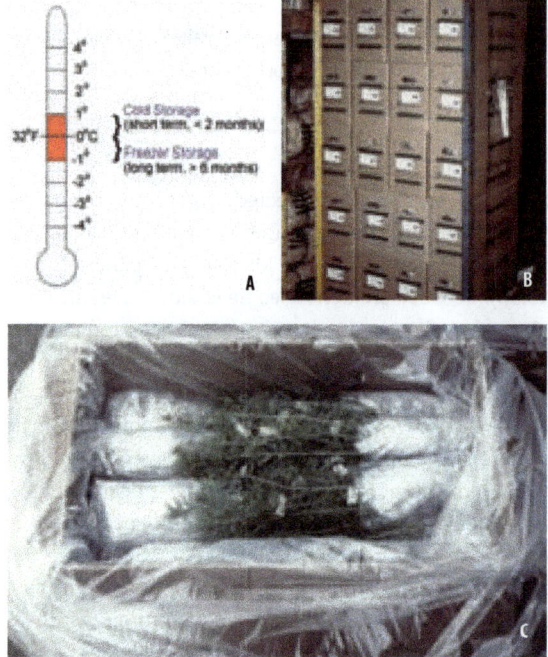

Figure 13.11—*(A) Although only a few degrees of temperature differentiate cooler and freezer storage, the duration of storage is very different. (B) Packaging is critical for freezer storage and most nurseries use waxed boxes that (C) are lined with a plastic bag to retard moisture loss.* Photos by Thomas D. Landis, illustration by Jim Marin.

The two types of refrigerated storage used in native plant nurseries are cooler and freezer storage, which are differentiated by their temperatures (figure 13.11A) and by type of packaging around the plants (table 13.3).

Refrigerated storage is effective for holding nursery stock because the lower temperatures suspend plant metabolic activity and conserve stored carbohydrates. The rule of thumb is to use cooler storage when plants are stored less than 3 months (or when shipments of plants occur throughout the storage period) and freezer storage when the duration is longer than 3 months because it significantly reduces incidence of storage molds. Because freezing converts all the free water in the storage container to ice, the development of pathogenic fungi such as gray mold (*Botrytis cinerea*) is retarded (figure 13.4F).

The type of packaging for refrigerated storage is very important. Most nurseries use special wax-impregnated boxes because they withstand moist conditions and are easy to handle and stack (figure 13.11B). Because long-term refrigeration withdraws moisture, boxes must be lined with a plastic liner (figure 13.11C).

Most conifers do well in freezer storage, while broadleaved trees and shrubs do better in cooler storage. Considerable variation exists among species, however, and there is no substitute for practical experience. Some species, such as black walnut and eastern dogwood, have serious problems with rots in refrigerated storage. Other dry desert species that retain their leaves all year, such as sagebrush and bitterbrush, are very susceptible to *Botrytis* and should not be stored under cooler storage. Very little information, unfortunately, has been published about the refrigerated storage of other native plants.

MONITORING STOCK QUALITY IN STORAGE

During overwinter storage, plants can be visualized as being in a state of "suspended animation"—plants are alive but their physiological functions have slowed to a minimum. The critical limiting factor that maintains dormancy during storage is temperature. Therefore, temperature should be rigorously monitored throughout the overwinter storage period. It is important to measure temperature at the levels of the plants and especially around the roots. Electronic thermometers with long probes are very useful for monitoring temperature in storage containers (figure 13.12A). Small and inexpensive temperature monitoring devices can monitor temperature over time, and the data can be downloaded to a computer (figure 13.12B). Any thermometer must be calibrated annually to make sure it is accurate. To calibrate, place the temperature probe in a mixture of ice and water; the temperature should read exactly 32 °F (0 °C) (figure 13.12C).

After temperature, the next most critical factor to monitor is moisture because even hardy, dormant others 1993). The best storage for these types of species is a waterproof structure that has good drainage. Removing leaves around the bud crowns prior to overwinter storage is also helpful.

Rodent damage is a major cause of loss because most overwintering systems are a perfect nesting site for rodents (figure 13.4G). Rodent damage is usually worse during years of deep snow accumulation because rodents are less able to forage for food and are protected by the snow from predators.

SPECIAL LIFTING WINDOWS

Just as with plants destined for overwinter storage, plants harvested for **summer outplanting** should undergo some hardening to prepare them for outplanting. Often, stock for summer planting is moved from the greenhouse to an outdoor area and the amount and frequency of fertigation is reduced for a period of at least a month before the plants are to be outplanted. Because these plants will still be actively growing, are relatively succulent, and have minimal or no cold hardiness, they should be held only in sheltered storage or should be cooler stored for a few days. The plants can be shipped directly to the outplanting site in their containers, or they can be graded and packed as described previously. If the latter, the plants must be outplanted promptly to avoid problems with desiccation and overheating.

Plants for **fall outplanting** usually have received a moderate amount of cold hardening but are not fully dormant when harvested. Nurseries either can ship plants in the container or pack them as for summer outplanting. If refrigerated storage is available at the nursery or on the outplanting site, only cooler storage is recommended because freezer storage may damage nonhardy tissue. The storage duration should be limited to a few days or weeks.

For either summer or fall outplanting, nursery location should be considered. Nurseries located close to the project area may be able to lift plants and deliver them quickly to the outplant site with little or no storage ("hot plant"). As the distance increases, however, some type of cooler storage facility is needed to preserve plant quality.

SHIPPING CONTAINER STOCK

Regardless of when the lifting window occurs, nursery plants have optimum quality when they are in the nurs-

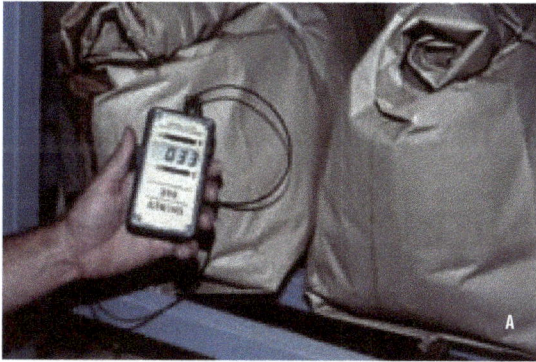

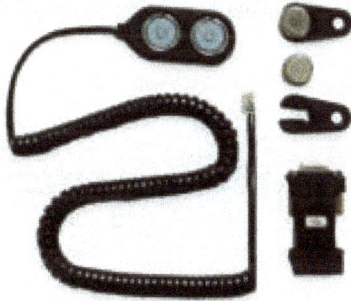

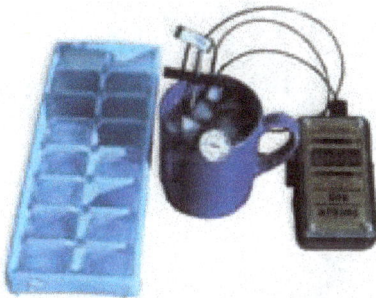

Figure 13.12—(A) Temperature can be monitored with long-stemmed electronic thermometers. (B) Small devices such as the iButton® can monitor temperature for weeks or months and the data can be downloaded to a computer. (C) Calibrate any thermometer in a water-ice mixture to make sure it is accurate. Photos A and C by Thomas D. Landis, B by David Steinfeld.

plants can dry out during overwinter storage. Although desiccation is more of a concern with evergreen species as they will transpire whenever exposed to heat and light, deciduous species can be damaged, too. Therefore, check plants routinely during the storage period and irrigate if necessary.

Some species, such as taprooted herbaceous dryland perennials, species with bulbs or fleshy roots, and some species of grasses, are especially susceptible to winter mortality when moisture is excessive (Iles and ery and are most vulnerable when they are shipped to the outplanting site. Therefore, it is important to be particularly cautious when handling nursery stock to make sure that it has the best chance for survival and growth. This is particularly true of plants not stored under refrigeration because they will be less dormant and hardy. The true impact of careless handling is not immediately apparent; it causes a degree of sublethal injury that will be observed only as a decrease in survival and growth after outplanting. Because all types of abuse or exposure are cumulative, think of nursery plant quality as a checking account. Plants are at 100 percent of quality when they are at the nursery, and all stresses are withdrawals from the account (figure 13.13). Note that it is impossible to make a deposit; nothing can be done to increase plant quality after a plant leaves the nursery.

Minimizing Exposure

Nursery plants should be handled with care and protected from direct exposure to sun and wind. Research has shown that desiccation is more of a concern than warm temperatures during shipping. A comprehensive evaluation of the various types of stresses affecting plants during storage, handling, and outplanting revealed that desiccation of the root system was the most damaging factor and that direct sunlight and high temperatures were significant only as they increased moisture stress. Special tarps are available that reflect sunlight and insulate nursery plants during the shipping process (figure 13.14A). Never place non-reflective tarps directly on seedling boxes; this practice will increase the temperature more than if the sun shines directly on them. If plants have to be carried long distances, some nurseries "jellyroll" plants to protect roots from desiccation. The process consists of aligning the plants with the roots folded in a wet cloth, rolling them into a bundle, and securing the bundle (figure 13.14B).

Physical Shock

Nursery stock should always be handled with care during shipping. Although research has shown that plants are relatively tolerant of vibration and dropping, all types of mishandling are cumulative and should be avoided. Placing container plants in cardboard boxes provides some protection and white boxes are best (fig-

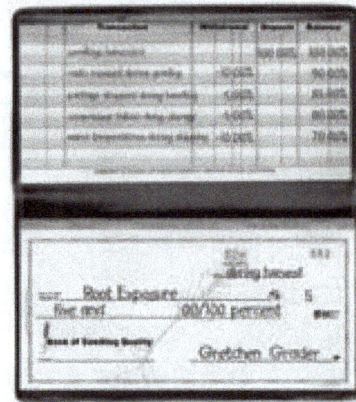

Figure 13.13—*It is useful to think of nursery plant quality as a checking account; all types of abuse or stress are cumulative and can only decrease quality.* Illustration by Jim Marin.

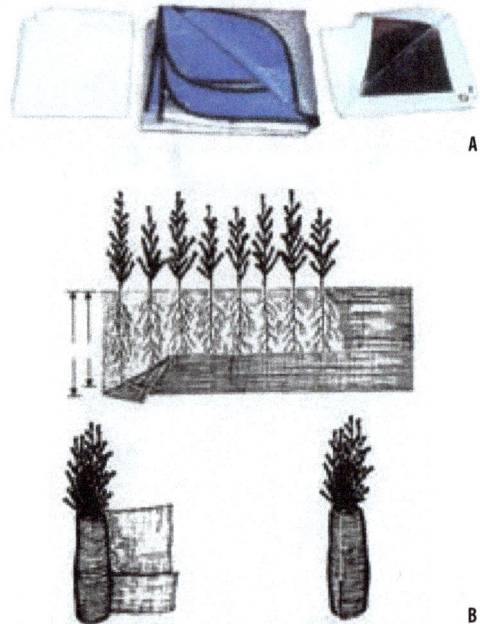

Figure 13.14—*(A) Special tarps protect nursery plants from direct exposure to sun and wind during shipping. (B) Jellyrolling has been used when plants have to be transported for long distances.* Photo by Thomas D. Landis, illustration by Jim Marin.

ure 13.15A). During shipping, plants should be stored under an insulating tarp or in a white, insulated truck box. For large shipments, truck boxes can be equipped with racks to protect the plants (figure 13.15B).

CONCLUSIONS AND RECOMMENDATIONS

The successful overwintering of container plants is one of the most challenging and important aspects of nursery management. To ensure maximum winter survival of container plants, growers should provide adequate nutrition during the growing season, properly harden stock during late summer and autumn, determine root hardiness by species, and provide the degree of protection needed by the species or stock type. Many types of overwintering systems can be employed, and, depending on location, climate, and the species grown, more than one system may need to be used.

Determining when it is safe to harvest plants so they maintain a high level of quality throughout the storage period is one of the most challenging parts of nursery management. Nursery managers must use a storage facility that compliments the condition of plants at harvest time and maintains their peak physiological condition until outplanted. Clients should work as a team with their nursery to maintain stock quality. The key concept to remember is that all stresses are both damaging and cumulative and that seedling quality cannot be improved after harvest.

Figure 13.15—(A) Placing container stock in boxes help protect them from physical shock and (B) they should be handled carefully during shipping. Photos by Thomas D. Landis.

LITERATURE CITED

Barnes, H.W. 1990. The use of bubble-pac for the overwintering of rooted cuttings. The International Plant Propagators' Society, Combined Proceedings. 40: 553-557.

Green, J.L.; Fuchigami, L.H. 1985. Special: overwintering container-grown plants. Ornamentals Northwest Newsletter. 9(2): 10-23.

Iles, J.K.; Agnew, N.H.; Taber, H.G.; Christians, N.E. 1993. Evaluations of structureless overwintering systems for container-grown herbaceous perennials. Journal of Environmental Horticulture. 11: 48-55.

Mandel, R.H. 2004. Container seedling handling and storage in the Rocky Mountain and Intermountain regions. In: National proceedings, forest and conservation nursery associations—2003. Proceedings. RMRS-P-33. Ft. Collins, CO: U.S. Department of Agriculture, Forest Service, Rocky Mountain Research Station: 8-9.

Mathers, H.M. 2003. Summary of temperature stress issues in nursery containers and current methods of production. HortTechnology. 13(4) 617-624.

Mathers, H.M. 2004. Personal communication. Assistant professor, extension specialist: nursery and landscape, Department of Crop and Soil Science, Ohio State University, Columbus, OH.

Perry, L.P. 1990. Overwintering container-grown herbaceous perennials in northern regions. Journal of Environmental Horticulture. 8: 135-138.

Whaley, R.E.; Buse, L.J. 1994. Overwintering black spruce container stock under a Styrofoam SM insulating blanket. Tree Planters' Notes. 45 (2): 47-52.

White, B. 2004. Container handling and storage in Eastern Canada. In: Riley, L.E.; Dumroese, R.K.; Landis, T.D., technical coordinators. National proceedings, forest and conservation nursery associations—2003. Proceedings. RMRS-P-33. Ft. Collins, CO: U.S. Department of Agriculture Forest Service, Rocky Mountain Research Station: 10-14.

ADDITIONAL READINGS

Davis, T. 1994. Mother nature knows best. Nursery Manager 10(9): 42-45.

Whitcomb, C.E. 2003. Plant production in containers II. Stillwater, OK: Lacebark Publications and Research. 1129 p.

APPENDIX 13.A. PLANTS MENTIONED IN THIS CHAPTER

bitterbrush, *Purshia* species

black walnut, *Juglans nigra*

cedars, *Thuja* species

eastern dogwood, *Cornus florida*

junipers, *Juniperus* species

sagebrush, *Artemisia* species

sedges, *Carex* species

Beneficial Microorganisms

Kim M. Wilkinson

14

The web of life depends on microorganisms, a vast network of small and unseen allies that permeate the soil, water, and air of our planet. For people who work with plants, the greatest interest in microorganisms is in the complex communities that are part of the soil. Beneficial microorganisms are naturally occurring bacteria, fungi, and other microbes that play a crucial role in plant productivity and health. Two types of beneficial microorganisms, mycorrhizal fungi (figure 14.1) and nitrogen-fixing bacteria (figure 14.2), are considered beneficial to plant health. Mycorrhizal fungi and nitrogen-fixing bacteria are called "microsymbionts" because they form a symbiotic (mutually beneficial) relationship with plants.

In natural ecosystems, the root systems of successful plants have microbial partnerships that allow plants to survive and grow even in harsh conditions. Without their microsymbiont partners, plants become stunted and often die. Frequently, these failures are attributed to poor nursery stock when the real problem was the lack of the proper microorganism. In the nursery, microsymbionts can be introduced by "inoculating" the root systems of the plants with the appropriate beneficial microorganism to form an effective partnership. Inoculation methods usually incorporate the microorganisms into the growing media.

WHY USE BENEFICIAL MICROORGANISMS IN THE NURSERY?

As discussed in Chapter 2, *The Target Plant Concept*, plants for land restoration may in some ways be considered a root crop. In natural ecosystems, the root

Ectomycorrhizae on pine by Thomas D. Landis.

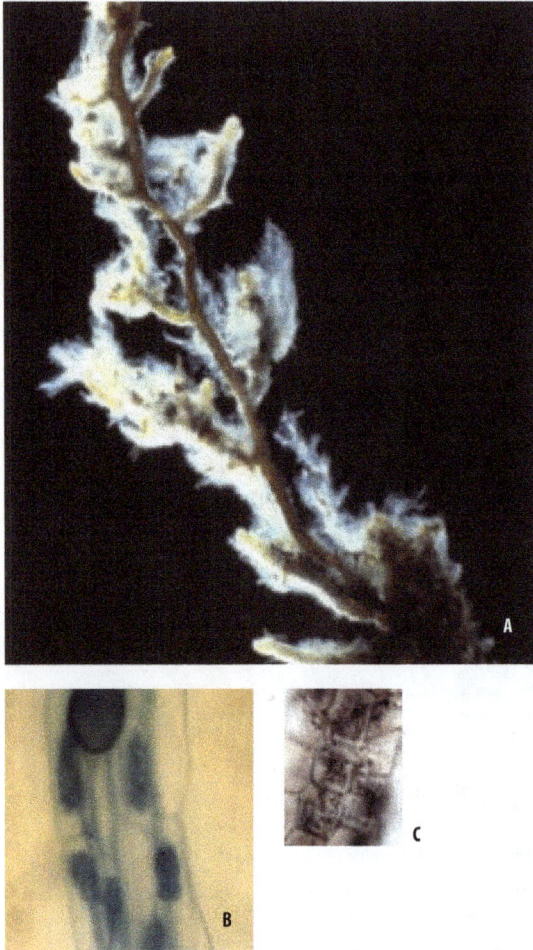

Figure 14.1—*The three types of mycorrhizal fungi are (A) ectomycorrhizal, which are visible to the naked eye, (B) arbuscular, and (C) ericoid, which are seen only with the aid of a microscope.* Photos A and B by Michael A. Castellano, C by Efren Cazares.

systems of successful plants have microbial partnerships with mycorrhizal fungi and, if applicable, with nitrogen-fixing bacteria. In the nursery, where plants have easy access to water and fertilizer, the benefits of these partnerships may not be apparent and their absence may go unnoticed. But, in the field, plants need every advantage. Plants that leave the nursery with microbial partnerships are better able to survive independently. Uninoculated plants have to "fend for themselves" to find microbial partners in the field and, while doing so, may grow poorly or die due to the stresses of outplanting or the harsher conditions of the outplanting site. In addition, many plantings take place

on deforested or degraded land. Because microsymbionts often do not survive in soil in the absence of their host plants, native populations of microsymbionts may be low or nonviable.

Inoculating plants in the nursery is an opportunity to introduce select microbial partners. Similar to using select seeds, the nursery manager can match plants with optimal microbial partners for site conditions. The use of select microsymbionts has been shown to greatly improve plant survival, productivity, and growth. For this reason, the presence of microsymbionts is often an important target plant characteristic.

Using microsymbionts in the nursery has the following benefits:

— Reduced fertilizer use in the nursery.
— Improved plant health and vigor.
— Improved resistance to disease during the germination and establishment phases.
— Provided the opportunity to introduce superior or selected partnerships to meet the needs of the site conditions.

Inoculated plants may establish more quickly with less external inputs such as fertilizers, thereby reducing costs. For situations in which revitalized ecosystem function is a project goal, using plants that are already fixing nitrogen or partnered with mycorrhizal fungi can help accelerate the process.

MYCORRHIZAL FUNGI

Mycorrhizal fungi form partnerships with most plant families and forest trees. "Myco" means "fungus" and "rhizae" means "root;" the word "mycorrhizae" means "fungus-roots." Many plants depend on their partnership with these fungi. The host plant roots provide a convenient substrate for the fungus and supply food in the form of simple carbohydrates. In exchange for this free "room-and-board," the mycorrhizal fungi offer benefits to the host plant:

1. Increased water and nutrient uptake—Beneficial fungi help plants absorb mineral nutrients, especially phosphorus and several micronutrients such as zinc and copper. Mycorrhizae increase the root surface area, and the fungal hyphae access water and nutrients beyond the roots. In the field, when plants lack

mycorrhizae, they become stunted and sometimes "chlorotic" (yellow) in appearance.

2. Stress and disease protection—Mycorrhizal fungi protect the plant host in several ways. With some fungi, the mantle completely covers fragile root tips and acts as a physical barrier from dryness, pests, and toxic soil contaminants. Other fungi produce antibiotics which provide chemical protection.

3. Increased vigor and growth—Plants with mycorrhizal roots survive and grow better after they are planted out on the project site. This effect is often difficult to demonstrate but can sometimes be seen in nurseries where soil fumigation has eliminated mycorrhizal fungi from the seedbed. After emergence, some plants become naturally inoculated by airborne spores and grow much larger and healthier than those that lack the fungal symbiont.

Mycorrhizal fungi and nitrogen-fixing bacteria are the two main microsymbionts of concern to nurseries. Mycorrhizal fungi form partnerships with most plant families and all forest trees. Most trees depend on their partnership with these fungi. The fungi, present on plant root systems, enhance plant uptake of nutrients (especially phosphorus) and water and protect the roots from soil pathogens. The following three types of mycorrhizal fungi are important to native plant nurseries:

— Ectomycorrhizae (ECM). These fungi form partnerships with many temperate forest plants, especially pines, oaks, beeches, spruces, and firs (figure 14.1A)
— Arbuscular Mycorrhizae (AM). These fungi are found on most wild and cultivated grasses and annual crops; most tropical plants; and some temperate tree species, including cedars, alders, and maples (figure 14.1B)
— Ericoid Mycorrhizae. These fungi form partnerships with plants in the families of heath (Epacridaceae); crowberry (Empetraceae); sedge (Cyperaceae); and most of the rhododendrons (Ericaceae), including the genera with blueberries, cranberries, crowberries, huckleberries, kinnikinnick, azaleas, and rhododendrons (figure 14.1C).

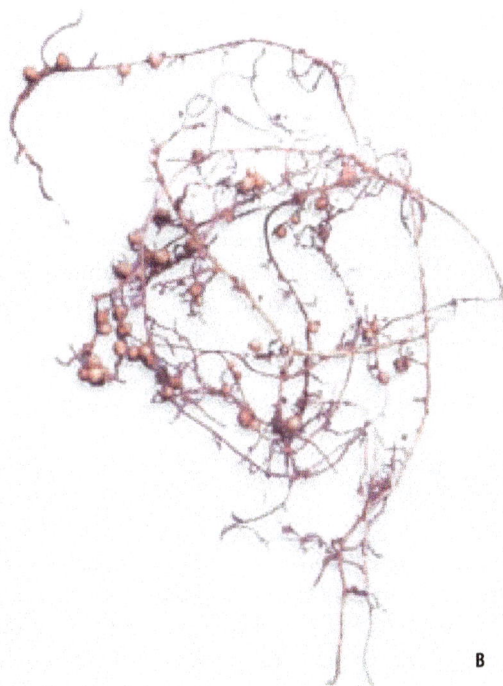

Figure 14.2—*Nitrogen-fixing bacteria include* Rhizobium *that form relationships with plants in the legume family. Rhizobium nodules on the roots of (A) leadplant and (B) acacia.* Photos by Tara Luna.

These fungi may be obtained from either commercial suppliers or from soil collected around a healthy host plant of the species being propagated. In all cases, inoculum must physically contact living roots of the plant in order to colonize effectively. Ways to acquire and successfully apply mycorrhizal fungi are explained in subsequent sections. While the fungi are similar in how they function and in their benefits to host plants, they appear differently on roots. Each mycorrhizal type has a unique application method that must be described independently. However, management practices in the nursery are similar and will be discussed together. The important thing to remember is that different plant species have specific fungal partners that must be matched appropriately to be effective (table 14.1).

ECTOMYCORRHIZAE (ECM) FUNGI

What are ECM fungi and what plants do they affect? Many familiar mushrooms, including puffballs and truffles are fruiting bodies of ECM fungi. These fruiting bodies are a small, aboveground portion of the total organism; underground, the fungus covering the short feeder roots of plants may be enormous. They form easily visible structures on roots and have great importance to many temperate forest species, especially evergreens and hardwoods in the beech and birch families. ECM fungi extend the volume of the feeding area of roots by many times and protectively coat the feeder roots.

SOURCES AND APPLICATION OF ECM FUNGI

Three common sources of ECM fungi inoculants are soil, spores, or pure-culture inoculant.

Soil

Historically, topsoil, humus, or duff from beneath ECM fungi host trees has been used to inoculate nursery plants. This practice is more common in bareroot nurseries than in container nurseries. Sometimes litter, humus, or screened pine straw from the forest floor has also been used. Because sterilization would kill these beneficial fungi, unsterilized soil and organic matter are incorporated into the growing media, up to 10 percent by volume. Three disadvantages of using soil as a source of inoculants are that (1) large quantities of soil are required, which can make the process labor intensive and have a detrimental effect on the natural ecosystem, (2) the quality and quantity of

Table 14.1—Plants and their mycorrhizal partners
(after Landis and 1989)

Ectomycorrhizae

birch	larch
Douglas-fir	oak
fir	pine
hemlock	spruce

Ectomycorrhizal and arbuscular mycorrhizal fungi

juniper
poplar
walnut

Arbuscular mycorrhizal fungi

ash	sweetgum
cherry	sycamore
maple	tuliptree
plum	western redcedar
redwood	

Ericoid mycorrhizal fungi

azalea	kinnikinnick
blueberry	pipsissewa
cranberry	rhododendron
crowberry	rushes
huckleberry	sedges

spores may be highly variable, and (3) pathogens may be introduced along with the inoculant. Therefore, using soil as a source of inoculant is discouraged unless spores or pure-culture inoculum is unavailable. If soil is used, inoculum should be collected from plant communities near the outplanting site. Small amounts should be collected from several different sites and care should be taken not to damage the host plants.

Spores

Nurseries can make their own inoculum from spores. Collected from the fruiting bodies of mushrooms, puffballs, and especially truffles (figure 14.3), these fruiting bodies, full of spores, are rinsed, sliced, and pulverized in a blender for several minutes. The resulting thick liquid is diluted with water and poured into the growing media of germinating seedlings or newly rooted cuttings. Plants are usually inoculated 6 to 12 weeks after sowing or striking (figure 14.4). Two applications 2 to 3 weeks apart are recommended to ensure even inoculation. Some spore suspensions are also available from commercial suppliers of mycorrhizal inoculants. The quality of commercial sources varies, however, so it is important to have this verified.

Pure-Culture Inoculum

Mycorrhizal fungi are also commercially available as pure cultures, usually in a peat-based carrier (figure 14.5). Most commercial sources contain several different species of ECM fungi. The quality of commercial sources can be variable; it is important to make sure a product with a verified high spore count is applied. Commercial inoculum can be purchased separately and mixed into the growing medium as per the instructions on the product and prior to filling containers, or purchased already mixed into bales of growing medium. Using commercial sources may be the easiest way to begin learning how to acquire and apply inoculant. Because the inoculum is from pure cultures, finding selected strains to match site needs may be difficult.

VERIFYING THE EFFECTIVENESS OF ECM FUNGI INOCULATION

With practice, nursery staff can learn to recognize ECM fungi on the root systems of plants—they are fairly easy to see. During the hardening phase, short feeder roots should be examined for a cottony white appearance on the roots (figure 14.6A) or a white or brightly colored mantle or sheath over the roots (figure 14.6B). Unlike pathogenic fungi, mycorrhizae will never show signs of root decay and the mycelia around the root will be visible. Sometimes, mushrooms or other fruiting bodies will occasionally appear in containers alongside their host plants (figure14.6C). While these structures are visible to the unaided eye, it is also recommended to send plant samples to a laboratory for verification. A local soil extension agent or university can likely assist with this process.

ARBUSCULAR MYCORRHIZAE (AM) FUNGI

What are AM fungi and what plants do they affect? More than 80 percent of the world's plant families are associated with AM fungi. AM fungi can be observed only under a microscope and are essential for many

Figure 14.3—*Fruiting bodies of ectomycorrhizal fungi, such as (A)* Boletus satanus *common under oaks, (B)* Cantharellus cibarius *(chantrelles) common under spruces, and (C)* Amanita muscaria *common under pines, can be collected and used as a source of inoculum for native plant nurseries.* Photo A by Dan Luoma, B by Thomas D. Landis, C by Michael A. Castellano.

Figure 14.4—*Inoculating tree seedlings with ectomycorrhizal fungi. Two applications 2 to 3 weeks apart are recommended to ensure even inoculation.* Photo by Michael A. Castellano.

Tablets for sowing

P-V Carrier for incorporation

Powder for root dip slurry

Packet for outplanting

Figure 14.5—*Forms of commercial inoculant available for nurseries include spores in a carrier to be used at the nursery or at the outplanting site.* Photo by Thomas D Landis.

temperate hardwood trees, most annual crops and grasses, and most tropical trees and plants. AM fungi mycelia penetrate plant roots and extend far into soil, taking up nutrients and water for the plant. Unlike ECM fungi, however, AM fungi on the roots of plants are not visible to the unaided eye, and their much larger spores are not easily disseminated by wind or water.

SOURCES AND APPLICATION OF AM FUNGI

Inoculant for AM fungi may be collected from soil underneath AM fungi host plants and incorporated into the growing media. As with ECM fungi, this method is discouraged because of damage to natural

ecosystems, variability of effectiveness, and risk of introducing pests and pathogens. Two main sources of AM fungi inoculant for nurseries are "pot culture," also known as "crude" inoculant, and commercially available pure cultures. Because AM spores are relatively large, it is critical to ensure that spores come in to direct contact with the root systems or seeds. Spores will not easily pass through irrigation injectors or nozzles, and do not move downward or through the soil with water. Therefore, thorough incorporation into the growing media is necessary.

Pot Culture

In pot culture inoculant, a specific AM fungi species is acquired either commercially or from a field site as a starter culture and then added to a sterile growing medium. A host plant such as corn, sorghum, clover, or an herbaceous native plant, is then grown in this substrate; as the host grows, the AM fungi multiply in the medium (figure 14.7). Shoots of host plants are removed and the substrate, now rich in roots, spores, and mycelium, is chopped up and incorporated into fresh growing medium before containers are filled and seeds are sown or cuttings stuck. This technique is highly effective for propagating AM fungi in the nursery. For details on how to use this method, consult the publications in the Literature Cited section of this chapter, particularly the book *Arbuscular Mycorrhizas: Producing and Applying Arbuscular Mycorrhizal Inoculum* (Habte and Osorio 2001).

Commercial Culture

Commercial sources of AM fungi inoculant are also available, usually containing several species or strains. These products are thoroughly incorporated into the growing medium before filling containers. Because AM fungi spores are so small and fragile, they are usually mixed with a carrier such as vermiculite or calcined clay to aid in application. Products are meant for incorporation into soil or growing media. Inoculation effectiveness has been shown to vary considerably between different products so it is wise to test before purchasing large quantities of a specific product. Laboratories can provide a live spore count per volume, which is the best measure of inoculum quality.

Verifying the Effectiveness of AM Fungi Inoculation

Unlike ECM fungi, AM fungi are not visible to the unaided eye. To verify the effectiveness of AM fungi inoculation, roots must be stained and examined under a microscope. This verification can often be done easily and inexpensively through a soil scientist at a local agricultural extension office.

ERICOID MYCORRHIZAL FUNGI

Plants that form partnerships with ericoid mycorrhizal fungi are able to grow in nutrient-poor soils and harsh conditions, including bogs, alpine meadows, tundra, and even in soils with high concentrations of certain toxic metals. Ericoid mycorrhizal fungi form partnerships in the plant order Ericales in the heath (Epacridaceae), crowberry (Empetraceae), and most of the rhododendron (Ericaceae) families. Similar to ECM fungi and AM fungi, ericoid mycorrhizal fungi must come into contact with the roots of host plants to form partnerships. Ericoid mycorrhizal inoculant is available as commercial cultures or from soil near healthy host plants. The product or soil is mixed into nursery growing medium. The fungus forms a net over the narrow "root hairs" of the plants, infecting the outer cells. The plant's cell membrane grows to envelop the fungal hyphae. Nutrients are shared through the membrane that forms the boundary between the fungus and plant roots. Laboratory confirmation is recommended to verify that successful inoculation has taken place.

MANAGEMENT CONSIDERATIONS FOR MYCORRHIZAL FUNGI

As with the introduction of any new practice, it takes time to learn how to work with mycorrhizal fungi. In some cases, working with a pure-culture product may be the easiest way to begin; the nursery can then expand into collecting and processing its own inoculant sources, if desired. It should be noted that some select or pure-culture inoculants may support high productivity in certain site conditions but may be less productive than native strains on other sites. If possible, the nursery can work with a specialist to help with the following tasks:

— Determining the best sources of inoculant and evaluating their effectiveness in the nursery.

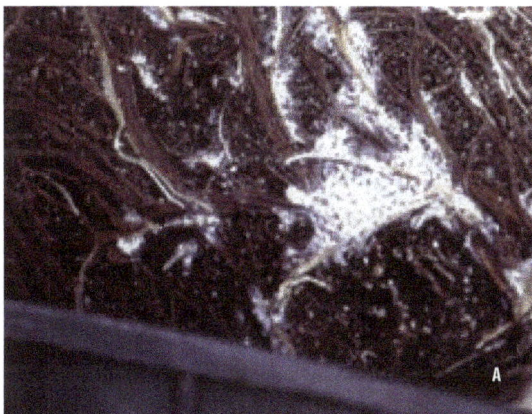

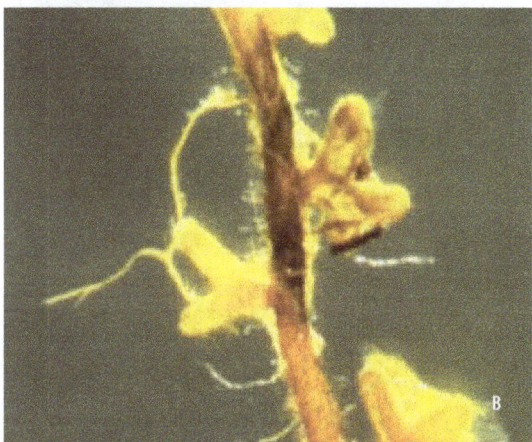

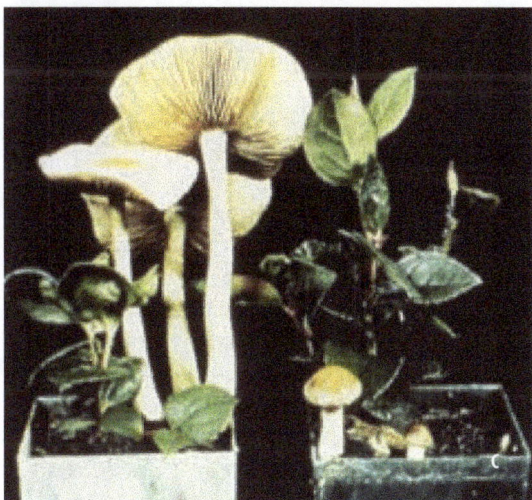

Figure 14.6—*Staff can learn to recognize the presence of ectomycorrhizal fungi (ECM) by examining plants during the hardening phase. (A) Spruce with and without cottony-white ectomycorrhizae on the roots, (B) brightly colored ectomycorrhizal fungi on the roots of lodgepole pine, and (C) fruiting bodies growing from containers.*
Photo A by William Sayward, B and C by Michael A. Castellano.

Figure 14.7—*In pot culture inoculant, a specific arbuscular mycorrhizal fungi species is acquired as a starter culture and added to a sterile potting medium with a fast-growing host plant. The shoots of host plants are removed, and the substrate, now rich in roots, spores, and mycelium, is chopped up and incorporated into the growing medium before containers are filled.* Photo by Tara Luna.

- Selecting optimal mycorrhizal partners for the species and outplanting sites.
- Designing outplanting trials to evaluate plant survival and mycorrhizal performance in the field and modifying the inoculant sources if improvements are needed.

In addition to learning how to effectively apply mycorrhizal fungi, some modifications of nursery management will be required to support the formation of mycorrhizal partnerships in the nursery. These modifications usually involve changes in fertilization and watering regimes and also likely will include changes in planting rates and scheduling.

Fertilization is probably the most significant adjustment. Mycorrhizal fungi extend the plant's root system and extract nutrients and water from the medium. High levels of soluble fertilizers in the growing medium inhibit formation of mycorrhizae. In some cases, the quantity of fertilizer used may be reduced by half or more due to the efficiency of nutrient uptake by mycorrhizal fungi. Fertilizer type and form is also important. An excessive amount of phosphorus in the fertilizer inhibits formation of the partnership; therefore phosphorus should be reduced. If nitrogen is applied, ammonium-nitrate is better used by the plant than is nitrate-nitrogen (Landis and others 1989). Generally, controlled-release fertilizers may be better than liquid fertilizers for inoculated plants because they release small doses of nutrients gradually rather than sudden higher doses. Water use is also affected; excessive or inadequate water will inhibit the presence of mycorrhizal fungi and the formation of the partnership, so watering schedules should be modified accordingly. Nursery staff who are willing to be observant and flexible as the nursery embarks on the use of mycorrhizal fungi will be the best decisionmakers in terms of modifying fertilization and watering regimes to support the microsymbionts.

Other adjustments may be necessary. Mycorrhizal inoculants may improve survival percentages of plants; these percentages will affect estimates and oversow rates. Scheduling may also be affected; inoculated plants may be ready for outplanting sooner than uninoculated plants. Applications of certain herbicides, insecticides, fungicides, and nematicides are detrimental to mycorrhizal fungi; susceptibility varies by species and pesticide applications should be assessed and adjusted.

When using mycorrhizal inoculants for the first time, it is recommended to start small and evaluate a few techniques and sources. Compare some trays or benches with and without mycorrhizae to determine how management and scheduling will be modified to support the presence of the fungi. Monitor the effectiveness of inoculation and keep records of crop development.

It should be noted that some plants form partnerships with both AM fungi and ECM fungi. These plants include juniper, poplar, and walnut (see table 14.1). In these cases, both kinds of mycorrhizal inoculant can be applied in the nursery for best results.

WHAT ARE NITROGEN-FIXING BACTERIA?

Nitrogen-fixing bacteria accumulate ("fix") nitrogen from the air. Many species live in root nodules and thereby share the nitrogen with their host plants. Unlike mycorrhizal fungi, which affect most trees and plants, only a fraction of native plants can form partnerships with nitrogen-fixing bacteria. The role of nitrogen-fixing bacteria and their partner plants in land restoration and ecosystem health is crucial. Nitrogen-fixing species are usually outplanted to help restore fertility and organic matter to the ecosystem. Soil at the outplanting site, however, may not contain a viable strain of bacteria to form a symbiotic partnership with the plant. Inoculating plants in the nursery ensures that nitrogen-fixing plants form an effective partnership to fix nitrogen. For nurseries that grow nitrogen-fixing plants, the introduction of nodule-forming bacteria to plant root systems is a key practice in meeting target plant requirements. Using these inoculants in the nursery can be an important part of accelerating rehabilitation of degraded land by enhancing plant survival and growth.

Without the bacterial partnership in place, plants are unable to make direct use of atmospheric nitrogen for their own fertility. Although plants that form this association are sometimes called "nitrogen-fixing plants" or "nitrogen-fixing trees," the plant itself is not able to accumulate nitrogen from the air. It is only through the partnership with nitrogen-fixing bacteria that these plants are able to obtain atmospheric nitrogen. It is a symbiotic partnership: the bacteria give nitrogen accumulated from the atmosphere to the plant, and in exchange the bacteria get energy in the form of carbohydrates from the plant.

Two types of nitrogen-fixing bacteria are used by plants: *Rhizobium* and *Frankia*. *Rhizobium* grow with some members of the legume family and plants of the elm family (Ulmaceae) (figure 14.8). *Frankia* are a kind of bacteria that form partnerships with about 200 different plant species distributed over eight families (figure 14.9). The species affected by *Frankia* are called "actinorhizal" plants (table 14.2).

Many, but not all, species of the legume family nodulate with *Rhizobium*. This family is made up of three subfamilies. Although most legume species in temperate regions of the world fix nitrogen, the subfamily Caesalpinioideae is mostly tropical in distribution and has fewer species that fix nitrogen.

HOW DOES BIOLOGICAL NITROGEN FIXATION WORK?

This symbiotic partnership between plants and their nitrogen-fixing microsymbionts consists of bacteria

Figure 14.8—*Legume species that form relationships with the nitrogen-fixing bacteria* Rhizobium *include (A) lupine, (B) American licorice, and (C) acacia. Because these species improve soil fertility on degraded lands, they are widely used for restoration projects* Photos by Tara Luna.

Figure 14.9—*Species that form relationships with nitrogen-fixing* Frankia *bacteria include (A) buffaloberry, (B) deerbrush, and (C) mountain-avens. As with legumes, these species improve soil fertility on degraded lands.* Photos by Tara Luna.

living in nodules on the roots of the plant. Each nodule contains millions of the bacteria that accumulate atmospheric nitrogen. The bacteria share this nitrogen with the plant and in exchange receive energy (in the form of carbohydrates produced by photosynthesis).

When the nitrogen-fixing plant sheds leaves, dies back, or dies, the nitrogen stored in the plant's tissues is cycled throughout the ecosystem. This process, part of the nitrogen cycle, is the major source of nitrogen fertility in most natural ecosystems (figure 14.10).

BENEFITS OF INOCULATING WITH NITROGEN-FIXING BACTERIA

Application of inoculants of nitrogen-fixing bacteria can have some direct benefits in the nursery. If an effective partnership is formed, most of the plant's needs for nitrogen will be met. As a result, the need to apply nitrogen fertilizer is reduced or eliminated. This partnership also reduces management decisions associated with nitrogen fertilizer application, such as considerations about when to fertilize and what type of nitrogen to use. The plant is essentially self-rationing; if a productive partnership is formed, the plant can usually take up as much nitrogen as it needs through the partnership. The reduction of soluble nitrogen applications also decreases the nursery's contribution to pollution from fertilizer runoff. Nitrogen-fixing nursery stock with nodulated root systems have exhibited faster early growth than seedlings that were not inoculated.

In the field, however, is where the benefits of the partnership are most apparent. Plants sent from the nursery with their root systems already nodulating have exhibited faster early growth than plants that were not inoculated. Nursery inoculation can reduce costs in establishment and maintenance; a few dollars worth of inoculant applied in the nursery can replace a lot of purchased nitrogen fertilizer over the life of the tree. Also, instead of providing spurts of soluble fertilizers (which may benefit surrounding weeds as well as the desired plant), the natural nitrogen fixation process provides a steady supply of nitrogen for the plant's growth. Faster growth early on can also lead to faster canopy closure, shading the soil and understory and reducing expenses of weed management. Early formation of the partnership with nitrogen-fixing bacteria also means a faster restoration of the natural nutrient cycling and fertility role of nitrogen-fixing species in

Table 14.2—What native plants form partnerships with nitrogen-fixing bacteria? (adapted from NFTA 1986)

Bacteria	Family	Subfamily (notes)	Common Names
Frankia	Birch (Beulaceae)		alder, birch
	Buckthorn (Rhamnaceae)		cascara, ceanothus, deerbrush
	Myrtle (Myricaceae)		myrtle
	Oleaster (Elaeaganaceae)		buffaloberry, silverberry
	Rose (Rosaceae)		bitterbrush, cliffrose, fernbush, mountain mahogany, mountain-avens
Rhizobium	Legume (Fabaceae)	Caesalpinioideae (about 1,900 species; about 23% fix nitrogen)	cassia, honeylocust, Kentucky coffeetree, nicker, redbud
		Mimosaideae (2,800 species; about 90% fix nitrogen)	acacia, albizia, mesquite, mimosa (2,800 species;
		Papilionoideae (about 12,300 species; about 97% fix nitrogen)	American licorice, black locust, clover, Indian breadroot, indigobush, leadplant, locoweed, lupine milkvetch, prairie clover, sesbania, sweetvetch, vetch

the ecosystem, which is usually the purpose of planting nitrogen-fixing species in the first place.

In many cases, uninoculated plants may eventually form a partnership with some kind of *Frankia* or *Rhizobium* strain after they are outplanted. These partnerships may not be with optimal or highly productive bacterial partners, and it may take months or even years for the partnerships to form if effective microsymbiont populations in the soil are low or inactive. Until the partnership forms, the plants are dependent on inputs of nitrogen fertilizers or the nitrogen available in the soil. Without fertilizer on poor sites, uninoculated plants will grow very slowly, sometimes outcompeted by weeds. Inoculating in the nursery ensures that plants form effective, productive partnerships in a timely fashion.

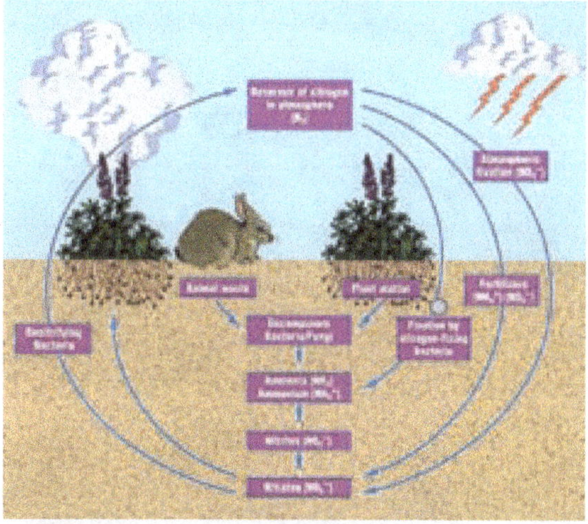

ACQUIRING INOCULANT FOR NITROGEN-FIXING BACTERIA

Inoculants for nitrogen-fixing bacteria tend to be very specialized. In other words, they are not "one size fits all." On the contrary, a different inoculant strain for each nitrogen-fixing species is usually necessary. Inoculants are live nitrogen-fixing bacteria cultures that are applied to seeds or young plants, imparting the beneficial bacteria to the plant's root system. Two forms of inoculant can be used in the nursery: pure-culture inoculant (figures 14.11A and B), and homemade (often called "crude")

Figure 14.10—*The nitrogen cycle plays a central role in fertility in natural ecosystems; beneficial microorganisms are critical to this process.* Illustration by Jim Marin.

inoculant (figure 14.11C). Cultured inoculant is purchased from commercial suppliers, seed banks, or sometimes, universities. Crude inoculant is made from nodules collected from the roots of nitrogen-fixing plants of the same species to be inoculated. Whichever form is used, care should be taken when handling nitrogen-fixing bacteria inoculants because they are very perishable. These soil bacteria live underground in moist, dark conditions with relatively stable, cool temperatures. Similar conditions should be maintained to

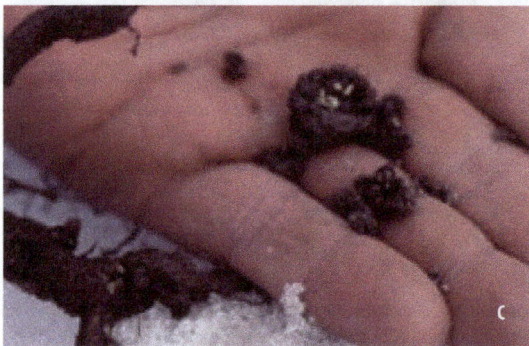

Figure 14.11—*Nitrogen-fixing bacteria are commercially available as (A) pure-culture inoculants, (B) often in a carrier. (C) They can also be prepared by collecting nodules from plants in the wild.* Photos A and C by Tara Luna, B by Mike Evans.

ensure the viability of inoculant during storage, handling, and application.

PREPARING PURE-CULTURE INOCULANT

Pure-culture inoculants of nitrogen-fixing bacteria usually come in small packets of finely ground peat moss. Some manufactured inoculants contain select strains that have been tested for forming optimally productive partnerships with their host species. Select-strain inoculants should be used if they can be obtained; these substances contain optimal partners for the species they were matched for, providing a good supply of nitrogen at a low cost to the plant. Superior strains can yield significant differences in the productivity and growth rate of the host plant; in some cases, they yield over 40 percent better growth (Schmidt 2000). Manufactured products usually come with application instructions; these directions should be followed. In general, about 100 g (3.5 oz) of cultured inoculant is sufficient to inoculate up to 3,000 plants. Because they contain living cultures of bacteria, these inoculants are perishable and should be kept in cool, dark conditions, such as inside a refrigerator.

Peat-based inoculants are added to chlorine-free water to create a slurry. (Allowing a bucket of water to stand uncovered for 24 hours is a good way to let chlorine evaporate.) If a blender is available, using it to blend some inoculant in water is a good practice to ensure that the bacteria will be evenly mixed in the solution. If a blender is not available, a mortar and pestle or a whisk can be used. Five to 10 g (about 0.2 to 0.4 oz) of manufactured inoculant can inoculate about 500 plants, usually exceeding the recommended 100,000 bacteria per plant. After plants begin to nodulate, nodules from their roots can serve as the basis for making crude inoculant, as described in the following paragraphs. This way, if desired, inoculant need be purchased only once for each plant species grown and thereafter perpetuated using nodules.

PREPARING CRUDE INOCULANT

Crude inoculant is made by using nodules, the small root structures that house the bacteria. Nodules can be seen on plant roots when nitrogen is being fixed. Each of the nodules can house millions of bacteria. For *Rhizobium*, a brown, pink, or red color inside is usually a good indicator that the millions of bacteria in the nodule are actively fixing nitrogen. For *Frankia*, desirable nodules will be white or yellow inside. Gray or green nodules should be avoided, because they are likely inactive.

To make crude inoculant, select healthy, vigorous plants of the same species as the plants to be inoculated. Expose some of the root system of a nodulating plant in the nursery or field. If available, choose plants that were inoculated with select bacteria, as described previously. Young roots often contain the most active nodules. Search for nodules with the proper color and pick them off cleanly. If possible, collect nodules from several plants. Put nodules in a plastic bag or container and place them in a cooler for protection from direct sunlight and heat. As soon as possible after collection (within a few hours), put the nodules in a blender with clean, chlorine-free water. About 50 to 100 nodules blended in about 1 qt (1 L) of water is enough to inoculate about 500 plants. This solution is a homemade liquid inoculant, ready to apply in the same method as cultured inoculant, as described in the following sections.

APPLYING INOCULANT
Inoculation should take place as early in the plant's life as possible, when the plant will most readily form the association. In nursery conditions, inoculant for nitrogen-fixing bacteria is commonly applied when seedlings are just emerging, usually within 2 weeks of sowing, or just after cuttings have formed roots (figure 14.12A). This helps ensure successful nodulation and maximizes the benefits of using inoculants. The quart (1 L) of liquefied inoculant made from either nodules or cultured inoculant as per the instructions in the previous paragraphs is diluted in more chlorine-free water. For 500 plants, about 1.3 gal (5 L) of water is used. This solution is then watered into the root system of each plant using a watering can.

VERIFYING THE NITROGEN-FIXING PARTNERSHIP
Allow 2 to 6 weeks for noticeable signs, listed below, that the plant has formed a symbiotic partnership with nitrogen-fixing bacteria.

— Plants begin to grow well and are deep green despite the absence of added nitrogen fertilizer (figure 14.12B).
— Root systems give off a faint but distinctive ammonia-like scent.
— Nodules are usually visible on the root system after about 4 to 6 weeks (figures 14.12C).

— Nodules are pink, red, or brown (for Rhizobium), or yellow or white (for Frankia).

MANAGEMENT CONSIDERATIONS
As with any nursery practice, becoming familiar with the application and management of nitrogen-fixing microsymbionts is a learning process. Several factors, listed below, are of primary concern to the nursery manager when using inoculants for nitrogen-fixing bacteria.

— **Timing.** The nursery manager should be mindful that inoculant is applied in a timely fashion, when seedlings are just emerging or cuttings are just forming new roots. This timing helps ensure successful nodulation and maximizes the nursery benefits of using inoculants.
— **Fertilization.** The use of nitrogen-fixing bacterial inoculant requires some adjustments in fertilization. Excessive nitrogen fertilizer will inhibit formation of the partnership. The application of nitrogen should be eliminated from nitrogen-fixing plants, and they might need to be isolated from non-nitrogen-fixing species to achieve this result.
— **Water Quality.** Excessive chlorine in water is detrimental to Rhizobium and Frankia. The water supply may need to be tested and a chlorine filter obtained if excessive chlorine is a problem in the water supply.
— **Micronutrients.** Some nutrients, including calcium, potassium, molybdenum, and iron, are necessary to facilitate nodulation. These nutrients should be incorporated into the growing media.
— **Sourcing Inoculants.** Locating appropriate sources of viable inoculants (either cultured or obtained as nodules) may require some research and time. The assistance of a specialist is invaluable. The benefits of successful inoculation are well worth the effort.

Most plants that form partnerships with nitrogen-fixing bacteria also require mycorrhizal partners; it is fine to inoculate plants with both microsymbionts as needed. Simply apply each inoculant separately, as described previously.

OTHER BENEFICIAL MICROORGANISMS
In natural soil, communities of bacteria, fungi, algae, protozoa, and other microorganisms make nutrients available to plants, create channels for water and air, maintain soil structure, and cycle nutrients and organic

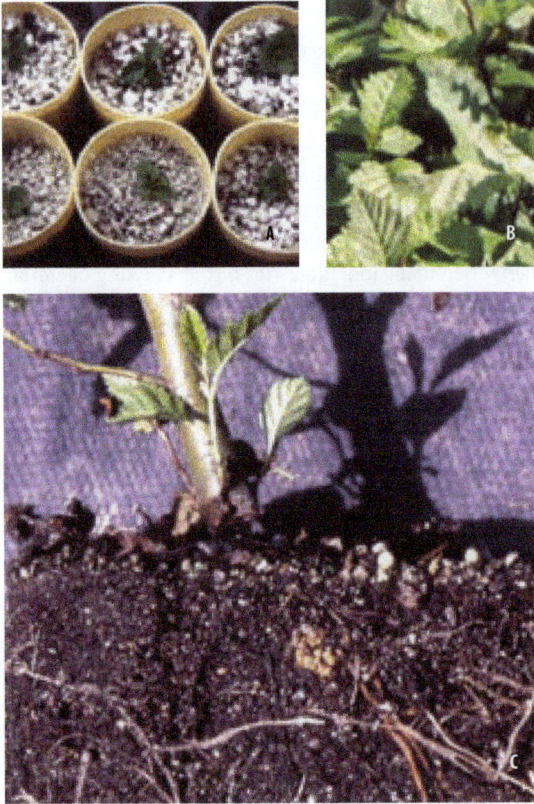

Figure 14.12—*(A) Alder seedlings, inoculated with* Frankia *at 4 weeks of age, grow well the next 2 months as (B) the partnership forms and the bacteria survive and multiply on the root system, (C) creating nodules that can be easily seen.* Photos by Tara Luna.

SUMMARY

Using mycorrhizal fungi and nitrogen-fixing bacteria in the nursery involves working with living organisms and adjusting the environment to foster them. In many cases, nurseries shifting from chemical fertilizer regimes into the use of beneficial microorganisms will have to make many adjustments to their management practices and schedule. The process of introducing microsymbionts does not involve a simple, direct replacement of chemical fertilizers. The results, however, in terms of stronger plants better able to survive independently makes taking up the challenge well worth the effort. In all, it is not as complicated as it may sound, and support can often be found through a local specialist or soil scientist.

By far, the greatest benefits of using microsymbionts in the nursery are seen in the field, with faster early growth and higher survival rates. The results in the field are the ones the nursery should focus on.

It is recommended to start small and try various sources of inoculants and options for their application. Work with a specialist or advisor if possible. Keep records and monitor results in the nursery. Verify that symbiotic partnerships are forming. After outplanting, follow up with clients to monitor the effectiveness of the microsymbionts. Also, teach clients about the importance of beneficial microorganisms: the more they understand the functions of these microsymbionts, the more they will appreciate the high quality of plant materials that have their microbial partnerships in place.

LITERATURE CITED

Habte, M.; Osorio, N.W. 2001. Arbuscular mycorrhizas: producing and applying arubscular mycorrhizal inoculum. Honolulu, HI: University of Hawai'i, College of Tropical Agriculture and Human Resources. 47 p.

Landis, T.D.; Tinus, R.W.; McDonald, S.E.; Barnett, J.P. 1989. The container tree nursery manual: volume 5, the biological component: nursery pests and mycorrhizae, Agriculture Handbook 674. Washington, DC: U.S. Department of Agriculture, Forest Service. 171 p.

Margulis, L.; Sagan, D.; Thomas, L. 1997. Microcosmos: four billion years of evolution from our microbial ancestors. Berkeley, CA: University of California Press 301 p.

NFTA. 1986. Actinorhizal trees useful in cool to cold regions. In: Roshetko, J., ed. NFTA (Nitrogen Fixing Tree Association) highlights: a quick guide to useful nitrogen-fixing trees from around the world. Morrilton, AR: Winrock International. NFTA 86-03: 1-2.

Schmidt, L. 2000. Guide to handling of tropical and subtropical forest seed. Humlebaek, Denmark: Danida Forest Seed Centre. 511 p.

matter. A healthy population of soil microorganisms can also maintain ecological balance, preventing the onset of major problems from viruses or other pathogens that reside in the soil. The practice of introducing beneficial microorganisms dates to ancient times. As a science, however, the use of beneficial microorganisms is in its infancy. Although thousands of species of microorganisms have been recognized and named, the number of unknown species is estimated to be in the millions. Almost every time microbiologists examine a soil sample, they discover a previously unknown species (Margulis and others 1997). Nursery staff should keep an eye on developments in this field and see how their plants can benefit from new insights into the roles of microorganisms. Working with these small and unseen allies is an important link to the greater web of life.

BENEFICIAL MICROORGANISMS

APPENDIX 14.A. PLANTS MENTIONED IN THIS CHAPTER

acacia, *Acacia* species

albizia, *Albizia* species

alder, *Alnus* species

American licorice, *Glycyrrhiza lepidota*

ash, *Fraxinus* species

azalea, *Rhododendron* species

beech, *Fagus* species

birch, *Betula* species

bitterbrush, *Purshia tridentata*

blacklocust, *Robinia pseudoacacia*

blueberry, *Vaccinium* species

buffaloberry, *Shepherdia* species

cascara buckthorn, *Frangula purshiana*

cassia, *Cassia* species

ceanothus, *Ceanothus* species

cherry, *Prunus* species

cliffrose, *Purshia* species

clover, *Trifolium* species

cranberry, *Vaccinium* species

crowberry, *Empetrum* species

Douglas-fir, *Pseudotsuga menziesii*

elm, *Ulmus* species

fernbush, *Chamaebatia* species

fir, *Abies* species

heath, *Erica* species

hemlock, *Tsuga* species

honeylocust, *Gleditsia triacanthos*

huckleberry, *Gaylussacia* species and *Vaccinium* species

Indian breadroot, *Pediomelum* species

indigobush, *Psorothamnus arborescens*

juniper, *Juniperus* species

Kentucky coffeetree, *Gymnocladus dioicus*

kinnikinnick, *Arctostaphylos uva-ursi*

larch, *Larix* species

leadplant, *Amorpha canescens*

locoweed, *Oxytropis* species

lodgepole pine, *Pinus contorta*

lupine, *Lupinus* species

maple, *Acer* species

mesquite, *Prosopis* species

milkvetch, *Astragalus* species

mimosa, *Mimosa* species

mountain mahogany, *Cercocarpus* species

mountain-avens, *Dryas* species

myrtle, *Myrtus communis*

nicker, *Caesalpinia* species

oak, *Quercus*

pine, *Pinus* species

pipsissewa, *Chimaphila umbellata*

plum, *Prunus* species

poplar, *Populus* species

prairie clover, *Dalea* species

redbud, *Cercis* species

redwood, *Sequoia* species

rhododendron, *Rhododendron* species

rushes, *Juncus* species

sedges, *Carex* species

sesbania, *Sesbania* species

silverberry, *Elaeagnus* species

spruce, *Picea* species

sweetgum, *Liquidambar styraciflua*

sweetvetch, *Hedysarum* species

sycamore, *Platanus* species

tuliptree, *Liriodendron tulipifera*

vetch, *Vicia* species

walnut, *Juglans* species

western redcedar, *Thuja plicata*

Holistic Pest Management

Thomas D. Landis, Tara Luna, and R. Kasten Dumroese

15

As any experienced grower knows only too well, nursery management is a continuous process of solving problems. Murphy's Law of "anything that can go wrong, will go wrong" sounds as if it were meant for native plant production.

One recurring problem is pests. Nursery managers have traditionally talked about "controlling" a pest. This approach usually involves waiting for an insect or disease to appear and then spraying some toxic chemical on the already dead or dying plants. Instead of a knee-jerk reaction to a specific problem, pest management should be a series of interrelating processes that are incorporated into the entire spectrum of nursery culture.

WHAT IS "HOLISTIC"?

Holistic is best described through an example. Holistic medicine, for instance, is a relatively recent concept in modern Western culture that emphasizes the need to perceive patients as whole persons, although it has always been a part of the traditional healing practices of Native Americans. It developed as a response to the increasing specialization in medical education that was producing physicians who treated organs rather than the body as a whole. Patients were shuttled from one doctor to another in search of a cure, but each was a specialist in one narrow field. Few could see the "big picture" or provide the patient with a comprehensive diagnosis. A related concern was the increasing dependence on drugs in the treatment of disease. Many patients

RaeAnne Jones of the Sioux Nation by Tara Luna.

felt that doctors were merely prescribing drugs to treat symptoms rather than trying to find out the real cause of the disease. Similarly, holistic pest management means looking at the big picture, not just observing symptoms but considering the overall health of the plant as well as the nursery environment when diagnosing a problem.

WHAT ARE DISEASES AND PESTS?

In native plant nurseries, a disease can be caused by biological stresses that we call pests or environmental stresses. Nurseries have many potential pests, including fungi (figures 15.1A–E), insects (figures 15.1F–J), nematodes, snails, and even larger animals such as mice (figure 15.1K). Other plant species, such as weeds and cryptogams (moss, algae, or liverworts) (figure 15.1L) can become pests when they compete with crop plants for growing space and light.

Plant disease can also be caused by abiotic (environmental) stresses, including frosts (figures 15.2A and B), heat (figures 15.2C and D), and chemicals (figures 15.2E and F). Sometimes, people are pests when they apply too much fertilizer, which causes chemical injury known as fertilizer burn (figure 15.2G).

THE "DISEASE TRIANGLE"

A useful concept to explain nursery pest problems is the "disease triangle," which illustrates the interrelationships among the pest, host, and environment (figure 15.3). All three factors are necessary to cause biotic disease. For example, a fungus or insect is able to survive inside the warm environment of a greenhouse and attack the host plant. Although many diseases may appear to involve only the host plant and the biological pest, environmental factors are always involved. Environmental stress may weaken the plant and predispose it to attack by the pest, or a particular environment may favor pest populations, enabling them to increase to harmful levels.

Abiotic disease can be visualized as a two-way relationship between the host plant and adverse environmental stress (figure 15.3). Abiotic diseases may develop suddenly as the result of a single injurious climatic incident, such as a freeze, or more gradually as a difficult-to-detect growth loss resulting from below-optimum environmental factors, such as a mineral nutrient deficiency.

APPLYING THE HOLISTIC APPROACH IN NURSERIES
The Conducive Nursery Environment

One basic tenet of holistic medicine is to recognize how much environment contributes to disease. We just discussed the disease triangle (figure 15.3), which stresses all three factors, but the environment takes an overriding role in native plant nurseries. The principle reason for raising plants in containers in a greenhouse is that all potentially growth-limiting factors can be controlled (table 15.1). Plants are sown at regular spacing in an artificial growing medium formulated for ideal pH and porosity. The atmospheric environment is automated to maintain ideal temperature and relative humidity both day and night. In some greenhouses, even carbon dioxide is supplied to accelerate photosynthesis.

Unfortunately, the ideal nursery environment is also a breeding ground for many pests. Fungus gnats are a good example. The larvae of these small flies can damage seeds or young germinants (figures 15.1G and H) but become a problem only when populations increase in greenhouses with excessively wet conditions. To make matters worse, greenhouses do not contain any natural parasites that normally control fungus gnats (table 15.1).

Recognizing that the nursery environment is well suited for disease problems, high-quality native plants can be best produced by a comprehensive approach using the following steps: (1) prevention through sanitation, (2) good crop scheduling, (3) keeping plants healthy, (4) daily monitoring and good recordkeeping, (5) accurate problem identification, (6) timely and appropriate control measures, and (7) encouraging beneficial organisms.

1. Disease Prevention through Sanitation

Keeping pests out of the nursery in the first place is critical. The container nursery environment is initially pest free, so the most logical approach to disease management is to prevent diseases by excluding pests from the growing area. All diseases are much easier to prevent than to cure.

Sanitation begins with nursery site selection. Any unnecessary vegetation should be removed because it provides cover for seed-eating rodents and birds and can harbor insect pests, including thrips, aphids, weevils, and the European crane fly. In existing nurseries,

Figure 15.1—*A collage of biotic nursery pests. Fungal pests include (A) damping-off fungi, (B) Phytophthora root canker, cedar-apple rust of (C) serviceberry leaves and (D) fruit, and (E) Botrytis blight. Insect pests include (F) aphids, larvae of fungus gnats eating (G) chokecherry seeds and (H) quaking aspen roots, (I) leaf-roller on green ash, and (J) microscopic spider mites on quaking aspen. (K) Mice and other rodents consume seeds. (L) Weeds, liverworts, and algae smother small plants.* Photos A–K by Thomas D. Landis, L by William Pink.

potential pests generally enter the growing area through the following sources:

— **Wind.** Airborne spores, seeds, or insects can be introduced through the ventilation system.
— **Water.** Fungus and cryptogam spores and weed seeds can be introduced through irrigation water.
— **Growing Media.** Most commercial mixes are considered "essentially sterile," but potentially harmful fungi have been isolated from some types of growing media or their components.
— **Containers.** Reusable containers may contain residual growing medium or plant roots that harbor pest propagules, moss, or algae from the previous crop.
— **Surfaces in the Growing Structure.** Floors, benches, and other surfaces in the growing area may harbor pests from the previous crop.
— **Propagation Materials.** Seeds, transplants, or cuttings are sometimes infected before they reach the nursery.
— **Transported Soil and Growing Media.** Infested materials can be carried into the growing area on tools, equipment, or the shoes of workers or visitors.
— **Mobile Pests.** Insects, birds, and rodents can enter the growing area directly.

2. Crop Scheduling Is Critical

Crop scheduling is an important component of holistic pest management. A typical native plant nursery will be growing a wide variety of plants with different growth rates. For example, most willows can be grown from seeds to shippable size in as little as 4 months whereas whitebark pine seedlings take 2 to 3 years to be large enough to ship. If you are growing a new species and are unsure of the growth rate, contact other nurseries, check the propagation protocols in volume 2 of this handbook, or check the Native Plant Network (http://www.nativeplantnetwork.org).

Slow-growing plants should be started first so they have maximum time to grow. Species that grow very quickly should be scheduled later in the season so that they do not become "top-heavy" and rootbound. Plants that have grown too large for their containers are easily stressed and prone to pest problems. Leafy, "leggy" overgrown plants can harbor insects and other pests can infect the rest of the crop.

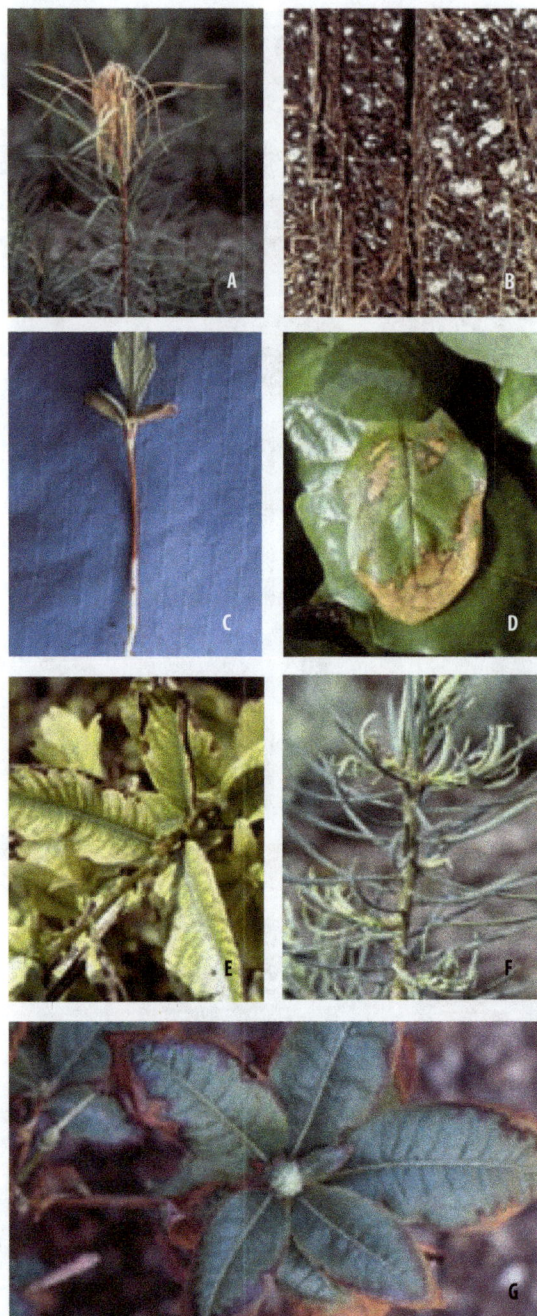

Figure 15.2—*Abiotic disease can be caused by any excessive environmental stress, including (A) frost injury to foliage and (B) especially roots, and (C) heat injury to germinants or (D) mature foliage. Chemical injuries can be caused by (E) the overliming of growing media or (F) the application of pesticides. (G) Fertilizer burn is an abiotic disease caused by improper fertilization.* Photos by Thomas D. Landis.

3. Keep Plants Healthy

A key component of any pest control system is to start with healthy crops; this practice is especially critical for holistic management. Healthy plants are more able to resist infection from fungi and attack from insects and other pests and can also tolerate environmental stresses better. Much of this resistance can be attributed to physical characteristics such as a thick, waxy cuticle on the foliage and some people believe that healthy plants have chemical defenses as well.

An important aspect of keeping plants healthy is to maintain conditions that are conducive for plant growth but do not favor pests or have damaging environmental extremes. For example, root diseases are common in all types of nurseries and are often brought on by environmental stresses. As an example, Swedish researchers studied the fungal pathogen *Cylindrocarpon destructans* in relation to root rot problems of pine seedlings in container nurseries. Suspected predisposing stress factors included excessive moisture, low light, and exposure to fungicides. They found that *C. destructans* does little harm to healthy seedlings but typically invades dead or dying roots. The fungus then uses these sites as a base for further invasion of healthy roots (Unestam and others 1989). On a practical basis, therefore, opportunistic pathogens such as *Fusarium* do not cause disease unless plants are under environmental stress (figure 15.4).

Marginal growing conditions for native plants often favor insect pests. For example, fungus gnats become a problem only under wet conditions and especially in locations where algae, moss, and liverworts have been allowed to develop. Often, these conditions exist under greenhouse benches, where water can puddle and excess fertilizer promotes their growth. This problem is particularly common where floors of greenhouses are covered by a weed barrier or gravel. Switching to concrete floors alone can often cure a fungus gnat problem (figure 15.5) because concrete dries faster and is easier to keep clean.

4. Daily Monitoring and Good Recordkeeping

Regular monitoring or "scouting" is a critical part of the holistic approach. A daily walk-through of the nursery will reveal developing pest outbreaks or horticultural problems that are conducive to pests while they are still minor and can be easily corrected. In small nurseries, the

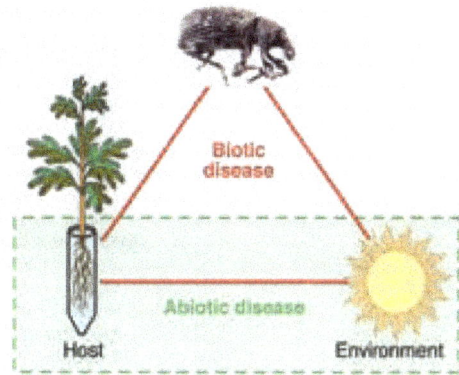

Figure 15.3—The "disease triangle" illustrates the concept that a host, a pest, and a conducive environment are necessary to cause biotic disease. Abiotic disease occurs when environmental factors, such as frost, injure the host plant. Illustration by Jim Marin.

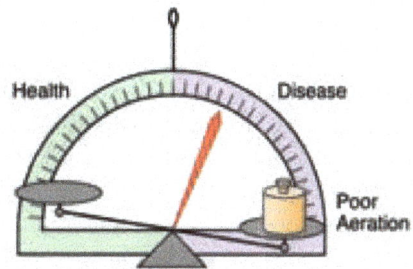

Figure 15.4—Many nursery diseases are caused by environmental stresses, such as poor aeration, which predispose plants to attack by opportunistic pests. Illustration by Jim Marin.

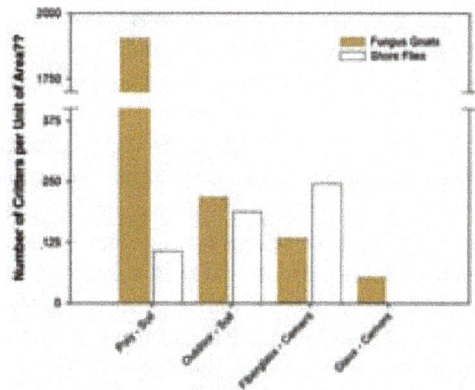

Figure 15.5—Many nursery pests, such as fungus gnats, thrive in nurseries with soil floors where they reproduce rapidly on algae, mosses, and liverworts to the point at which populations can cause significant damage to crops. Illustration by Jim Marin.

Table 15.1—The ideal propagation environment of a greenhouse consists of edaphic (pertaining to the soil) and atmospheric (pertaining to the air) factors that have advantages and disadvantages

	Advantages	Disadvantages	
	No Growth-limiting Factors in Physical Environment	Ideal Conditions for Some Pests	No Beneficial Organisms
Edaphic	Water Mineral nutrients	Fungus gnats, *Bradysia* spp.	Mycorrhizal fungi
Atmospheric	Light Humidity Carbon dioxide Temperature	Botryis blight, *Botrytis cinerea*	Insect parasites

grower or nursery manager should do the inspections, but, in larger facilities, one person should be designated as crop monitor and pest scout. This person should have nursery experience and be familiar with all the plant species at all growth stages. It is essential to know what a healthy plant looks like before you can notice any problem (figure 15.6). The person should also be inquisitive and have good observation skills. Often, irrigators serve as crop monitors because they are regularly in the nursery, checking whether plants need watering.

Crop monitors should carefully inspect each species being grown, record the temperature and environmental controls of the greenhouse, and make other observations. It is important to establish a monitoring and recordkeeping system for all areas of the nursery, including the greenhouse, rooting chambers, and outdoor nursery. These records are invaluable for avoiding future pest problems when planning the next crop. In addition to looking at crop plants, scouts should check all greenhouse equipment and monitor the growing environment. Pest scouts should carry a 10X to 20X hand lens for close inspection and a notebook or tape recorder to record observations. When a problem is noted, a camera with a closeup lens is an excellent way to document the problem (figure 15.7). All observations should be recorded in a daily log (figure 15.7), and any suspicious problems should be immediately reported to the nursery manager. Inspections should occur on a daily basis after sowing the crop and during the establishment phase, when plants are most susceptible to diseases such as damping-off. When problems are detected early, plants can be treated or isolated from the rest of the nursery or greenhouse crop.

Use yellow sticky cards (figure 15.8) to detect white flies, aphids, fungus gnats, and shore flies. Place one to four cards every 1,000 square ft (90 square m) and space them evenly in a grid pattern, with extra cards placed near the vents and doorways. Inspect these cards each week to detect and monitor these pests. Record the information and replace the cards as needed to keep track of population trends. Use the blue sticky cards, which are more attractive to thrips, around plants susceptible to this pest.

5. Accurate Problem Identification

Nursery managers and disease scouts must be able to identify problems quickly and accurately before the problems can inflict significant damage. Although biological pests such as fungi and insects are always present, abiotic stresses typically cause more problems. Disease diagnosis requires a certain degree of experience and training, and nursery workers should be trained to quickly spot new problems as well as incidents of abiotic injury. Workers who are in the growing area daily have the best chance to spot potential problems before they can intensify or spread.

Unfortunately, no single reference is available for the diseases and pests of native plant species. Some common nursery pests and causes of abiotic injury were provided in a previous section. Volume 5 of the *Container Tree Nursery Manual* (Landis and others 1989) series features identification keys and color photographs of many diseases, insect pests, and horticultural problems of conifer crops (figure 15.9). Although pest scouts should make a tentative diagnosis of disease and pest problems, they should confirm their conclusions with the nursery manager and a trained nursery pest specialist.

Many diseases and insect pests damage a wide variety of host plants. For example, damping-off fungi (see figure 15.1A) affect all species during germination and emergence and Phytophthora root rot (figure 15.1B) can attack many species of larger plants. Other pests are host specific, however, and disease scouts must understand their basic biology and life cycle for accurate diagnosis. Rust fungi are very specialized pests and attack only one group of plants, but some rusts alternate between host plants. Cedar-apple rust is caused by a fungus (*Gymnosporangium juniperi-virginianae*) and needs two hosts to complete its life cycle: junipers and certain plants of the rose family (Rosaceae) such as serviceberry or hawthorn. On junipers, the disease appears as woody, spherical galls. In the spring (early May), brown, horn-like projections called "telia" grow out of the woody galls (see figure 15.1D). During wet weather, the telia absorb water, become orange and gelatinous, and emit spores that infect the rose family plants. On this second host, the fungus infects leaves to cause bright-orange spots (see figure 15.1C), which eventually produce their "aecia" on fruit and the underside of leaves. In late summer, the aecia produce spores that reinfect junipers. The disease must pass from junipers to rose family plants to junipers again; it cannot spread between rose family plants. You can see that understanding the life cycle of the pest is essential for accurate diagnosis and management.

Some of the more common pests on native plants, ways to monitor them, and their diagnostic symptoms are included in table 15.2.

6. Timely and Appropriate Control Measures

One key concept of holistic pest management is the minimal use of chemicals. Chemical pesticides should be used only after all other environmental and horticultural controls have been considered. For example, you can reduce the use of pesticides by excluding pests via screens or barriers that can exclude small pests, such as insects, to big pests, such as mice and deer (figure 15.10). Most native plants can be grown with few fungicides and insecticides, but, because the nursery environment is so favorable for many pests, chemicals are sometimes needed. The important thing is to choose a pesticide that is targeted for the specific pest and is safest to use.

Many people believe that any natural or organic pesticide is always safe to use, but this belief is incorrect. A number of registered botanical insecticides can be toxic to applicators or the environment. The relative

WHAT IS THE MOST DANGEROUS CHEMICAL APPLIED IN THE NURSERY?

By far, the chemical used most frequently in the nursery that causes the most damage is water. Too much water on plants encourages damping-off, root disease, fungus gnats, moss and liverwort growth, excessive leaching of mineral nutrients and potential groundwater contamination, and foliar diseases such as Botrytis blight and may cause plants to grow rapidly and spindly, making them more susceptible to environmental stresses such as excessive heat, wind, or cold. Too little water on plants may cause damage to root systems through salt damage or desiccation, allowing entry points for root disease pests. Plants under severe moisture stress have lower resistance to stresses associated with heat, cold, and wind.

One of the best ways to limit pest problems is to irrigate plants properly.

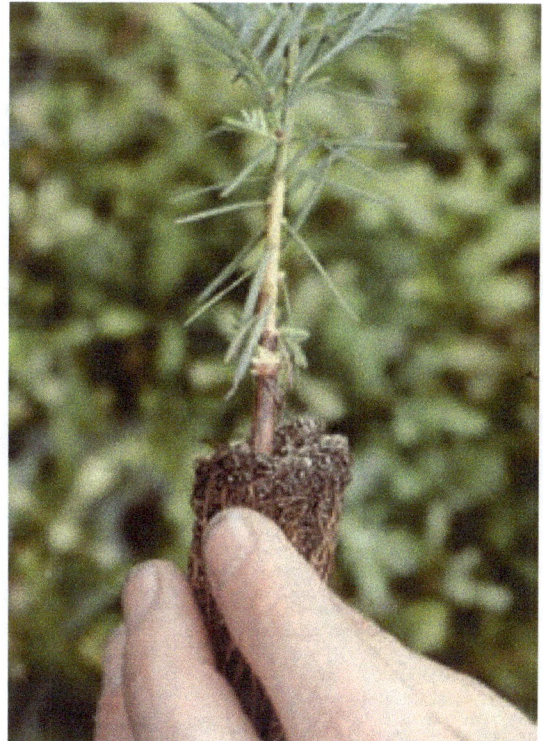

Figure 15.6—*Disease scouts must be familiar with all crop species at all stages of development. The burl on the stem is normal for this redwood seedling but could be a problem for other plants.* Photo by Thomas D. Landis.

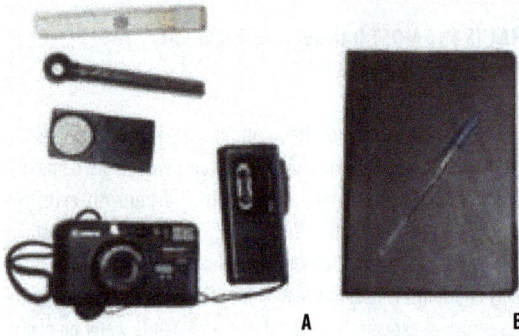

Figure 15.7—(A) Crop monitors and disease scouts should carry a hand lens, notebook or tape recorder, and camera to identify and document problems. (B) All potential problems should be noted in the daily log and reported to the nursery manager. Photos by Thomas D. Landis.

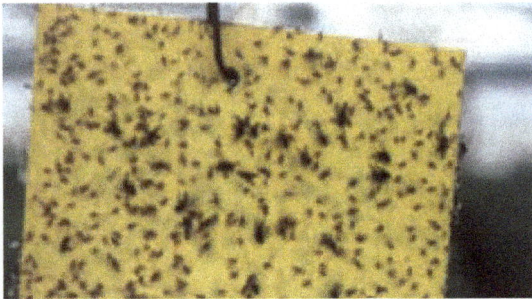

Figure 15.8—Yellow sticky cards are essential for monitoring the types and population levels of insects pests. Photo by Thomas D. Landis.

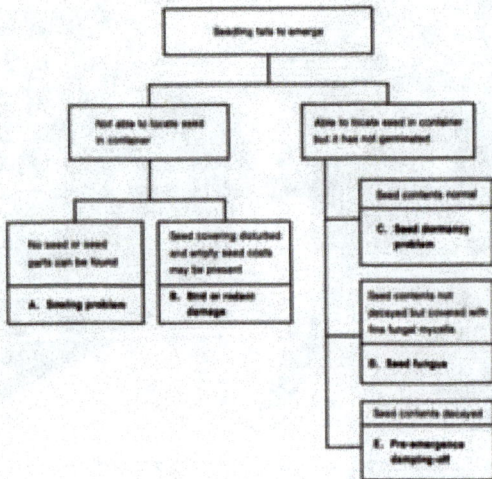

Figure 15.9—Pest identification keys are very useful in helping disease scouts follow a logical and critical procedure and make the correct diagnosis. Illustration by Jim Marin.

toxicity rating for any chemical is known as the LD50, which indicates the lethal dose that is required to kill 50 percent of a population of test animals. As can be seen in table 15.3, insecticides from natural sources can be just as hazardous as or even more hazardous than chemical pesticides.

Still, we recommend that you always consider natural pesticides first because they degrade faster than synthetics. Faster degradation reduces their impact on nontarget organisms; they are generally less toxic to humans, mammals, and beneficial insects such as bees. Most natural pesticides are less harmful to host plants when applied according to label directions.

The drawbacks of natural pesticides are that they must be applied more frequently because of their rapid degradation, they are often more expensive, may be more difficult to obtain, and little data exist on their effectiveness and long-term toxicity.

7. Encourage Beneficial Organisms

Holistic management involves using beneficial fungi, insects, and other organisms to help prevent pest impacts on crop plants. (table 15.2). Most people working in a nursery have heard about mycorrhizal fungi and realize that they have many benefits to the host plant including protection against root pathogens. The fungi that form mycorrhizae are often very host specific but there is little published information for many native plants (see Chapter 14, *Beneficial Microorganisms*).

SUMMARY

Your efforts to provide optimum growing conditions for the plants that you grow will go a long way toward preventing disease and insect problems. This practice, in conjunction with maintaining sanitation in the greenhouse, is the most useful and least expensive tool for preventing diseases and pests. Incorporate a holistic approach to diagnosing and treating problems by giving as much attention to the environment as to the pest. Many problems are triggered by environmental stresses that can be easily avoided or corrected by good horticultural practices. Pest control measures may not be needed if everyone at the nursery makes a concerted effort to prevent problems. By having regular, frequent monitoring of the crop and by keeping records, your nursery will be able to detect problems early and develop measures to prevent problems with future crops.

Table 15.2—Common diseases and pests, symptoms, prevention, and treatment measures for native plants

Pest	How To Monitor	Prevention	Signs and Symptoms	Biological and Organic Control Options
Aphids	Monitor 2 times weekly. Look on underside of leaves and on tips of new stems.	Shoot prune vigorous tender growth as needed. Watch for outbreaks in early summer months.	Plants have distorted new growth, sticky honeydew, and sooty mold.	Use aphid midges, aphid parasites, lady bugs, Safer soap applied every 6 days, pyrethrins.
Bacterial diseases	Inspect new plants coming into greenhouse. Look for water-soaked, dark brown to black leaf spots on leaves and wilted stem tips. Confirm diagnosis with a laboratory.	Increase spacing between containers as crop grows larger. Water only in the morning or use subirrigation. Keep humidity low (see appendix 15.A).	Plants are stunted with swollen or misshapen leaves. Look for water-soaked leaf spots or angular lesions on the stems.	Remove infected leaves as soon as problem is detected. Isolate infected trays of plants from the rest of the crop.
Botrytis blight	Concentrate monitoring where crop is closely spaced and air circulation is poor, and on tender leafy species. Look for dieback, stem cankers, and powdery gray mold on leaves.	Increase spacing between containers as crops grows larger. Water only in the morning or use sub-irrigation. Keep humidity low (see appendix 15.A).	Plants have leaf blights, stem cankers, gray mold.	Apply *Trichoderma harzianum* (PlantShield), *Streptomyces griseoviridis* (Mycostop).
Caterpillars	If moths are seen in the greenhouse, look for caterpillars. Look for fecal droppings, bites taken out of leaves and webbing (tent caterpillars).	Screening.	If damage is seen, look for caterpillars under containers or in growing media. Many caterpillars feed at night and hide during the day.	Apply BT *Bacillus thuringiensis* ssp. *kurstaki* (Pro DF) as needed, pyrethrins.
Crown and root rots	Monitor weekly for wilted, off-colored plants with discolored root systems. Pay attention to media that stays wet. Check soluble salt levels.	Do not overwater crop. Increasing spacing between containers as crop grows larger. Keep humidity low. (see appendix 15.A).	Plants are stunted, wilted, and off-color. Roots are discolored and turn brown or black. Main stem becomes weak and water soaked in appearance.	Apply *Trichoderma harzianum* (PlantShield, RootShield), *Trichoderma virens* (Soil Guard), *Streptomyces griseoviridis* (Mycostop).
Damping-off disease	Monitor daily during germination and establishment phases. Look for seeds that do not germinate and seedlings that collapse at soil line just after emergence. Discard infected containers immediately.	Cleanse seeds and growing area. Use sterile media and containers. Avoid over-sowing, crowding of seedlings, or planting seeds too deeply. Keep green-house and media temperatures warm during germination and establishment. Keep humidity low (see appendix 15.A).	Seeds do not germinate; seedlings collapse at soil line just after emergence. Dark dead spots appear on stems at soil line of emerged seedlings. Infected plants may later develop crown and root rot.	Use *Trichoderma harzianum* (PlantShield, RootShield), *Trichoderma virens* (Soil Guard).

Pest	How To Monitor	Prevention	Signs and Symptoms	Biological and Organic Control Options
Fungal leaf spots	Monitor weekly for leaf spots. With a hand lens, look for small fungal fruiting bodies. Confirm problem with a laboratory.	Use mesh benches to encourage airflow. irrigation. Keep greenhouse floor clean and free of pooled water. Water only in the morning or use subirrigation. Do not overwater crop. Keep humidity low (see appendix 15.A). Increase spacing between containers as crop grows larger.	*Alternaria* leaf spots are usually brown or black with a yellow border. *Septoria* leaf spots are small gray to brown with a dark brown edge.	Apply *Trichoderma harzianum* (Plant Shield). Remove infected leaves as soon as problem is detected. Isolate infected trays of plants from the rest of the crop.
Fungus gnats	Monitor every other day, especially during germination and establishment phases. Look for tiny winged flies near growing media surface. Use yellow sticky cards to detect adults.	Keep greenhouse floor clean and free of pooled water and algae, do not overwater crop. Use yellow sticky cards and a good seed mulch.	Plants have weak or stunted growth, seeds that do not germinate, root damage on seedlings.	Apply BT *Bacillus thuringiensis* ssp. *israeliensis* (Gnatrol) applied every 7 days as a drench, mite predators, parasitic nematodes.
Fusarium wilt	Look for downward bending leaves or "cupping" of leaf margins. Can be confused with water stress, root rot. Send sample to laboratory to confirm.	Use mesh benches to encourage airflow. Do not overwater crop. Keep humidity low (see appendix 15.A).	Leaves cup downward or stems bend in a crook. In later stages, brown streaks can be seen on the leaves. Orange spores may be on stem.	Apply *Trichoderma harzianum* (Plant Shield, Root Shield), *Streptomyces griseoviridis* (Mycostop) as a soil drench. Remove and isolate infected plants as soon as problem is detected.
Mealybugs	Look for small, oval, soft-bodied insects covered with a white, wax-like layer on the underside of leaves.		Plants may have white cottony residue. Sticky honeydew on leaves and sooty mold may develop.	Use predatory beetles, parasitic wasps; pyrethrins.
Powdery mildew	Monitor weekly. Inspect susceptible species. Look in areas near vents or any location with a sharp change between day and night temperatures. Use a hand lens to see white, powdery threads and spores.	Place susceptible species where drastic changes in temperatures do not occur. Water only in the morning or use subirrigation. Keep humidity low (see appendix 15.A). Increase spacing between containers as crop grows larger.	Plants may have white powdery fungal growth on upper or lower leaf surfaces. If severe, white coating can be seen on foliage.	Remove infected leaves as soon as detected. Move infected plants to structure with more constant temperatures. Treat with Neem oil, horticultural oil, Safer soap. Try test tray first. Can also use sulfur fungicide as an organic fungicide. Some plants are sensitive to sulfur injury so use lowest rate recommended. Do not apply within 2 weeks of an oil spray treatment.

Table 15.2.—(continued) Common diseases and pests, symptoms, prevention, and treatment measures for native plants

Pest	How To Monitor	Prevention	Signs and Symptoms	Biological and Organic Control Options
Rhizoctonia web blight	Monitor leafy herbaceous plants, especially where they are closely spaced. Look for cob-webby growth that mats leaves together.	Use mesh benches to encourage airflow. Place susceptible crops near vents and fans. Increase spacing between containers as crop grows larger. Keep humidity low (see appendix 15.A).	Stems and leaves may collapse and turn to mush with fine, web like fungal strands on the plant tissue and at soil line.	Use *Trichoderma harzianum* (Plant Shield, Root Shield).
Rusts	Look for yellow and rusty orange spots on the upper and lower leaf surface.	Group susceptible species where temperature and humidity can be easily controlled. Increase spacing between containers as crop grows larger. Keep humidity low (see appendix 15.A).	Rust brown spots or stripes may be seen on lower and upper leaf surface.	Isolate plants immediately.
Slugs	Look for chewed holes on leaves and trails of slime. Slugs hide under dense foliage and under containers and benches.	Keep plants on raised benches or pallets. Space containers as needed so that slugs can be detected easily.	Plants may have chewed holes on leaves with smooth edges and slime that dries into silvery trails on foliage.	Pick slugs off plants. Keep containers on benches. Use saucers filled with beer to attract slugs away from plants.
Soft scales	Look for yellow brown to dark brown scale insects along veins and stems.		Honeydew and sooty mold develop if scales are present.	Use parasitic wasps, Safer soap, pyrethrins.
Spider mites	Look on undersides of leaves especially along veins. Use a hand lens to look for webbing, egg clusters, and red adult mites. Look in areas of that are hot and dry, near the heaters and vents.	Lower greenhouse temperatures and raise humidity levels, especially in the south and west edges of the greenhouse and near vents and furnaces.	Plants may have light-yellow flecking of leaves, discolored foliage. Leaf drop and webbing occur during outbreaks and severe infestation.	Use predatory mites, predatory midges. Apply Safer soap every 6 days.
Thrips	Use blue or yellow sticky cards placed just above canopy foliage for detection.	Increase container spacing on leafy crops as needed to detect problems early.	Plants may have distortion of new leaves, buds, and shoot tips. White scars on expanded leaves.	Use predatory mites, pirate bugs, lacewings, Safer soap, and pyrethrins.

Table 15.2—(continued) Common diseases and pests, symptoms, prevention, and treatment measures for native plants

Pest	How To Monitor	Prevention	Signs and Symptoms	Biological and Organic Control Options
Viruses	Monitor weekly. Inspect all incoming plants. Send sample to laboratory to confirm.	Usually not a problem with native plants; can be a problem on cultivated varieties, ornamentals, plants grown by tissue culture.	Look for mosaic patterns on foliage, leaf crinkle or distortion, streaking, chlorotic spots and distinct yellowing of veins and stunted plants.	None. Remove and discard all infected plants immediately. Thoroughly clean area of greenhouse where infected plants were growing.
White flies	Use yellow sticky cards to detect adults. Look for adults on the uppermost tender leaves. Immature larvae are found on the underside of leaves.		Plants may have distorted new shoot and leaf growth.	Use predatory beetles, whitefly parasites, Safer soap applied every 7 days; pyrethrins.

Figure 15.10—*Native oaks can be especially vulnerable to pests such as mice and deer after they are sown in containers. Simple measures such as caging newly seeded containers and young seedlings can be used to exclude a variety of pests from insects to mice and deer.* Photo by Tara Luna.

Table 15.3—Comparative safety of common botanical and synthetic insecticides.

Insecticide	Class	Toxicity Rating (Oral LD50 in mg/kg)	Label Warning*
Nicotine	Botanical	50 to 60	Danger
Sevin	Synthetic	850	Warning/Caution
Malathion	Synthetic	885 to 2,800	Caution
Pyrethrin	Botanical	1,200 to 1,500	Caution
Neem	Botanical	13,000	Caution

Modified from Cloyd (2004); * danger is most toxic, caution is least toxic

LITERATURE CITED

Cloyd, R.A. 2004. Natural instincts. American Nurseryman 200(2): 38-41.

Landis, T.D.; Tinus, R.W.; McDonald, S.E.; Barnett, J.P. 1989. The container tree nursery manual: volume 5, the biological component: nursery pests and mycorrhizae, Agriculture Handbook 674. Washington, DC: U.S. Department of Agriculture, Forest Service. 171 p.

Unestam, T.; Beyer-Ericson, L.; Strand, M. 1989. Involvement of *Cylindrocarpon destructans* in root death of *Pinus sylvestris* seedlings: pathogenic behaviour and predisposing factors. Scandinavian Journal of Forest Research 4(4): 521-535.

ADDITIONAL READINGS

Dumroese, R.K.; Wenny, D.L.; Quick, K.E. 1990. Reducing pesticide use without reducing yield. Tree Planters' Notes 41(4): 28-32.

Heiskanen, J. 1993. Favourable water and aeration conditions for growth media used in containerized tree seedling production: a review. Scandinavian Journal of Forest Research 8(3): 337-358.

Heiskanen, J. 1997. Air-filled porosity of eight growing media based on Sphagnum peat during drying from container capacity. Acta Horticulturae 450: 277-286.

Holden, J.M.; Thomas, G.W.; Jackson, R.M. 1983. Effect of mycorrhizal inocula on the growth of Sitka spruce seedlings in different soils. Plant and Soil 71: 313-317.

James, R.L.; Dumroese, R.K.; Gilligan, C.J.; Wenny, D.L. 1989. Pathogenicity of *Fusarium* isolates from Douglas-fir seed and container-grown seedlings. Bulletin 52. Moscow, ID: University of Idaho, College of Forestry, Wildlife and Range Sciences. 10 p.

Juzwik, J.; Gust, K.M.; Allmaras, R.R. 1998. Influence of cultural practices on edaphic factors related to root disease in *Pinus* nursery seedlings. Plant and Soil 207(2): 195-208.

Keates, S.E.; Sturrock, R.N.; Sutherland, J.R. 1989. Populations of adult fungus gnats and shore flies in British Columbia container nurseries as related to nursery environment, and incidence of fungi on the insects. New Forests 3(1): 1-9.

Landis, T.D. 1984. The critical role of environment in nursery pathology. In: Dubreuil, S.H., comp. 31st Western International Forest Disease Work Conference, proceedings. Missoula, MT: USDA Forest Service, Cooperative Forestry and Pest Management: 27-31.

Landis, T.D.; Tinus, R.W.; McDonald, S.E.; Barnett, J.P. 1989. The container tree nursery manual: volume 4, seedling nutrition and irrigation. Agriculture Handbook 674. Washington, DC: U.S. Department of Agriculture, Forest Service. 119 p.

Landis, T.D.; Tinus, R.W.; McDonald, S.E.; Barnett, J.P. 1990. The container tree nursery manual: volume 2, containers and growing media. Agriculture Handbook 674. Washington, DC: U.S. Department of Agriculture, Forest Service. 88 p.

Lilja, A.; Lilja, S.; Kurkela, T.; Rikala, R. 1997. Nursery practices and management of fungal diseases in forest nurseries in Finland: a review. Silva Fennica 31(1): 547-556.

Olkowski, W.; Daar, S.; Olkowski, H. 1991. Common-sense pest control. Newtown, CT: Taunton Press. 715 p.

Thornton, I. 1996. A holistic approach to pest management. Nursery Management and Production 12(6): 47-49.

APPENDIX 15.A. CHECKLIST FOR PREVENTING DISEASES AND PESTS

1. Start with clean seeds. Seeds can be treated with a very mild diluted bleach or hydrogen peroxide solution before stratification or sowing to help prevent seed and seedling diseases.

2. Remove all plant debris before sowing the crop. Also, clean tables, aisles, side walls, and floors with a mild bleach or soap solution before sowing.

3. Vigilantly remove all weeds growing under benches and in the crop.

4. Use containers that have been cleaned (see Chapter 6, *Containers*).

5. Use a heat-pasteurized growing medium.

6. Prevent algae from forming on the floors and benches by ensuring proper and rapid drainage of excess irrigation water and by properly managing irrigation frequency. Algae and pools of water provide a breeding ground for fungus gnats and shore flies.

7. Use hooks to keep the hose nozzles off the floor and disinfect planting tools; dirty hose nozzles and planting tools can infect growing media with pathogens.

8. During sowing and the establishment phase of the crop, carefully manage the greenhouse environment to keep humidity levels and condensation problems low by venting the greenhouse frequently; avoid cool temperatures that delay germination. Do not overwater germinating seeds and seedlings. Remove dead and dying plants; make sure they are disposed of away from the nursery to prevent reinfection.

9. During the active growth stage, reduce humidity within the leaf canopy to prevent the development of many foliar diseases. Reducing humidity can be accomplished by improving air circulation by increasing distance between plants, increasing the frequency of ventilation in the greenhouse, and pruning shoots as necessary. Remove any plant debris on the floor on a regular basis. Remove dead and dying plants; make sure they are disposed of away from the nursery to prevent reinfection.

10. Water only in the morning, never later in the day. Favorable environmental conditions for several fungal diseases include a film of moisture for 8 to 12 hours, high relative humidity, and temperatures between 55 to 65° F (13 to 18°C). By watering early, rising daytime temperatures will cause water to evaporate from the leaf surfaces and reduce favorable conditions.

11. Use separate propagation structures for growing plants with very different environmental and horticultural requirements, or, if you have a single growing structure, group plants with similar growing requirements together and take advantage of microenvironments within the greenhouse. For example, the south side of the greenhouse is usually warmer and drier than the north or east section of the greenhouse. Plants requiring cool temperatures or those requiring more frequent irrigation should be grouped together on the north and east sides, while plants requiring drier conditions should be grouped on the south and west sides of the greenhouse.

12. Reduce humidity; high relative humidity encourages the development of many foliar diseases, including *Botrytis*, powdery mildew, and *Rhizoctonia*. Improve air circulation and reduce humidity and condensation with fans that produce horizontal air flow. Relative humidity and condensation can also be reduced by heating and venting moist greenhouse air; heat and vent two to three times per hour in the early evening after the sun sets and again during early morning. Many growers use oversized vent fans and louvers to increase air flow in the greenhouse.

APPENDIX 15.B. PLANTS MENTIONED IN THIS CHAPTER

chokecherry, *Prunus virginiana*

green ash, *Fraxinus pennsylvanica*

hawthorn, *Crataegus* species

juniper, *Juniperus* species

oaks, *Quercus* species

quaking aspen, *Populous tremuloides*

redwood, *Sequoia sempervirens*

serviceberry, *Amelanchier alnifolia*

whitebark pine, *Pinus albicaulis*

willows, *Salix* species

SEED
TREATMENT

16

Nursery Management

Kim M. Wilkinson

This handbook provides an overview of the factors that go into starting and operating a native plant nursery. Management includes all aspects of working with plants in all their phases of growth as described in Chapter 3, *Crop Planning and Developing Propagation Protocols*. Management also includes working with the community; organizing materials and infrastructure; planning educational activities and outreach; maintenance activities, such as watering and pest management; and much more (figure 16.1). Each of these elements will become part of the day-to-day and year-to-year aspects of managing the nursery. At the outset, the variety and complexity of tasks may seem overwhelming. This chapter provides a broad overview of the essential aspects of managing a native plant nursery.

WHO IS RESPONSIBLE?

For small nurseries, one person may be the sole operator who takes care of everything. In these cases, it is essential to have at least one backup person who understands crop status, knows everything that needs to be done, and knows how to do these things. The backup person can keep the nursery running and the plants healthy in case the primary person becomes unavailable.

Personalities and management styles vary widely among effective nursery managers. Some general characteristics, however, are important to good nursery management. If you are contemplating becoming a nursery manager, look at how to cultivate the following attributes in yourself:

Joanie Hall and Marie Crawford of the Confederated Tribes of the Umatilla Indian Reservation and James Randall of the Yakima Nation by Tara Luna.

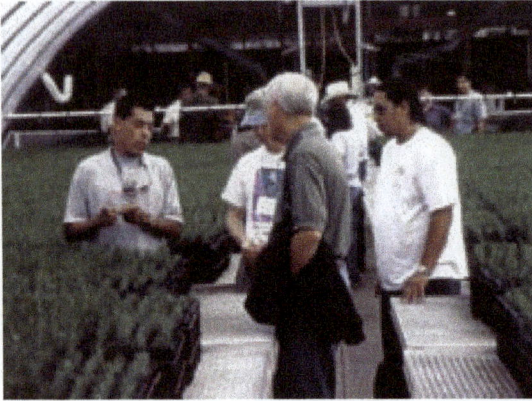

Figure 16.1—*Managing a nursery, such as the Southern Ute Forestry Nursery in Colorado shown here, involves more than producing plants. It includes working with the community; organizing materials, time, and infrastructure; carrying out trials and experiments; and relating with customers and the public.* Photo by Thomas D. Landis.

Figure 16.2—*Keen observation skills and the ability to "think like a plant" are key attributes of an effective nursery manager, such as Roy Tyler.* Photo by Charles Matherne.

Figure 16.3—*Effective managers focus on plant production but never lose sight of the bigger picture. Joy Hosokawa of the USDI National Park Service checks her seed supply.* Photo by Kim M. Wilkinson.

— Keen observation skills.
— Flexible management style (scheduling must not be rigid but rather adaptable to the shifting needs of living, growing plants).
— Ability to "think like a plant" (managing plants effectively is an art as well as a science; someone who has a "feel" for the crops will likely do better as a manager than someone who approaches crops strictly from an engineering perspective).
— Willingness to be responsible for plants in the nursery.

As the nursery grows in size, the manager may oversee the same tasks but delegate many of them to other staff members. Naturally, all staff should be committed to the success of the nursery. The manager understands both the "how" and the "why" of tasks to be done, and each person on staff should gain that knowledge as well. Every staff member is an important part of the team, and they should be clear about how their role fits into the big picture. Even for large nurseries, however, just one person, the manager, needs to take responsibility for the crop and understand the requirements for its management (figure 16.2). Nursery operations cannot be conducted "by committee" because too many variables are involved in working with a living, growing crop to risk confusion or irresponsibility.

WHAT NEEDS TO BE DONE?

At first, the number and complexity of tasks involved in operating a nursery may seem daunting. A checklist can provide an overview of what needs to be done. No blueprint or schedule will apply to all nurseries. Many tasks, however, are applicable to most nurseries. The example overview checklist in appendix 16.A can be modified and customized to a specific nursery's management needs. As determined by the nursery's situation, activities may be divided into daily, weekly, monthly, and seasonal tasks (figure 16.3).

Each of the tasks in appendix 16.A is discussed in this chapter. Only a few required tasks must happen each day; these are the "essential" tasks that keep the crop alive and healthy and the nursery functioning on a daily basis. These essential tasks include watering, keeping daily records, and monitoring and managing crops as they go through the establishment, rapid growth, and hardening phases. See Chapter 3, *Crop*

Planning and Developing Propagation Protocols, for more detail on crop production activities. It is important, however, not to get so caught up in these daily activities that other essential but less pressing tasks go undone. The other tasks must be prioritized and scheduled as well.

PLANNING

Schedule an overview and planning session on a weekly basis to take a "bird's-eye view" of the nursery, pressing activities, and long-term goals. This is an opportunity to decide priority tasks for the coming week and month. The needs of the plants, environmental conditions, and many other factors require flexibility and responsiveness in management style. Crops usually will conform poorly to an exact schedule and may perform differently in different years, which is why weekly assessment and planning is so important. Attempting to make rigid schedules (such as "weed every Tuesday") are often far less effective than planning the week to respond to the observed needs and conditions of the crop.

The observation skills of the nursery manager and staff are the greatest assets to effective planning. Taking time on a weekly basis to review the daily log, plant development record, and other observations by the manager and staff will help with prioritizing the work to be done. What is happening with the crop? What growth phase is it in: establishment, rapid growth, or hardening? Is it on schedule? What needs to be done next: transplanting, moving to a new structure, altering fertilization rates? Are we observing anything that might indicate a potential problem, such as the presence of a potential pest? Do clients need to be updated on the progress of their crop and the schedule? Is a new crop ready to be sown? Once the list is complete, prioritize the list from most to least important activities, assigning roles and tasks to the appropriate staff members.

Some planning time should also be spent in a proactive state of mind, focusing not just on what is urgent or most pressing. Planning should include scheduling some activities that are important to the nursery's larger mission but may fall by the wayside if the manager gets too engrossed in the day-to-day details of running the nursery. Updating plant protocols is an essential activity that should take place after each crop has been shipped, based on information in the daily log and plant development records. Other important but seemingly less pressing activities include public relations or educational activities, deciding what trials or experiments to conduct, and making time to attend a conference or training. These valuable activities should be put in the schedule so they are not neglected.

ROUTINE DAY-TO-DAY TASKS

Daily activities include the essentials of keeping the plants alive and healthy. At a minimum, these activities usually include daily watering of some portion of the crop. Other tasks, such as weeding, monitoring the atmospheric environment, fertilizing, or pest management, are carried out as necessary.

Daily observation of the crop and the nursery environment is an essential part of good nursery management. Crops in the nursery are living, growing plants, and observing them is an important way to relate with them and understand their needs. Keeping simple but systematic written records on a daily basis facilitates this process and also hones good observational skills. Writing down observations also creates valuable records that can be used in the future. Two daily records are used: a daily log or journal and, for each crop, a plant development record.

Keeping a **daily log** is an essential nursery practice. The log does not need to be elaborate. To begin with, the log can be as simple as writing the day's date and jotting down some details in a notebook about observations and activities at the end of each day. Keep this notebook easily accessible at the manager's workstation and make a habit of entering something in it each day, even if the observations seem unimportant at the time. What was done today? Were any supplies purchased? How many hours of labor were spent on which crop? How much time was spent on management activities (for example, watering)? What crops are coming up? What observations can be made about the crop or the nursery environment? What plants were sold and for how much? As the nursery grows in size and complexity, entering this information into a computer (even simple spreadsheets) will make the information easier to track. Events each day create small amounts of vital information that will contribute immeasurably to improving nursery management and productivity over time (figure 16.4). This information is invaluable for making many decisions, including the following:

- Budgeting funds.
- Estimating schedules to produce future crops.
- Determining what labor-saving equipment might give the most benefit for the cost.
- Analyzing nursery expenses.
- Improving profits or production.
- Replicating successful crops.

The **plant development record** is another key recordkeeping tool that heightens awareness and hones good observation skills. It is a simple form; an example is provided in this chapter and is discussed in Chapter 3, *Crop Planning and Developing Propagation Protocols*. Again, the plant development record should be kept in an easily accessible place and a few notes should be jotted down as changes occur with the crop.

The manager or a designated "crop monitor" should observe the crop every day (figure 16.5). This task can be done once daily as a formal practice and can also be integrated into quick "walk-throughs" at other times of the day. Sometimes the observations are casual; occasionally measurements will be taken. The person monitoring the crop should understand what "normal" should look like for that crop and the environment (based on experiences reflected in the protocol and crop schedule) and be highly sensitive to any deviations from that norm. Experience and the use of all five senses will make this job easier with time. A checklist can be developed for the primary crop monitor and all the nursery staff to train them in what to look for. These observations can catch potential problems long before they become emergencies. Observations may include the following.

Appearance. How does the crop look? Is the shoot-to-root ratio proper for the stage of growth? Are signs of nutrition or disease problems visible on the roots or foliage? Inspect closely for any insect pests. If diagnosed early, pest problems can be handled quickly and effectively. Are beneficial microsymbionts visible on the root systems? Is the crop developing as predicted and is it on schedule for outplanting?

Smells. Some problems such as gray mold may be discernable to the experienced grower, and a major outbreak can be averted if it is caught early enough. Overheating motors, broken fans, furnace problems, and other factors can also be detected by the sense of smell.

Figure 16.4—*The timing of routine tasks, such as thinning pine seedlings at the nursery operated by the Confederated Tribes of the Colville Reservation in Washington State, should be recorded in a daily log. These records will help you plan, budget, and schedule for future crops.* Photo by Kim Wilkinson.

Figure 16.5—*Making regular observations of the crop, and staying alert and aware are key to avoiding problems.* Photo by Tara Luna.

Noises. Does a motor need oiling? Is an engine pulsating when it should be running steadily? Is water running when or where it shouldn't?

Feel. Is the temperature and humidity in the proper range? Are fans running and is the air moving through the nursery as it should? Are plant root systems moist but not overly wet?

Although one designated crop monitor is responsible for this task, all staff should understand that observation and being alert and aware are key to heading off problems. Staff should be given every opportunity to share their observations with the nursery manager. Depending on the personality of the staff member, this information may be more comfortable to record in a written staff log that the manager reads daily or it may be easier to share verbally. The manager should welcome and encourage staff to share their observations, as this practice builds observation skills and greater crop awareness.

Observations and notes made in the daily journal and plant development records will later be used to improve and develop propagation protocols (see "Recordkeeping" in the following paragraphs).

CROP PLANNING AND PRODUCTION TASKS

The details of planning crops are discussed in Chapter 3, *Crop Planning and Developing Propagation Protocols*. The process of crop production includes:

— Understanding the three growth phases crops go through (establishment, rapid growth, and hardening) and the distinct requirements for each phase.
— Making growing schedules for crop production from seed procurement through outplanting and detailing changes as the growing cycle progresses.
— Listing space, labor, equipment, and supplies required to support the crop during the three stages of growth.
— Keeping written records, including a daily log and plant development record.
— Developing accurate propagation protocols so the successes of this crop can be replicated next time.

The work to produce a crop consists of managing the plants through each phase of development so that plants receive what they need and are as strong and healthy as possible for outplanting. Once a schedule has been made showing what plants need to be sown and by when, the tasks of preparing growing media, filling containers, and sowing can be scheduled accordingly (figure 16.6). In the establishment phase, plants begin to germinate, and thinning, transplanting, and inoculation with microsymbionts will take place. As the plants move from the germinant phase to the rapid growth phase and later to the hardening phase, their needs will change.

For some nurseries, plants will be physically moved from a germination or rooting area to a more open environment (figure 16.7). For other nurseries, climate control might produce the same effect. Fertilization and watering regimes are changed for each of the three phases. When the crop is ready, it will be harvested and shipped as described in Chapter 13, *Harvesting, Storing, and Shipping*. Again, daily observations and weekly assessment and planning will enable the manager to schedule appropriately and produce a successful crop.

RECORDKEEPING

Two main kinds of records are kept for nurseries: horticultural and financial.

Horticultural Records

Good horticultural records are essential to keeping production on track and precluding serious problems. Horticultural records include the following:

— A daily log.
— Plant development records for each crop.
— Crop-growing schedules and facilities schedules.
— Plant protocols (regularly updated and revised).
— Inventory assessment (so you know what crops will be ready when).

The daily log and plant development records provide the ability to update and revise propagation protocols, records that show how to produce that crop successfully in your nursery. The protocols provide guidance for each new crop in developing the production plan and listing needed materials and supplies. The schedules are essential to meet targets and serve clients.

Keeping an inventory enables crops to be tracked, particularly by seed source, client, and date of availability. A crop inventory should include:

— All plants in the nursery by bench or structure number.
— Current developmental stage of the crop.
— Details of delivery (site, name of client, seed source, and anticipated delivery date).

Having these facts on hand is an important part of working with clients. Records of inventories and supplies can be kept in the computer and updated weekly or as stock changes.

Financial Records

Keeping financial records is a key activity if the nursery is to thrive in the long term. The daily journal and other records should track labor spent on various activities, money spent for materials, and overhead costs such as rent and utilities. These numbers can be factored in to accurately estimate the cost of each crop, allowing the manager to correctly budget time and funds to produce crops. It also is essential in determining what the manager must charge for various plant materials sold.

In keeping financial records, be sure to note the following factors:

— Size of stock.
— Time to grow.
— Labor (in person-hours) required through all phases.
— Materials required and their cost (for example, seeds, growing media).
— Need for custom culture (for example, special containers, extra labor).
— Overhead costs (for example, utilities).
— Cost inflation over time.
— Typical losses (percentage of crop discarded).

To keep up with maintenance and production, a manager must have the necessary materials on hand when they are needed. These materials include supplies for production such as seeds, growing media, containers, and so forth. For nurseries in remote areas, obtaining some supplies such as specialized containers or spare parts for equipment may require a lot of waiting time. In these cases, extras of essential items should be kept on hand. If spare parts are used for repairs, they should be replaced right away.

SEASONAL CLEANUP

Usually, some time is available between crops or at the end of each season that lends itself to some "deep cleaning" and maintenance. Cleanliness is essential to avoid disease problems and also to maintain a professional, appealing image for the nursery. A clean environment builds customer confidence and staff morale. Every 2 to 6 months, just after shipping out a large order, or at the end of each season, is an opportunity to perform the following tasks:

Figure 16.6—*The management of nurseries, such as the native plant nursery operated by the Confederated Salish and Kootenai Tribes in Montana, involves planning the timing of essential production tasks, such as filling containers, as well as ensuring that all the necessary supplies (for example, containers, growing media, seeds, and so on) are available.* Photo by Dawn Thomas.

Figure 16.7—*At the Santa Ana Pueblo Nursery in New Mexico, plants are moved from one structure to another as they go through their three phases of growth. (A) Good planning and management maintains open paths and an easy flow of work between structures and (B) between sections of plants separated and identified by seed source.* Photos by Tara Luna.

— Dispose of any holdover stock.

— Clean and hose down the floors and tables (applying dilute bleach or other cleanser if no plants are present).

— Clean and sterilize containers (see Chapter 6, *Containers*).

— Flush out irrigation system and run a cup test. See Chapter 10, *Water Quality and Irrigation*, for more details.

— Conduct other equipment checks and make any repairs.

— Replace roof plastic, if necessary.

This is also a good time to step back from the nursery activities and reassess long-term goals and strategies. A staff meeting and some "big picture" planning is valuable to make sure the nursery stays on track with its mission.

TRAINING, DISCOVERY, AND PROBLEMSOLVING

Working with plants and nature is an ongoing educational process. There is always more to learn. Exchanging information with other growers in person or through reading is a great way to increase knowledge, as is learning directly from the plants through observation, research, and experimentation. Problem solving is also an essential aspect of running a nursery and leads to a better understanding of the plants and their needs.

Training

Training and ongoing education is of great value to the nursery and staff. The more growers understand their work and the effects of their activities, the more they will be able to relate to the crop. Attending training sessions and conferences and reading published literature on topics of interest are important investments in the nursery's growth (figure 16.8). A chance to learn from other growers has no substitute. Visiting other nurseries and hosting field days, at which growers can visit your nursery, can be an important part of cultivating supportive, informative relationships (figure 16.9). These events can be wonderful times to step out of the world of your nursery and gain a broader perspective. It is also an opportunity to cultivate relationships with other growers with whom information can be shared later.

Discovery

Daily observations, keeping records in the daily journal, maintaining plant development records, and up-

Figure 16.8—*Exchanging information with other growers through associations such as the Intertribal Nursery Council (shown here at the 2003 meeting) is a very fruitful way to learn more about plants and nursery management.* Photo by Kim Wilkinson.

dating protocols are the foundation for understanding how crops grow and develop in your nursery environment. Inevitably, more information will be sought. Goals in research include improving productivity and health, increasing survival rates, or experimenting with growing new species yet to be cultivated in your nursery. For native plant growers, making discoveries through simple trials and experimentation is often a key aspect of successful nursery development, because this is the best way to learn how to cultivate a species. On a monthly or seasonal basis, the nursery staff might meet to determine some of the most pressing questions facing the nursery. These questions will shape priorities for trials and study. For example, at the very beginning stages of nursery development, when production levels are small, the nursery may decide to try different kinds of containers to determine the best ones for the crops grown or to experiment with different seed treatment techniques for a new species. Later in nursery development, other pressing questions will arise. What problems are recurring that might be preventable with better understanding? What could improve efficiency? What could produce a stronger crop? The nursery might decide to test improved seed sources and assess their performance in the nursery and the field to improve target plant objectives. Ways to design, carry out, and assess these experiments and trials are detailed in Chapter 17, *Discovering Ways to Improve Crop Production and Plant Quality*.

Problemsolving

Good management, staff training, monitoring, and planning will minimize emergency situations in the

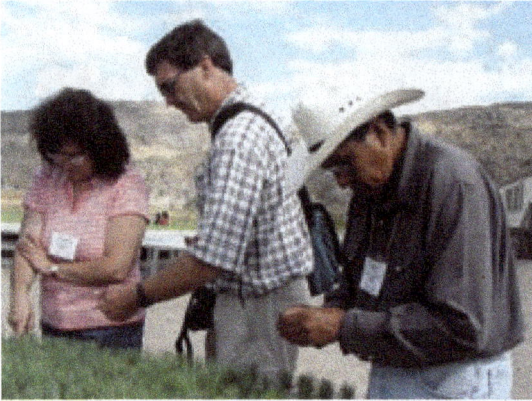

Figure 16.9—*Hosting visits to your nursery and attending tours of other native plant nurseries is a great way to cultivate supportive relationships.* Photo by Kim M. Wilkinson.

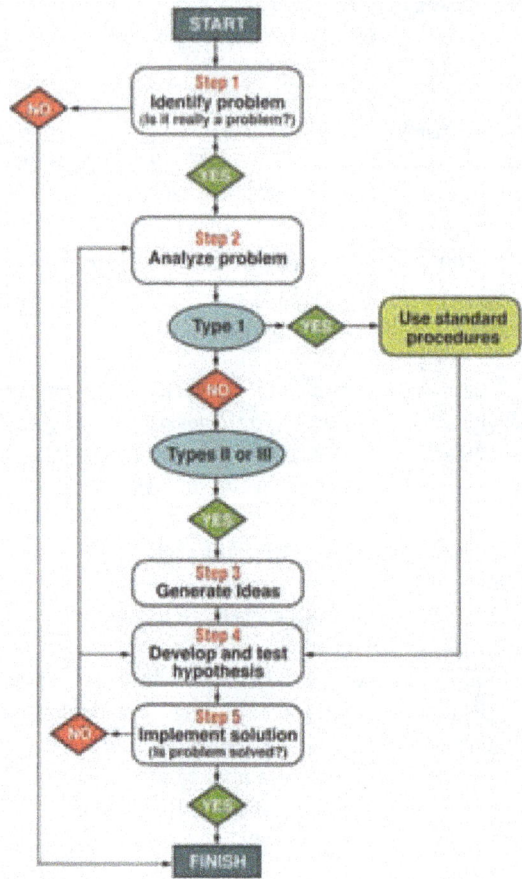

Figure 16.10—*Problem-solving matrix.* Illustration by Jim Marin based on Landis (1984).

nursery. Even the best manager, however, cannot avoid problems entirely. Perhaps a crop does not develop as expected, or an unknown pest problem arises. Some problems, such as difficulties with the irrigation system, appear suddenly and must be handled instantly. Others require a longer term approach. With experience, troubleshooting problems may become easier. This five-step systematic approach can be helpful when approaching long-term challenges (figure 16.10):

— **Identify Problem.** Is it really a problem? What seems to be wrong?
— **Analyze Problem.** What happened exactly? When did it start?
— **Generate Ideas.** Identify potential sources of the problem. Consulting literature, other nurseries, staff members, or outside sources of help such as extension agents or specialists can aid in gathering information.
— **Develop and Test Hypothesis.** At some point, a conclusion about the source of the problem must be decided and acted upon.
— **Implement a Solution.** Decide on a way to solve the problem. Observe the results. If the problem is not solved, start again with step 2.

Do not be reluctant to reach out to colleagues, other nursery managers, or other professionals. Everyone faces obstacles once in a while, and we can all help each other learn more about plants as we share our experiences.

WORKING WITH STAFF

Working with staff requires special skills. All staff should be trained to relate to the crops, observe and detect potential problems, and understand and carry out their direct responsibilities. Each staff member should understand the important role he or she plays in the big picture of the nursery's success. Some education in horticulture is very helpful to allow staff to "think like a plant." Training in safety is also essential for staff members. Be sure everyone knows how to properly use all equipment on the premises. Staff should be encouraged to stay curious and learn more about the plants and their production.

The manager's task in working with staff is to assign roles, goals, and tasks and then follow up to make sure tasks were completed. The manager should be open to

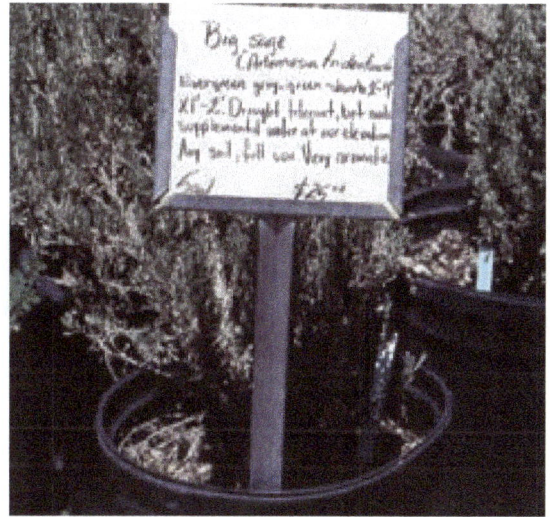

Figure 16.12—*Customers and the public benefit from learning more about how to plant native species. The Santa Ana Pueblo Nursery in New Mexico uses signs to educate visitors about native plants, empowering clients to cultivate them successfully.* Photo by Tara Luna.

Figure 16.11—*Sharing knowledge of native plants and their uses to future generations is a goal shared by many native plant nurseries, including (A) the Browning High School on the Blackfeet Reservation in Montana and (B) Moencopi Day School on the Hopi Reservation in Arizona.* Photo A by Tara Luna, B by Thomas D. Landis.

receiving feedback from staff about how to make their work more efficient, and the manager should also provide feedback so staff members can optimize the value of their work. Sharing the nursery's mission and overall goals fosters a greater sense of purpose among staff—improving morale and avoiding the "it's-just-a-job" attitudes.

Clear communication is a central part of good relations among managers and staff. Although meetings can seem to be an inconvenience, brief staff sessions are invaluable in linking the day-to-day tasks with the big picture for the nursery. Meetings can be scheduled on Monday morning to do some strategic planning and prioritize activities for the week. At the end of the week, another brief meeting can be held to assess and evaluate progress, and to identify priorities for the coming week.

COMMUNITY EDUCATION AND OUTREACH

Community education is an important activity for many native plant nurseries (figure 16.11). The perpetuation of native plants and cultural traditions are often a central part of the nursery's mission, and the nursery's efforts are lost if the community is not ready, willing, and able to accept plants the nursery will produce. Hosting workshops or field days, writing educational materials or articles, and attending local fairs or trade shows can educate the community about the nursery's mission and the plant materials being produced (figure 16.12).

Many native plant nurseries work with school groups and environmental or cultural education activities. Connecting with living, growing plants can be a wonderful activity for school and youth groups and a meaningful way to pass on the knowledge of plants and their cultural uses to younger generations.

INTERACTING WITH CLIENTS AND THE PUBLIC

Although generating new clients is essential, keeping and building the trust of existing clients is even more crucial. Word-of-mouth referrals are a very important part of a thriving nursery practice. For contract growing, create a clear agreement in writing. In some circumstances, such as growing for neighbors or close community members, a written agreement may seem unnecessary; however, the written agreement is invaluable for creating clear expectations, enhancing communication, and making everything go more smoothly for everyone concerned. It is far better to experience a little awkwardness up front than have a major miscommunication on your hands at delivery time! Both the client and a nursery representative should sign the agreement, and each should keep a signed copy. The terms of the agreement should include the following:

— A description of the plant materials to be provided (for example, species, container type, plant size).
— The anticipated schedule.
— The quantity of plant materials to be provided.
— The price per unit and the total price for the order.
— When and how payment will be made
— What will happen in certain situations, such as the following:
  – If a payment from the client is late.
  – If the client picks up their plants late or fails to pick them up.
  – If the nursery is unable to deliver the plants as described.

The sample contract in appendix 16.B is for demonstration purposes only. It should be tailored to meet the needs of specific nurseries. Underlines indicate fill-in-the-blank parts. It is best to consult with a legal expert to make sure the contract protects the nursery and conforms to local legal statutes.

Aside from clear agreements up front, another way to build good relationships with existing clients is to maintain frequent communication during the development of their crop. Customers appreciate, and often enjoy, staying informed about their crop's progress. Sending e-mail updates (perhaps including digital photos) or occasionally having a phone conversation or inviting the client up to visit their crop can go a long way in keeping customers involved in crop production and committed to the schedule (figure 16.13). Here are a few suggestions for interacting with customers.

If the client is an organization with several staff members, ask them to assign one sole contact person for the crop. In turn, the nursery should have just one contact person for that client. This one-on-one arrangement precludes many kinds of potential misunderstandings and also helps to develop long-term relationships and trust.

The Target Plant Concept (see Chapter 2) is a useful way to communicate expectations. No one enjoys an unpleasant surprise at plant delivery time. Clients should be clear on what size plants they will receive before they place an order.

Crop production may vary slightly from year to year. If possible, when the order is placed, agree on a window of time for plant delivery that spans a few weeks rather than setting an exact date. Based on the state of the crop, the exact date for delivery can be determined closer to the tentative delivery date.

See things from the client's point of view. Often, tremendous effort and expense go into planning a project and preparing land for outplanting. Acquiring plants is a central part of this process but, in terms of expense, may be a small percentage of the total project cost. Nursery staff should do everything in their power to meet set schedules. If any problems or delays are anticipated with the crop, clients must be updated immediately so they can modify their plans accordingly.

Set up a feedback system with customers after the order is complete. Give the client an opportunity to communicate about their experience in obtaining plants from the nursery. The nursery should also follow up about plant performance. Ideally the nursery contact person can visit the outplanting site and check on the progress of the crop over time. Observations can be used to improve target plant specifications for that outplanting environment. Feedback from the client can also be used to improve customer service in the future.

Re: Update on your order for 500 *Arctostaphylos uva-ursi*
Date: May 10, 2005

Dear Jane, I hope this note finds you well. I'm writing to let you know your 500 bearberry are doing great—we moved them into the final hardening stage of production last week, so they are getting toughened up for outplanting! We're right on schedule, delivery should be in three weeks, definitely not later than June 1. Let's arrange a pick-up day for that week. Please write back to confirm you got this message when you get a chance. Look forward to hearing from you!

—Gloria, Manager, New Meadows Nursery

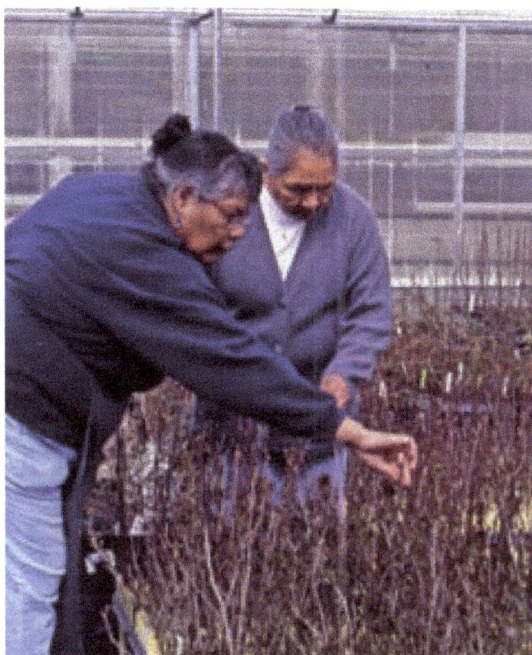

Figure 16.13—*Having clients visit the nursery is an important part of building good relationships.* Photo by Tara Luna.

SUMMARY

Expensive equipment, a big nursery, and a large staff are no guarantee of a successful nursery. On the contrary, a small nursery that is well planned and well managed can produce excellent plant materials and serve many community needs (figure 16.14). Management involves an understanding of scientific, technical, interpersonal, and economic aspects of the nursery. Managing a nursery, however, is an art as well as a science. The art aspect will be learned through experience. Observational skills, a flexible management style, and a willingness to be responsible for the crops are key attributes of a successful manager.

An essential aspect of good management is to have a structured organization with clearly defined responsibilities and one manager willing to assume responsibility for the crop. Feedback from clients and staff, information gathered from trials and daily records, and continuing research and education will consistently build on understanding and improving plant production over time. A strong desire to better understand nature and plants makes the work meaningful, productive, and satisfying for all concerned. As vital information is gathered, it should be recorded in plant protocols so this valuable knowledge may be passed on for the future.

Figure 16.14—*Well-planned, well-managed facilities, such as the Santa Ana Pueblo Nursery and Garden Center in New Mexico, can produce excellent plant materials and serve many community needs.* Photo by Tara Luna.

LITERATURE CITED

Landis, T.D. 1984. Problem solving in forest-tree nurseries with emphasis on site problems. In: Duryea, M.L., Landis, T.D., eds. Forest nursery manual: production of bareroot seedlings. Martinus Nijhoff/ Dr W. Junk Publishers. p. 307-314.

APPENDIX 16.A. ACTIVITIES

Planning (Weekly)

Strategic planning for the week and upcoming months includes the following tasks:

Get an overview of what needs to be done.

Look at crop growing schedules, facilities schedules, and deadlines.

Assess crop development and maintenance required.

Assess potential problems.

Make a schedule for the week and month.

Prioritize tasks.

Delegate tasks.

Follow up to make sure tasks were done.

Get another overview to plan next action items.

Conduct long-term planning (vision and goals for future and their steps).

Routine Tasks (Daily)

Watering.

Maintenance (for example, weed or pest control, fertilizing).

Monitoring and observing the crops.

Noting crop progress on a plant development record (daily or weekly).

Recording general observations and activities in daily journal.

Crop Production Tasks (Weekly or seasonal)

Consulting growth schedules and facilities schedules of what needs to be sown, moved, fertilized, shipped, and so on.

Establishment tasks (for example, making growing media, sowing seeds, inoculating with microsymbionts).

Rapid early growth phase tasks (for example, fertilizing, monitoring).

Hardening phase tasks (for example, changing fertilization and light regimes).

Updating clients about crop development.

Packing and shipping.

Culling and purging.

Recordkeeping

Horticultural records

Maintaining daily log (for example, environmental conditions, labor, activities for the day).

Making notes in the plant development records for each crop.

Creating and updating crop growing schedules and facilities schedules.

Updating and revising plant protocols.

Conducting crop inventory assessment and updating.

Financial records

Labor and time spent.

Money for materials.

Overhead costs (for example, utilities).

Cost estimating (so can budget and/or charge appropriately).

Income monitoring.

Inventory of supplies for production and maintenance (for example, growing media, fertilizers, containers and trays, irrigation parts).

Seasonal Cleanup (Seasonal, or between crops)

Purge holdover stock.

Clean floors, tables, and so on.

Clean and sterilize containers.

Check and repair equipment and infrastructure.

Learning More, Researching, and Problemsolving

Learning more

Attend trainings and conferences.

Learn from other nurseries; host and attend field days and visits.

Read published literature (for example, Native Plants Journal).

Research

Determine the most pressing questions to prioritize trials and research.

Design and conduct experiments and trials.

Assess findings from data and records.

Pursue troubleshooting and problemsolving:

Approach problems systematically.

Identify problem.

Analyze problem.

Know who to call for help (for example, another nursery, soil scientist, pest person, irrigation specialist).

Generate ideas.

Develop and test hypothesis.

Implement a solution.

Working with Staff

Provide staff education and training.

Assign roles, goals, and tasks.

Follow-up on tasks.

Give and receive feedback and input (observations and improvement suggestions).

Plan meetings, safety awareness, and so on.

Working with Clients and Potential Customers

Update clients about crop development.

Get feedback from existing and former clients.

Follow up on field performance of plants.

Offer public education and outreach.

APPENDIX 16.B. SAMPLE CONTRACT

CONTRACT
With New Meadows Nursery

This agreement, entered into this _____ day of _____ (month), _____ (year), by and between New Meadows Nursery and _____ (hereinafter referred to as "Client") witnesses as follows:

Whereas New Meadows Nursery is organized to provide plant materials for outplanting; whereas Client is interested in purchasing plant materials from New Meadows Nursery, it is agreed between the parties as follows:

I. Plant Materials Provided by New Meadows Nursery

In time for the spring planting window (May 1-June 30) New Meadows Nursery will provide 500 plants grown in Ray Leach 98 Cone-tainers™, with stems 8 to 10 inches long and having firm root plugs of the following species:

Arctostaphylos uva-ursi   (kinnikinnick)   Price: $3.00 ea

II. Fees

Client agrees to pay New Meadows Nursery $1500 for the 500 plants listed above. Payment shall be made in the following way: an initial fee of $750 (50% of the total for plant materials) is required to begin propagation, with the balance of $750 to be paid prior to dispatch of the plant materials. Other fees, such as shipping/delivery charges if applicable, will be billed separately and are also to be paid in full prior to dispatch of the plant materials.

    If any payment as per the above schedule remains overdue for more than 60 days, Client acknowledges that New Meadows Nursery may take legal action to collect the overdue amount. In such event, Client will be responsible for all reasonable litigation expenses incurred by New Meadows Nursery, including, but not limited to, court costs and attorney fees.

III. General Conditions

a) New Meadows Nursery agrees to use its best efforts to provide the plant materials listed in Section I above;

b) Client understands and acknowledges that New Meadows Nursery shall in no way bear liability for results produced in use of plant materials. New Meadow's maximum liability is limited in amount to the amount paid by Client to New Meadows Nursery for the purchase of the plant materials under all circumstances and regardless of the nature, cause, or extent of any loss.

c) In the event that Client cancels the order for plant materials in whole or in part, Client agrees to pay the balance due for the full amount for plant materials as listed in Section I.

d) New Meadows Nursery reserves the right to prorate or cancel any order, in whole or in part, because of natural disaster, disease, casualty, or other circumstances beyond our control. In the event that New Meadows Nursery is unable to provide the plant materials listed in Section I above by June 30, the initial fee paid by Client may be applied to another purchase, credited to a future order, or refunded, as requested by Client. In any other event, the initial fee is nonrefundable and the entire balance is due.

e) Client agrees that all plant materials ordered must be dispatched (picked up, shipped, or delivered) within 30 days of notification of readiness as determined by New Meadows Nursery. Plant materials not dispatched within 30 days are subject to a storage fee of $0.01 cents per plant per day; plant materials not claimed within 45 days of notification of readiness are forfeit.

IV. Conclusion

This agreement, executed in duplicate, sets forth the entire contract between the parties and may be canceled, modified, or amended only by a written instrument executed by each of the parties thereto.

    This agreement shall be construed as a State of _____ contract.

    Witness the hands and seals of the parties hereto, each duly authorized, the day and year first written above.

_____

Client, Tall Mountain Wildlife Sanctuary   Date

_____

Gloria Greenthumb, New Meadows Nursery   Date

Discovering Ways To Improve Crop Production and Plant Quality

Kim M. Wilkinson

Working with plants is a process of discovery. Being curious and aware, paying close attention, and staying open and adaptive are important practices. Books and people can help us learn about plants in the nursery, but the very best teachers are the plants themselves. "Research" is simply paying close attention, tracking what is happening and what is causing it to happen, asking questions, and seeking answers. In other words, research is something many growers already do. Scientific research is simply "the testing (systematic, controlled, empirical, and critical investigation) of ideas (hypothetical propositions about presumed relations among natural phenomena) generated by intuition." If research is done well, the process can yield useful, accurate information. If done ineffectively, the process can waste time or yield meaningless or inaccurate conclusions (Dumroese and Wenny 2003). The purpose of this chapter is to help design easy trials and experiments to discover useful, meaningful ways to improve crop production and quality.

Asking questions and finding answers is the essence of learning. Some people may be lucky enough to have an elder or mentor in their life who is always asking them questions and pressing them to explore and discover more. Carrying out research, asking questions, and keeping records are ways to "self-mentor." A systematic approach supports the making of accurate observations, honing senses and awareness so that discoveries otherwise missed can be made and shared. It is widely acknowledged that working with plants is an art as well as a science: observation, senses, emotions, empathy, and intuition

Fertilizer experiment by R. Kasten Dumroese.

Figure 17.1—*Even simple experiments, such as this seed treatment trial for penstemon, can teach us a lot about how to grow plants.* Photo by Tara Luna.

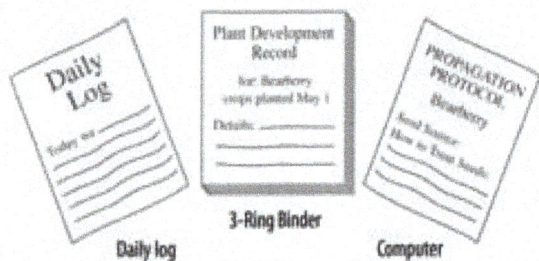

Figure 17.2—*The daily log, plant development record, and propagation protocol are the three basic kinds of records that form the foundation for learning more.* Illustration by Jim Marin.

play important roles. When a question comes up that is important to try to answer, it is time to think about doing a trial or experiment. Because growers are often working with plants for which minimal literature or outside information is available, many questions about optimal techniques will come up. The observations may be subjective or objective; research projects can be simple or complex. Learning how to effectively carry out experiments to evaluate new plant production techniques is essential to discovering relevant and applicable answers.

WHY TRY NEW THINGS IN THE NURSERY?

Often, the tendency is to take the "path of least resistance" and use known or established production techniques. Especially for species that are new to the nursery, the first technique that was tried and that produced an adequate plant may have become the established protocol. The original technique, however, may be more costly or inefficient than alternate methods, and may not produce the best quality plant for the outplanting site. A few modifications could go a long way to improve production, plant quality, and, ultimately, plant survival and growth after outplanting. Simple experiments allow the nursery to try out new techniques, ideas, and problem solving strategies (figure 17.1).

The tasks of keeping up with day-to-day nursery management may feel like more than enough to fill the schedule, and time for experimentation may seem a low priority. In truth, however, most nurseries already engage in investigations on a regular basis. Experimentation is taking place every time a new idea is tried out, a question leads to alternative strategies, or a problem is analyzed and solved. Taking a little care to be systematic and to follow a few guidelines will dramatically increase the benefits of these activities as well as provide greater confidence in conclusions drawn from them.

Often, simple evaluations can be carried out simultaneously with producing an order for plants. For example, in most cases, if the nursery has an order for plants, it can produce them using established methods as described in the protocol and also grow some additional plants for research at the same time. One variable can be altered for the additional plants, such as using a modified growing medium, a new type of container, or a different mycorrhizal inoculant. In this way, each new crop represents an opportunity to try something new on a small scale. The discoveries could greatly improve production efficiency and seedling quality over time. Because of these potential benefits, it is worth putting a little effort into trying new things.

Conducting simple experiments and trials in the nursery can accomplish the following:

— Produce better plants.
— Speed up production.
— Save money, labor, and materials.
— Improve outplanting survival and performance.
— Contribute to greater knowledge of the plants.

MAKING OBSERVATIONS AND KEEPING RECORDS

A foundation for improving plant health and quality is a good understanding of current production practices and how plants respond to them. The basic recordkeeping discussed in Chapter 3, *Crop Planning and Developing Propagation Protocols*, provides this foundation. The following three types of records are most important:

— A daily log of general conditions and activities.
— Development records for each crop that are filled out as the plants develop.
— Propagation protocols that describe from start to finish how the plants are currently grown (figure 17.2).

These records prevent the nursery from wasting time repeating strategies that do not work while providing a plan to help duplicate successful crops. This information also establishes how crops normally perform, and it can be used to recognize problems or gauge the impact of production methods and alternatives. Perhaps most important, keeping data and cultural records in a written format creates information that can be passed on to future nursery staff or others in the community. Without these records, valuable information (perhaps gleaned from a lifetime of nursery work) may be lost, and the new manager will have to start over. Knowing in a measurable way what is "normal" helps the nursery experiment with modifications that can improve crops and productivity over time.

SOLVING PROBLEMS THROUGH OBSERVATION

"Seeds of candle yucca did not germinate well the first year we tried growing it in the nursery. After spending time in the habitat and observing the plants in their natural setting, I observed that yucca seedlings germinated in the crown of the mother plant within the dried cane stalks. I collected a dried cane stalk and cut it into a block, and inserted seeds into it. The seeds germinated very well…

Photos by William Pink

…Next, I applied this natural germination pattern on a larger scale by cutting yucca cane blocks into small pieces, inserting one seed into each and placing them in a plug tray.

The sugars in the cane blocks must help with the germination of the seeds, and I transplant them into individual containers once the root and shoot is well developed."

William Pink, Pechanga Band of the Luiseno Indians

WHERE DO IDEAS FOR EXPERIMENTS AND TRIALS COME FROM?

Ideas for experiments may come from many places, including the following:

— A casual observation the nursery wishes to verify.
— A pressing question that seems to recur in the daily log.
— An informal trial in which a difference is observed.
— Something that can be improved (for example, percentage of germination).
— A desire to work with a new species or a new seed source.
— A desire to try out a new technique or idea.

Because staff time is limited, ideas for experiments will need to be prioritized in terms of their potential positive impact and importance to production. Nevertheless, it is good to keep a list of any potential experiments or trials that staff think would be beneficial, and keep these ideas on hand to try out as time allows. Some types of experiments may be easy and fast to set up and conclude, and they can be done efficiently as part of crop production.

Experiments and trials can be conducted to improve nursery production, learn about new species, or to try out different strategies for growing plants. Many topics lend themselves well to experimentation, including the following:

— Developing seed treatments and germination techniques.
— Researching microsymbiont sources or application methods.
— Testing out new seed sources.
— Altering watering regimes.
— Trying new container types.
— Changing an aspect of management, such as timing for moving crops from one environment to another.

GATHERING INFORMATION AND SPECIFYING A QUESTION

At this point, some background research is in order. The person initiating the experiment may look back on their own experiences and observations to search for ideas. Checking the existing plant development records, plant protocol, and daily log is essential to track

EXAMPLE: STEP 1— AN IDEA FOR AN EXPERIMENT

For example, perhaps one native species that is grown in the nursery consistently has low percentage germination. Less than 25 percent of the seeds germinate after they are planted, and germination is sporadic, taking place over 4 weeks. These results occur despite good seed sterilization and handling practices, and a seed test sent to a laboratory indicated that a higher percentage of the seeds are viable. The nursery wishes to try to increase the percentage germination. The first step is to look at the options for experimentation. What could be contributing to the low germination?

what has happened and what has been tried in the past. Information can also be gathered from nursery staff, specialists, extension agents, and associates from other nurseries. Look up the subject in journals, books, and electronic sources of information. A little background research like this can help narrow the focus of the experiment to a question that can be answered.

Note that the question must be limited in scope, pertaining to just one aspect of production. It would not work to focus on multiple variables at the same time. For example, if many aspects of the seed treatment process (for example, season harvested, collection and processing method, sowing times and methods, and media used during the germination

EXAMPLE: STEP 2— GATHERING INFORMATION

Perhaps not much information is available in the literature that discusses this particular species directly. According to publications and other nursery growers consulted, however, several closely related species have a hard seedcoat that is usually scarified prior to planting. According to literature and other growers, mechanical scarification (nicking by hand with a nail clipper) is sometimes used, and other nurseries use a 20-second hot water treatment to scarify their seeds. Perhaps lack of proper scarification is causing the poor germination. The question that the experiment will address is formulated: How does scarifying the seeds affect germination?

phase) were all modified at the same time, how would the experimenter know which modification caused a difference? Pose a focused question that can be answered by experimentation.

THE HYPOTHESIS

The hypothesis is the proposed answer to the focused question or problem posed.

EXAMPLE: STEP 3—THE HYPOTHESIS

The hypothesis might be "Mechanical scarification of the seeds by hand-nicking the seedcoat with a small nail clipper will result in improved germination." If time allows, it may be desirable to pose more than one treatment to answer the same question. Perhaps a second treatment to be tested separately for this example would be "Hot water scarification for 20 seconds will result in improved germination." These two treatments (mechanical and hot water scarification) will be compared against a third treatment that is not modified in any way. In other words, if the seeds are usually not scarified according to the protocol and the plant development records, the third set of seeds will not be scarified. This third treatment is the control for the experiment.

THE VARIABLES OF THE EXPERIMENT

All experiments have the following three essential parts:

— The control (the way things are usually done).
— The independent variable (the one thing that is changed for the experiment).
— The dependent variable (the thing being observed as it is affected by manipulating the independent variable).

The Control

An essential aspect of experimenting is to have a control. For experiments in plant production, the control is simply the way the nursery usually does things according to established protocols. Nothing is changed. For the scarification example, growing medium, seed source, fertilizer, watering regime, light, container type, spacing, and all other factors would be carried out as usual. The performance of control plants grown in the

Figure 17.3—*Seed scarification and/or stratification requirements are often discovered through experimentation and trials. Careful labeling of the control and each treatment is essential.* Photo by Tara Luna.

usual way will be measured against the performance of plants grown in a modified way. Plants in the control and the experiment should be started at the same time and kept in the same areas of the nursery to eliminate other potential differences between them, thereby isolating the factor being evaluated (figure 17.3). If possible, it is a good idea to have several identical controls.

The Independent Variable

The independent variable is the one factor purposely changed for the sake of the experiment. In the example, the scarification of seeds is the only aspect of production that is manipulated. Everything else must stay the same. The seed source, collection, and cleaning; growing medium; light and watering regime during germination; and all other factors would not be modified. Only the independent variable is altered. If more than one variable is manipulated, isolating which variable caused a change is impossible.

The Dependent Variable

The dependent variable is the variable being observed. In the example, the percentage germination is being observed.

Repetition and Blocks

Research is largely about isolating the independent variable and eliminating the possibility that any other factors could be contributing to observed differences.

Repetition is an essential aspect of this process (figure 17.4). If too few plants are used for the control or the treatment, any differences observed may simply be a coincidence. For this reason, it is valuable to have as many control and treatment plants as is reasonable for the experiment.

Variability in the nursery environment can affect experiments, and this variability may go unnoticed. For example, two benches thought to be equal may have subtle differences in irrigation or light received. Because of environmental variability, it is best to place treatments and controls right next to each other, on the same bench or even in the same tray. When repetitions are carried out (with multiple blocks), have the block of each kind of treatment (including the control) grouped together, but place each of the repetitions in several different parts of the nursery if possible. In other words, if you have four blocks, each with a control, treatment 1, and treatment 2, place each of these blocks in a different part of the nursery. That way, if a difference is observed, the researcher can be more confident in the results because the possibility of a fluke is reduced.

A few tips for having good repetition include the following:

— Have at least 30 plants per treatment and 30 plants per control, or grow at least three trays each of treatment and control.

Figure 17.4—*Repetition need not be large-scale, but it is important to reduce the possibility that outside factors could be affecting observed differences in the experiment.* Photo by Tara Luna.

— If possible, replicate treatments in three or four different locations within the nursery.
— Situate treatments and the control immediately adjacent to each other to reduce variation in microclimate (also consider splitting the block or tray [half treatment and half control]).

Because each growing season is different, the experiment may be repeated one or two more times on subsequent batches of plants. After the first experiment, however, the results may indicate strongly that the new method led to better germination or higher quality plants. If so, for subsequent crops, this new way can become the protocol (the way plants are grown at the next production time), and, if desired, a small batch of seedlings using the "old" method can also be grown for comparison purposes and to verify that the new method works better.

CARRYING OUT THE EXPERIMENT

If the researcher feels unsure about the validity of the proposed elements of an experiment, find an ally in the local university or agricultural extension system to discuss the plan briefly. It usually takes only a few minutes to have someone verify that the experiment is soundly designed. This is a wise investment of time, saving the trouble of investing in research that does not succeed in answering the question posed.

After the hypothesis is posed and the treatments and repetitions decided on, it is time to plan when and how the experiment will actually be carried out. Unless the problem addressed is urgent (that is, interferes with production), it may be most economical to wait until the nursery has an order to produce the species that is part of the research. The production of these plants carried out according to the usual protocol will be the control. Extra plants can be planted at the same time from the same seed source and on the same day, with only the independent variable manipulated.

A few tips for starting and carrying out the experiment include the following:

— Have one person in charge of the experiment, making observations and collecting all data. This person should be involved in the research from start to finish. Having one person collect data eliminates the possibility that two people are measuring or making observations in different ways. It also ensures that one person is responsible for keeping records.

— If special materials are needed for the experiment (for example, a different microbial inoculant, a new seed source, a special growing medium ingredient), be sure to have them on hand before the experiment starts.

— Mark treatments clearly with durable, easy-to-read labels. Nothing is worse than discovering a group of plants performing outstandingly but with no record of what was done differently. Mark the control group as well.

— Do not count on experimental treatments to produce marketable plants. Use established protocols to meet client requirements. Plants devoted to research should be above the count required for the order. If the experimental subjects turn out to be of high quality and saleable, that will be a side benefit. Experimental plants, however, will likely be different in size or performance from the bulk of the order. If growing on contract, the client may be interested in accepting research plants to continue the trial in the field. Agreements should be clarified in advance regarding experimental plants.

— Take careful notes and keep a journal documenting every step of the experiment as it is carried out. Changes may occur rapidly and go unnoticed if care is not taken to record them. Sometimes the independent variable will affect one brief but important stage of plant development. Keep data organized, ideally entered into a computer spreadsheet just after taking measurements and observations.

— Be prepared to carry out the experiment more than once.

MAKING OBSERVATIONS AND COLLECTING DATA

The focus of the experiment dictates the types of information to be collected and observations to be made. When gathering data, keep the process as simple and straightforward as possible, and reduce risks of non-applicable or meaningless results. For example, the wet weight of live plants will vary considerably depending on irrigation and time of day; therefore, weighing live plants does not usually generate meaningful data for experiments. For small experiments, have just one person take measurements and gather "hard" data in order to reduce variations in the way data are collected. For larger exper-

EXAMPLE: STEP 4— CONDUCTING THE EXPERIMENT

Let's say this experiment on seed scarification will be carried out simultaneously with growing an order of plants for that species. The nursery should not count on being able to use any of the seeds from the experiment (the new method may increase, decrease, or have no effect on germination percentages). So, the procedures produce the correct number of seedlings to meet the order should be carried out as usual. Seeds should be collected and processed as described in the protocol and records for that species. If an order for 100 plants of the species with the usual expected 25 percent germination is received, according to the established protocol, the nursery will probably need to sow about 450 seeds to compensate for the low germination and other losses. At the same time, additional seeds from the same seedlot would be scarified using mechanical scarification. A minimum of 30 seeds should be treated and sown, although, if the supply of seeds is abundant, the researcher might choose to treat more seeds with the new method for a good-sized sample. All these seeds are planted on the same day and placed in the same environment as the control treatment. If they are in separate trays, they should be placed side by side to eliminate any other variables (such as differences in light or humidity) that might affect the germination rate. For a larger experiment, replicating the trial in three or four different locations in the nursery helps eliminate outside variables. The experimental set is clearly marked with exactly what is different about it. All seeds will be treated identically otherwise. If a third treatment (20 seconds with hot-water scarification) is used on a third set of seeds, it, too, should be handled identically otherwise and clearly marked. The control, scarified, and hot-water treated seeds are placed side by side in the same environment or in replicated blocks, as the control. Now, it is time to watch, observe, and collect data.

iments, however, it is a good idea for several people to take data on all treatments, including the control, in order to cancel out bias in data collection. If several people will be collecting data, make sure each person is trained to measure using the same procedures.

Although information on how to keep records of crop progress is provided in Chapter 3, *Planning Crops and Developing Propagation Protocols*, data collection for the experiment should focus on the dependent variable for that experiment (in the example, percentage of germination). Other data and observations, if available, however, may be collected as well if time allows. Even if they are not quantified, observations about the appearance and vitality of plants can be especially useful for many experiments.

The best timing for data collection also varies depending on what is being studied. Although any period of rapid change for the crop can be a useful time to gather data, in general, the most meaningful results tend to be gathered:

- During germination (as in the example in this chapter).
- At the beginning and end of the rapid growth phase.
- At the end of the hardening phase (just before shipping).
- After outplanting (usually after the first 3 to 12 months in the field, up to 5 years).

Measurements may include the following but not every experiment will require all these measurements— only the ones relevant to the study. Stick to simple measurements and observations that are meaningful.

Germination Timing and Percentages. A percentage of germination can be determined by comparing the total number of seeds planted with the number of healthy germinants that emerge for each seed treatment. Timing is very important to monitor: sometimes the percentage of germination will ultimately be the same but one treatment may facilitate uniform and rapid germination while another treatment may be uneven or delayed. Encourage daily or weekly measurements to capture differences in rate of germination.

Plant Height and Root-Collar Diameter. These measurements are useful to compare changes to the timing and development of plants to the control and to previous crops described in the plant development records and propagation protocol. Root-collar diameter measurements are often taken about 0.25 in (0.5 cm) above the medium (figure 17.5A). Height can be measured from the growing medium surface to the top of the growing point on the stem (not the top of the leaf) (figure 17.5B).

Shoot-to-Root Ratios. Shoot-to-root ratios are taken only periodically and usually only as small samples, because these measurements destroy the plants sampled. They are based on oven-dry weight. Carefully remove any medium from the roots and dry the plant samples for 72 hours at 150 °F (66 °C). A convenient way to handle plants is to put them into paper lunch bags before placing them into an oven. The treatment can be written directly on the bag to avoid confusion. After the plants are dry, cut the sample at the place where the stem meets the roots (the root collar; often a change of color occurs here) and weigh the shoots and roots separately to get the ratio.

Plant Vigor. Plants can be subjectively rated at the beginning and end of each of the growth phases using a numeric rating system, such as 1 to 5. Clear guidelines must be developed for the numerical scale to give a relative estimate of plant vigor. For example, 1 = no vigor, plant appears on verge of death; 2 = poor and slow growth; 3 = some growth, some vigor; 4 = plant looking vigorous; 5 = plant appears to be thriving and very vigorous.

Insect and Disease Analysis. Noting, and perhaps quantifying, the presence of disease or insects can be done regularly. Good times to make notes are during emergence and at the ends of the establishment, rapid growth, and hardening phases. Samples of pests or diseases can be sent to the local agricultural extension office for identification, if necessary.

Outplanting Survival and Growth. Field survival and performance after outplanting can also be evaluated using most of the previously described techniques.

RECORDING, ASSESSING, AND SHARING FINDINGS

Keeping good records is a key part of successful experimentation. Entering observations and measurements into a computer or project journal is a very good practice (figure 17.6). A simple tabular format is fine for most types of data and makes capturing and assessing the data easier. If feasible, only one person should be taking measurements and recording data in the journal in order to eliminate variable styles of measurements (figure 17.7). Others may contribute to subjective evaluations, however, and the person in charge of the research project may solicit the observations of other staff members and enter these observations in the journal as well.

Some types of experiments may focus on just one phase of growth, such as the germination phase. Many others will follow plants through all phases. Regardless, when the final phase is complete, it is time to step back and assess the data and observations collected. Data must be organized to be interpreted. If percentages were used, the data can be converted into a graph or chart to visually show differences between the control and the experiment. Any results should be shared with other staff and entered into the records. If the experiment was focused on producing one species, the results should be entered into the protocol for that species, even if no difference was observed. If a difference was observed, and the experimental seedlings had better germination, survival, or quality than the control, the experiment should be repeated at least once or twice more to verify the results. In the interim, however, the "new" production technique can tentatively become the new protocol, with the old technique repeated on the smaller scale for comparison purposes. If, after a few repetitions, the same results are found, the old method can be retired and the new one adopted as the official protocol.

If no difference was observed, or if the experimental treatment performed more poorly than the control, that is still very valuable information. As Thomas Edison said, "Every wrong attempt discarded is another step forward." The observation should be entered in the plant protocol notes. (That is, "A trial to scarify this species using a 20-second hot-water (200 °F [93 °C]) treatment resulted in 0 percent germination.") Now that it is known that 20 seconds in 200 °F hot water decreased germination (and likely killed the seeds with too much heat), other scarification techniques (lower temperature, shorter time in hot water, or mechanical methods) can be tried. Noting ideas for future experiments is an important part of concluding the experiment.

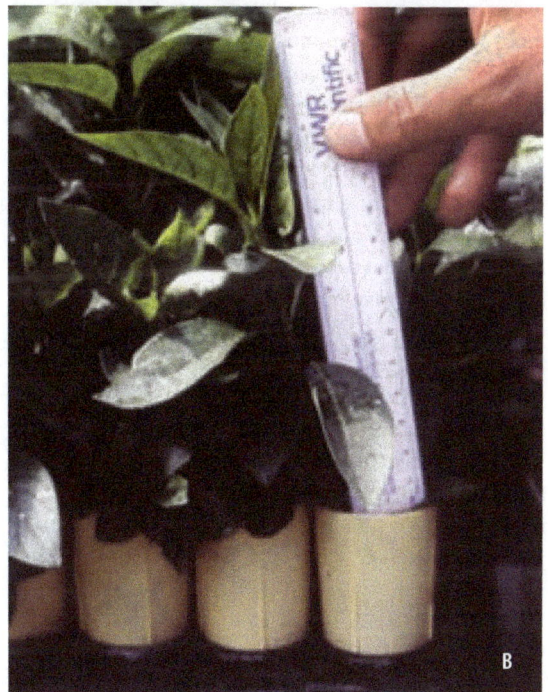

Figure 17.5—(A) Root-collar diameter measurements can be taken at the medium line or elsewhere along the stem as long as the place where the measurements are taken is consistent from plant to plant. (B) Height can be measured from the medium surface to the top of the growing point on the stem (not the top of the leaf). Photos by Tara Luna.

EXAMPLE: STEP 5—
OBSERVATION AND DATA COLLECTION

According to the protocol, germination usually takes place sporadically over a period of 4 weeks in order to achieve 25 percent germination. Therefore, the experiment will run 4 weeks. Each tray of seeds is monitored daily and the number of germinants recorded. If the emergents are to be transplanted into larger containers before the end of the experiment, it is critical to make sure the counts are accurate prior to transplanting. Ideally, germinants from each treatment are transplanted in separate trays and/or carefully marked, even in their new containers. At the end of the 4 weeks, the experiment will be complete.

In the early years of the nursery, especially one growing native and culturally important plants, the improvement rate resulting from experiments and trials can be very high. Many new techniques that are tried will lead to better quality plants and more efficient production. As the nursery gains expertise in the most effective strategies for each species, the focus of experiments may shift in later years to bring about more subtle improvements.

Again, keeping good records and making brief notes of experimental results in the protocols have tremendous long-term value. For example, perhaps the example species will be mechanically scarified by hand using a small nail clipper for some years after this experiment. Then, one day, the nursery receives a very large order for that species and the staff begins to wonder if they can save the labor of hand scarification by using hot water instead. They will be very glad to find results and suggestions from the first experiment, and they will be able to build on that knowledge. Also, consider sharing results at nursery meetings, by submitting short papers to professional publications such as *Native Plants Journal*, or by uploading the information into the Native Plant Network (http://www.native plantnetwork.org).

SUMMARY

When a nursery is just starting out, many aspects of plant production are waiting to be discovered. Being aware and "connected" to plants will make nursery employees better observers and better growers. Having a "green thumb" is possible, and it can be cultivated by

EXAMPLE: STEP 6—
RESULTS

The final count of germinants from the control treatment is tallied: 114 germinants out of 450 seeds, slightly over the usual 25-percent germination. A graph could be made to show their cumulative germination on each day—in the example cumulative germination was increasing at a constant rate throughout the 4 weeks of the trial and it appeared germination would continue to occur even though the experiment had concluded. For the seeds that were mechanically scarified, 79 out of the 100 seeds germinated: 79 percent. If a graph were made of the number of germinants emerging per day, it would show that most of the germinants emerged 3 to 8 days after sowing, with germination rates tapering off dramatically after the eighth day. For the third treatment, in which seeds were treated with hot water for 20 seconds as a scarification technique, no germinants had emerged by the end of 4 weeks. Therefore, mechanical scarification yielded high germination percentages that occurred rapidly—germination characteristics desired in a nursery. The protocol will be revised to call for mechanical scarification as the method of seed treatment.

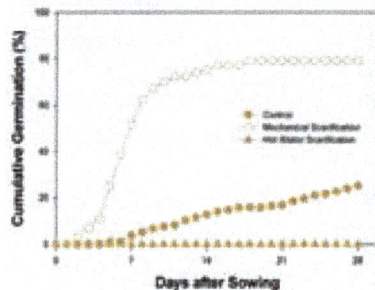

asking questions. Research is a way of exploring and learning from plants through direct experiences. Experimenting is a key part of nursery development because it leads to an increased understanding of the plant's needs and, potentially, to increased productivity.

Keeping some baseline records is invaluable for nursery productivity and development. Daily logs, plant development records, and propagation protocols form the foundation of future research and trials to improve plant quality. Nurseries that are interested in

EXAMPLE: STEP 7 —
NOTING FAILURES AND SUCCESSES
FROM EXPERIMENTS

Mechanical scarification worked very well, resulting in improved germination percentages and more uniform timing for germination. Mechanical scarification will become part of the standard protocol for seed treatments. As far as the hot water treatments, even though a closely related species is known to respond well to hot water, the 20-second treatment in 200 °F (93 °C) hot water on these seeds resulted in 0 percent germination. It is likely that the 20-second exposure was too long for this species and "cooked" the seeds. A note should be made of this result and of the questions that arise: Perhaps a shorter time in the same temperature water or a lower water temperature could scarify the seeds effectively without harming them? Next time, maybe just 10 seconds in 200 °F (93 °C) hot water should be tried or 20 seconds in 170 °F (77 °C) hot water. If a large order for this species is received in the future, scarification will be very labor-intensive to do mechanically by hand. The notes about the hot water option, what did not work and what might work, will be a key piece of information for future discoveries.

Figure 17.6—*A few basic measuring tools and a good way to keep records is all you need for many kinds of experiments.* Photo by Tara Luna.

engaging in larger scale or more intensive research may be able to get assistance from a local college or university.

Remember the following essential aspects of experimentation:

— Always have a control treatment (that is, the way the nursery usually does things).
— Keep it simple; define the problem and hypothesis to alter just one variable.
— Eliminate other variables (amount of water or light, for example, must be equal).
— Carry out several repetitions on groups or blocks of plants.
— Take data accurately and precisely.
— Keep records updated.
— Share important findings with staff and others.
— Stay curious and open to learning more: working with plants is an endless process of discovery.

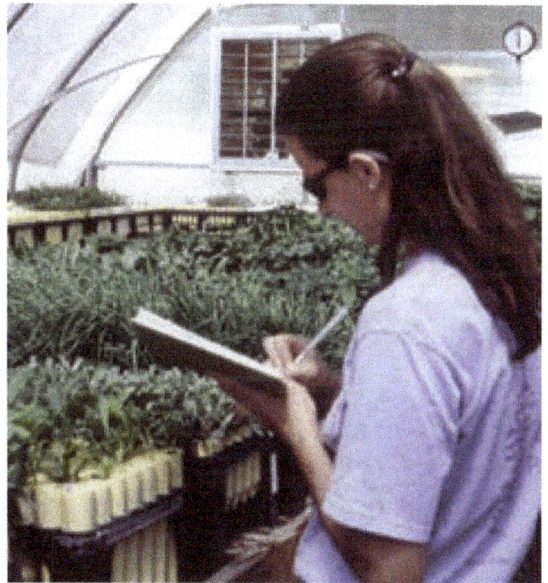

Figure 17.7—*Keeping good records is a key part of successful experimenting.* Photo by Tara Luna.

LITERATURE CITED

Dumroese, R.K.; Wenny, D.L. 2003. Installing a practical research project and interpreting research results. Tree Planters' Notes 50(1):18–22.

ADDITIONAL READINGS

Landis, T.D.; Tinus R.W.; McDonald, S.E.; Barnett, J.P. 1994. The container tree nursery manual: volume 1, nursery planning, development, and management. Agriculture Handbook 674. Washington, DC: U.S. Department of Agriculture, Forest Service. 188 p.

Wightman, K.E. 1999. Good tree nursery practices: practical guidelines for community nurseries. Nairobi, Kenya: International Centre for Research in Agroforestry. 95 p.

APPENDIX 17.A.
PLANT MENTIONED IN THIS CHAPTER

Candle yucca, *Yucca schidigera*

www.ingramcontent.com/pod-product-compliance
Lightning Source LLC
Chambersburg PA
CBHW080514220326
41599CB00032B/6083